INTRODUCTION *to* GAME THEORY *in* BUSINESS *and* ECONOMICS

T0313087

THOMAS J. WEBSTER

M.E.Sharpe
Armonk, New York
London, England

To my sons, Andrew Nicholas and Adam Thomas—two beautiful minds.

Library of Congress Cataloging-in-Publication Data

Webster, Thomas J.
 An introduction to game theory in business and economics / Thomas J. Webster.
    p. cm.
 Includes bibliographical references and index.
 ISBN 978-0-7656-2237-2 (pbk. : alk. paper)
 1. Game theory. 2. Economics--Psychological aspects. 3. Strategic planning. I. Title.

HB144.W43 2007
330.01'5193--dc22   2007043943

Printed in the United States of America

The paper used in this publication meets the minimum requirements of
American National Standard for Information Sciences
Permanence of Paper for Printed Library Materials,
ANSI Z 39.48-1984.

BM (p)   10   9   8   7   6   5   4   3   2

Contents

Acknowledgments

I would like to express my heartfelt appreciation to my colleagues and students at Pace University who provided many invaluable insights and suggestions during the preparation of this text. I would also like to express my appreciation to those individuals who were gracious enough to review earlier drafts of this text. Although their feedback was not always positive, or flattering, it was always constructive and invaluable. In particular, I would like to extend a special thanks to Professors Walter Antognini, Roy Girasa, and Art Magaldi of Pace University for their legal expertise, Professor John Teall of Rensselaer Polytechnic Institute for his insights and encouragement, and Assistant Dean Germaine Hodges of Pace University for her friendship, forbearance, and honesty. I would like to express my appreciation to Professor Udayan Roy of Long Island University for his painstaking review of early drafts of several chapters of this text. His critical comments, especially in the area of mixed strategies, were invaluable. I would like to thank my son, Adam Webster, a student at the University of Arizona, who reviewed early drafts of this text and contributed a case study on global warming. Finally, I would like to express a very special thanks to Lynn Taylor, Nicole Cirino, Angela Piliouras, and Eileen Chetti at M.E. Sharpe, who managed to suffer through my incessant nagging with much grace and aplomb.

A Note to Students and Instructors

TO THE STUDENT

More than in most disciplines, students of business and economics come to appreciate very early in their education the importance of thinking strategically. Strategic interaction between and among rivals is the essence of competition in the marketplace. A business manager who attempts to anticipate how rivals will react to changes in product pricing, the introduction of a new product line, an increase in advertising expenditures, expanding into overseas markets, or the naming of a new chief operating officer is thinking strategically. Sometimes this game is played just once, but more often than not this game is played repeatedly. Rivals' decisions can be made simultaneously or sequentially. If I do this, what will you do? And, if you do that, how will I react, and so on and so forth. The essence of game theory is to make sense of this back-and-forth (circular) reasoning. By putting ourselves in the shoes of rivals, we turn the game to our advantage. The objective of the study of game theory is to understand strategic interaction—move and countermove. Indeed, no graduate or undergraduate program in business and economics is complete without a rudimentary understanding of the principles of game theory that extend beyond the prisoner's dilemma paradigm, which can be found in most fundamentals of microeconomics textbooks.

This book is designed to appeal to both graduate and advanced undergraduate students in business, economics, and other social sciences with varied educational backgrounds. Most of the material presented in this text can be understood by students with a basic understanding of the principles of microeconomics. For the most part, mathematical training beyond elementary algebra is unnecessary, although a one-semester course in differential calculus is necessary to master the material dealing with continuous strategies and best-response functions. This is especially true when we discuss models of imperfect competition and strategic trade policy. The appendices to Chapters 1 and 13 present a review of the basic mathematical and statistical concepts necessary to master the material presented in this text.

TO THE INSTRUCTOR

This text originated from a series of lectures that were prepared for beginning, part-time MBA students with diverse backgrounds. The objective was to convey to these students an appreciation of benefits of strategic thinking in a casual and nonthreatening manner. These lectures constitute the first three chapters of this text. As interest among students developed, these lectures were extended to include more esoteric topics, which were presented in a dedicated course on basic strategy. As the topic coverage was expanded, it became necessary to elevate somewhat the level of mathematical rigor, but not to raise it so high as to alienate most students.

The first three chapters of the text deal with one-time, noncooperative, pure-strategy, static (simultaneous-move) games with complete information. Chapters 4 and 5 consider repeated, pure-strategy, static games in which cooperation between and among the players is possible, but not guaranteed even when it is in their collective best interest to do so. In these chapters, basic rules are developed to assess the stability of cartel arrangements in static games that are infinitely repeated, or are repeated a finite number of times where the end of the game is certain and uncertain. These chapters also discuss the use of trigger strategies and enforcement mechanisms for binding players to an agreement. Chapter 6 introduces the idea that mixing pure strategies may itself constitute a strategy. The objective of this chapter is to demonstrate that, unlike many games in pure strategies, mixing strategies almost always results in a unique outcome.

The first six chapters of the text require that the players choose from among a discrete number of strategies. By contrast, Chapters 7 through 10 extend the analysis of static games to include choices from a continuous spectrum of strategies. Beginning with Chapter 11, the focus shifts from static to dynamic (sequential-move) games. The concepts developed in Chapter 11 are used to analyze basic bargaining scenarios in Chapter 12. In Chapter 13, pure strategies involving uncertain payoffs are introduced and applied to an examination of torts and contracts in Chapter 14. Chapter 15 applies the ideas developed in earlier chapters to an analysis of auctions. The final chapter of the text presents a brief review of the special problems associated with dynamic games in which the players have less than complete information about their rivals' characteristics, intentions, and payoffs.

This text includes several case studies, applications, and demonstration problems that are designed to enhance the student's formal and intuitive appreciation of the underlying logic and principles of game theory. For instructors who adopt the text for classroom use, an online Test Bank and Instructor's Manual are available. The Test Bank consists of hundreds of multiple-choice questions, while the Instructor's Manual provides suggested answers to selected end-of-chapter exercises.

CHAPTER

1

INTRODUCTION TO GAME THEORY

In this chapter we will:

- *Discuss the underlying rationale and objectives of studying game theory;*
- *Review the history of the development of game theory;*
- *Explore the concept of strategic behavior, and how game theory can be used to analyze situations involving move and countermove;*
- *Introduce the basic vernacular of game theory;*
- *Explore the most basic of game theoretic scenarios—the prisoner's dilemma;*
- *Introduce the concept of a strictly dominant strategy.*

INTRODUCTION

Most of us think of a game as an activity involving two or more players engaged in a recreational parlor activity, such as chess or gin rummy, or a sporting event, such as football or baseball. The objective is to win the game because, as the saying goes, "to the winner go the spoils." Sometimes the spoils are little more than "bragging rights," often symbolized by a memento of some sort, such as a trophy, ring, blue ribbon, or medallion. Sometimes the spoils are monetary in nature, such as winning the pot in a game of poker. Sometimes the winner is rewarded with both cash and trinkets. For example, the owner of the winning team in the Super Bowl is presented with the sterling silver Vince Lombardi Trophy, and each player receives cash and a gold and diamond ring. After winning Super Bowl XXXVIII, for example, each member of the New England Patriots received $68,000, while each player for the losing Carolina Panthers received $36,500. In business, we tend to think of the winning "team" as the firm earning the biggest profits, or capturing the largest market share, or achieving some other desirable objective more successfully than its rivals. However, unlike football games, in which the rivals are fierce competitors, it is often in companies' best interest to cooperate in order to achieve a mutually advantageous outcome.

In general, the game of life is made up of a series of decisions, which affect, and are affected by, the decisions of others. How well we play this game is based on our strategic skills, understanding of the rules, knowledge of our rivals' intentions, ability to predict the repercussions of our actions, and so on. It would be foolish, for example, for a chess player to move without first considering the prior play, and likely countermoves, of an opponent. It would be foolhardy for the owner of a gasoline station to set the price of gasoline without first considering the prices charged by other gasoline station owners in the neighborhood, and the most likely response by rivals to a change in the *status quo.* It would be negligent for a national leader to ignore the possible consequences of a foreign policy that risks war. Strategic interaction, move and countermove, is at the heart of all decision-making processes. Moreover, to be successful we must consider not only the possibility of conflict, but also of cooperation. That is to say, strategic interaction may be either cooperative or noncooperative. In short, **game theory** is the study of how individuals interact in situations

1

involving move and countermove in which the objective is mutually exclusive or mutually beneficial. Game theory is the study of strategic behavior.

> **Game theory** The study of how rivals make decisions in situations involving strategic interaction, i.e., move and countermove.

A SHORT HISTORY OF GAME THEORY

Game theory is the formal study of strategic behavior. Strategic behavior involves any situation in which the decisions of competing individuals or groups are mutually interdependent. One of the earliest attempts to analyze strategic behavior may be attributed to the work of French economist Augustin Cournot (1838), who attempted to explain the interdependent production decisions of duopolies. His analysis of strategic behavior was modified five decades later by Joseph Bertrand (1883), whose work emphasized the role of strategic behavior in product pricing. In the twentieth century, Cournot's and Bertrand's analyses of imperfectly competitive market structures were extended by Edward Chamberlin (1933), Heinrich von Stackelberg (1934), Paul Sweezy (1939), and others. However, these efforts to provide insights into the strategic behavior of oligopolistic firms were handicapped by their reliance on traditional optimization techniques.

In 1921, the French mathematician Émile Borel published several papers on the theory of games. Examining the role of bluffing and second-guessing in poker games with imperfect information, Borel aimed to establish the existence of dominant strategies, which he believed would have military and economic applications. Although he is credited with the first modern formulation of mixed strategies and although he suggested a minimax equilibrium for two-player games with three or five possible strategies, he did not develop his ideas very far. Because of this, most historians credit Hungarian-born mathematician John von Neumann with developing and popularizing game theory.

As a junior faculty member at Princeton University, von Neumann began his work on game theory by attempting mathematically to model the behavior of poker players. In 1928 he published his first paper on game theory in which he proved the minimax theorem and introduced the concept of the extensive form game. Recognizing the importance of game theory to economics, von Neumann teamed up with German-born economist Oskar Morgenstern to develop his ideas more fully. Modern game theory finds its origins in their 1944 classic, *Theory of Games and Economic Behavior.*

Although von Neumann and Morgenstern emphasized the role of cooperation in economic behavior, Princeton mathematician John Forbes Nash believed that strategic behavior between and among economic agents was essentially rivalrous. In his twenty-seven-page doctoral dissertation and several papers published in the early 1950s, Nash demonstrated that noncooperative games with a known end may have what he called a "fixed point" equilibrium. In such games, a rational player will adopt a strategy that is the best response to the strategies adopted by rivals. In his honor, the outcome of such games is usually referred to as a Nash equilibrium. According to Nash, business managers, for example, will cooperate with competitors when it suits them, but will violate implicit or explicit agreements whenever it is in their best interests to do so. The ideas developed by Nash have since become the central focus of game theory. Although written primarily for economists, the applications of game theory to such diverse fields as sociology, politics, warfare, and diplomacy (see, for example, Schelling [1960]) became immediately apparent. Game theory has even been applied to problems in evolutionary biology.

Since the 1970s, game theory has fueled a revolution in economic thought. In the 1990s, game theory captured the public's imagination when it was used to design an auction to allocate licenses for the "narrow-band" frequencies used by mobile telecommunication services such as cellular phones, pagers, and other wireless devices. These auctions both efficiently allocated Federal

Communications Commission licences and generated billions of dollars for U.S. government coffers.

STRATEGIC BEHAVIOR

Strategic behavior involves situations in which the actions of an individual affect, and are affected by, the actions of other individuals. In other words, the decision-making process is mutually interdependent. Much of economic theory is predicated on the assumption that decision makers attempt to achieve the best possible outcome subject to constraints. Consumers, for example, are assumed to allocate their scarce income and wealth to maximize their level of utility or satisfaction. Firms are often assumed to maximize profits subject to a fixed operating budget. In each case, the decisions made by consumers and firms are simplified by assuming that their actions have no measurable effect on the behavior of others. In the real world, however, these situations are much more complicated because they involve strategic behavior. Optimal decisions made by a single consumer or firm may very well depend on the decisions made by other consumers or firms.

Business managers have much to learn from game theory. Managers who are able to put themselves in their competitors' shoes will be in a better position to predict how their rivals will react to changes in market strategy. Although game theory is not a cookbook that provides a recipe for every strategic contingency, it does suggest an approach to analyzing possible future outcomes in situations involving move and countermove.

> **Strategic behavior** When the actions of an individual or group affect, and are affected by, the actions of other individuals or groups.

In many respects, running a business is like playing a game of football or chess. The object of the game is to achieve an optimal outcome. But unlike these games, the "best" outcome does not always mean that your opponent loses. In many cases, the best outcome often results from cooperation among players. When cooperation is impossible or illegal, the objective is to win the game. But, victory does not always go to the strongest, fastest, or most talented. Victory often belongs to the player who best understands the rules of the game and has a winning strategy. The purpose of this book is to introduce the reader to the basic principles of game theory, and to demonstrate what it takes to make correct decisions in situations involving strategic behavior.

CASE STUDY 1.1: GLAXO AND ZANTAC

GlaxoSmithKlein, which began in Great Britain as Glaxo Holdings PLC, is the world's largest pharmaceutical company. Its global reach includes more than forty manufacturing facilities worldwide. The company's transformation from a moderately sized health care conglomerate into a fast-growing pharmaceutical colossus stemmed largely from its development of Zantac. Zantac represented a major competitive challenge to SmithKline's Tagamet, which was then the preeminent anti-ulcer medication and the world's best-selling drug.

Launched successfully in Europe in the late 1970s, Zantac was shortly thereafter introduced to the U.S. market. By 1984, Zantac had captured 25 percent of the new prescription market. By the end of the decade, Zantac controlled more than half of the global market for anti-ulcer remedies, and was the largest-selling prescription drug in the world. From 1980 to 1988, Glaxo's sales nearly tripled, with total sales amounting to nearly $2 billion. By 1989, Glaxo had become the second-largest pharmaceutical company in the world.

Glaxo's tremendous success can be explained both by market power and by a unique marketing strategy involving a series of strategic partnerships with drug companies around the world. Market power is the ability of a firm to set the price of its product above its marginal cost of production. It cost the company relatively little to produce Zantac, but its price margin and monopoly profits were very high. Glaxo's initial ability to charge a high price for Zantac resulted from patent protection, without which prices would have been driven down to more competitive levels.

Over the next few years, however, Glaxo's hold on the market began to weaken. A wave of drug company mergers in 1989 left the company fourth in worldwide sales. Zantac's dominance of the ulcer treatment market ended in April 1997 when the U.S. division of Novopharm announced its intention to market a generic version of Zantac in the United States and Puerto Rico. The announcement followed a federal court ruling that Novopharm could produce the drug without infringing on one of Glaxo Wellcome's two patents, which was set to expire the following July. In July 1999, the discount drug seller RxUSA sold a 30-tablet, 250-mg box of Zantac for $85.95, and a 300-tablet, 300-mg generic version for $95. In addition to competition from Novopharm, clinical trials demonstrated little difference in the treatment success rates of Zantac and several of its close substitutes, including SmithKline's Tagamet. Additionally, industry analysts doubted Glaxo's ability to maintain its lead in new drug introductions.

Surprisingly, despite the growing number of competitors, Glaxo's prices and profits were not significantly affected. There are a number of possible explanations. In 1990, Glaxo announced worldwide regulatory trials for its new anti-migraine drug, Imigran (sumatriptan). Moreover, in early 1991 the United States Food and Drug Administration approved Zofran Injection, an important new treatment for the prevention of nausea and vomiting in cancer patients receiving chemotherapy. In just over a year, the drug was available in most of the world's markets and registered sales of more than $400 million.

Glaxo also tried to capture some of the entrepreneurial energy of smaller companies through a series of joint ventures. In 1991, in exchange for about $20 million, Glaxo purchased an equity stake in Gilead Sciences, Inc., a company with the potential for creating anti-cancer drugs. The company also sold its interests outside of prescription drugs and increased its research and development allocations. One such development was ceftazidime, an injectable antibiotic that received a strong market reception in Japan.

Another factor that explains the absence of more intense price competition is the creation of production synergies arising from Glaxo's 1995 merger with Burroughs Wellcome & Company around the time that it introduced Zantac. The merger created important economies of scale by linking similar avenues of research. Even more importantly, Glaxo Wellcome announced its intention in January 2000 to merge with SmithKline Beecham. Today, GlaxoSmithKline annual sales exceed $40 billion, of which Zantac's sales constitute about $3 billion.

The Glaxo saga underscores several issues that concern game theorists. Its ability to exercise market power to earn above-normal rates of return on its Zantac investment stemmed not only from patent protection, but from a research and development strategy that allowed it to enter markets that were dominated by SmithKline. Moreover, Glaxo's aggressive marketing strategy in the early 1980s gave the company a competitive edge in several overseas markets. Its advertising strategy successfully differentiated the company's product from those of a growing number of competitors. Finally, Glaxo's merger strategy allowed the company to increase its market share and enhance its ability to price above marginal cost.

WHAT IS GAME THEORY?

Game theory is the formal study of how rivals make decisions in situations involving strategic interaction. A game is any situation in which the final outcome depends on decision makers' strategic choices. In game theory, these decision makers are referred to as **players**, which may be individuals, companies, common interest groups, or any combination thereof. In game theory, players are assumed to behave rationally. By **rational behavior** we mean that players endeavor to optimize their outcomes, which are referred to as **payoffs**. Sometimes the objective is to maximize these payoffs, as in the case of profit-maximizing firms and utility-maximizing consumers. At other times, the objective is to minimize payoffs, such as in the prisoner's dilemma (discussed below), in which the players' objective is to spend as little time in jail as possible.

> **Player** A decision maker in a game.
>
> **Rational behavior** When players endeavor to optimize their payoffs.
>
> **Payoff** The gain or loss to a player at the conclusion of a game.

A number of elements are common to all games. To begin with, all games have rules defining the order of play (i.e., the sequence of moves by each player). Players' moves are based on strategies. A **strategy** defines a player's moves in situations where choices need to be made. In a one-time game (that is, a game involving only one move), a player's strategy and move are synonymous. In a game involving multiple moves, a strategy is a player's game plan. It is a contingency plan of action based upon the behavior of rivals. In game theory, we often refer to such contingency plans as a **pure strategy**. A pure strategy is a complete and nonrandom game plan. Knowledge of a player's strategy allows us to predict his or her next move when confronted with choices. By contrast, a strategy may be a game plan that involves randomly mixing different strategies. A **mixed strategy** is a game plan that involves randomly mixing pure strategies.

> **Strategy** A decision rule that defines a player's moves. It is a complete description of a player's decisions at each stage of a game.
>
> **Pure strategy** A complete and nonrandom game plan.
>
> **Mixed strategy** A game plan that involves randomly mixing pure strategies.

The collection of all players' strategies is called a **strategy profile**. In this text, strategy profiles will be summarized within curly brackets. For example, suppose that in a one-time game the managers of two rival firms, A and B, are considering a *high price* or a *low price* strategy. This game has four possible strategy profiles: {*High price, High price*}, {*High price, Low price*}, {*Low price, High price*}, and {*Low price, Low price*}. The first entry within the curly brackets is the strategy adopted by firm A and the second entry is the strategy adopted by firm B. Each strategy profile defines the outcome of the game and the *payoffs* to each player. In this text, the payoffs to each player will be summarized in parentheses. For example, suppose that each firm earns a profit of $1 million by adopting a *high price strategy.* The payoffs to firm A and firm B would be written as ($1 million, $1 million).

> **Strategy profile** The collection of all players' strategies.

There are two basic types of games: **simultaneous-move** and **sequential-move games**. In simultaneous-move games, also referred to as **static games**, players move at the same time. The distinguishing characteristic of a static game is that neither player is aware of the decisions of the other players until all moves have been made. In a two-player game, player *A* is unaware of the decisions of player *B,* and vice versa, until both have moved.

> **Simultaneous-move game** A game in which the players move at the same time.

> **Sequential-move game** A game in which the players take turns.

> **Static game** A game in which the players are ignorant of their rivals' decisions until all moves have been made. A simultaneous-move game is an example of a static game.

An example of a simultaneous-move game is the children's game "rock, paper, scissors." In this game, two players in unison recite the words "rock, paper, scissors" before simultaneously showing either a rock (a closed fist), paper (an open hand), or scissors (a separated index and middle finger). The outcome of this game depends on the combination of moves. If one player shows rock and the other player shows scissors, rock wins because "rock breaks scissors." If one player shows rock and the other player shows paper, paper wins because "paper covers rock." Finally, if one player shows scissors and the other player shows paper, scissors wins because "scissors cut paper." If both players make the same move, the game results in a tie. The important thing about this game is that neither player is aware of the other's intentions until both players have moved.

In a static (simultaneous-move) game the players are not actually required to move at the same time. "Rock, paper, scissors" could also be played, for example, by isolating the players in separate rooms to prevent communications between them. A third individual, whom we will call the referee, asks each player to privately reveal his or her move, after which a winner is declared. The essential element of this game is that each player must move without prior knowledge of the move made by the other player.

In a sequential-move game, on the other hand, the players take turns. Sequential-move games are sometimes referred to as **multistage** or **dynamic games**. In a two-player game, player *A* moves first, followed by player *B,* followed again by player *A,* and so on. Unlike a simultaneous-move game, player *B*'s move is based upon the knowledge of how player *A* has already moved. Moreover, player *A*'s next move will be based on the knowledge of how player *B* moved in response to player *A*'s last move, and so on. Examples of sequential-move games include board games such as chess, checkers, and Monopoly. The models developed by Cournot, Bertrand, and Stackelberg attempted to model the strategic interaction between duopolies as sequential-move games.

> **Dynamic (multistage) game** The same thing as a sequential-move game.

In addition to rules, games are also defined by the number of times the game is played. **One-time games**, for example, are played only once. **Repeated games** are played more than once. If you and a friend agree to play just one game of backgammon, you are playing a one-time game. If, on the other hand, the winner is the player who wins two out of three games, you are playing a repeated game.

> **One-time game** A game that is played just once.

> **Repeated game** A game that is played more than once.

Game theory is perhaps the most important tool in the economist's analytical kit for analyzing strategic behavior. Game theory represents a significant improvement over earlier attempts to model strategic interaction because it can be applied to the behavior of firms in any competitive environment.

THE PRISONER'S DILEMMA

The basic elements of a game may be illustrated with what is perhaps the best known of all game theoretic scenarios—the **prisoner's dilemma**. Surprisingly, this important contribution to the development of game theory did not first appear in a scholarly journal, but in the classroom. In 1950, the well-known mathematician Albert J. Tucker was a visiting professor at Stanford University. While attempting to illustrate to a group of psychologists the difficulty of analyzing certain kinds of games, Tucker developed the prisoner's dilemma. S. J. Hagenmayer (1995) wrote that "Mr. Tucker's simple explanation has since given rise to a vast body of literature in subjects as diverse as philosophy, ethics, biology, sociology, political science, economics, and, of course, game theory."

The prisoner's dilemma is an example of a two-player, noncooperative, simultaneous-move, one-time game in which both players have a **strictly dominant strategy**. A player has a strictly dominant strategy if it results in the best payoff regardless of the strategies adopted by other players. This game is noncooperative in the sense that both players are unable or unwilling to cooperate to achieve a mutually advantageous outcome.

> **Strictly dominant strategy** A strategy that strictly dominates every other strategy. It is a strategy that results in the best payoff given the strategies adopted by the other players.

To illustrate the prisoner's dilemma, consider the following situation, which is described in Dixit and Nalebuff (1991), Schotter (1985), Luce and Raiffa (1957), and elsewhere. Two individuals are taken into custody by the police following a robbery, but after the thieves have disposed of the booty. Although the police believe them to be guilty, the authorities do not have enough evidence to convict them of the crime. In an effort to extract a confession, both suspects are taken to separate rooms and interrogated. If neither suspect confesses, the only thing that either suspect can be convicted of is loitering at the scene of the crime, which carries a penalty of six months in jail. On the other hand, if one suspect confesses and turns state's evidence against the other suspect, that suspect will go free by a grant of immunity, while the other suspect receives ten years in jail. Finally, if both suspects confess they will be convicted, but because each has agreed to cooperate the penalty is five years of jail time on the lesser charge of "breaking and entering."

The entries in the cells of the payoff matrix in Figure 1.1 summarize the jail time (in years) for each suspect from each combination of strategies. The payoffs are negative to highlight the fact that more prison time results in greater unhappiness (disutility). We will adopt the convention that the first entry in the parentheses refers to the payoff to the player identified on the left-side of the payoff matrix, while the second entry refers to the payoff to the player identified at the top of the payoff matrix. The setup depicted in Figure 1.1 is also called a **normal-form game**.

> **Normal-form game** Summarizes the players and payoffs from alternative strategies in a static game.

The game depicted in Figure 1.1 has four possible strategy profiles: {*Confess, Confess*}, {*Confess, Silent*}, {*Silent, Confess*}, {*Silent, Silent*}. In the situation depicted in Figure 1.1, the worst outcome is for a suspect not to confess (*Silent*) while the other suspect confesses (*Confess*). To see this, consider the lower, left-hand cell of the payoff matrix, which represents the decision by suspect *A* to confess and the decision by suspect *B* not to confess. The result of the strategy profile {*Confess, Silent*} is that suspect *A* is set free, while suspect *B* goes to jail for ten years. Since the payoff matrix is symmetric, the strategy profile {*Silent, Confess*} results in the opposite outcome.

FIGURE 1.1

Prisoner's Dilemma

Suspect B

		Silent	Confess
Suspect A	*Silent*	(−½, −½)	(−10, 0)
	Confess	(0, −10)	**(−5, −5)**

Payoffs in years: (Suspect *A*, Suspect *B*)

> **Prisoner's dilemma** A game in which it is in the best interest of all players to cooperate, but where each player has an incentive to adopt his or her dominant strategy.

It should be remembered that the prisoner's dilemma is an example of a noncooperative game. Neither suspect will know the other's intentions until after all moves have been made. Since both suspects are being held incommunicado, they are unable to cooperate. Under the circumstances, if both suspects are rational, the decision of each suspect (i.e., the move that results in the largest payoff) will be to confess. Why? Consider the problem from suspect *A*'s perspective. If suspect *B* remains silent, it is in suspect *A*'s best interest to confess since this will result in no jail time as opposed to six months in jail by not confessing. On the other hand, if suspect *B* confesses, it is once again suspect *A*'s best move to confess, since this would result in five years in jail, compared with ten years by not confessing. In other words, suspect *A*'s best strategy is to confess, regardless of the strategy adopted by suspect *B*. Since the payoff matrix is symmetrical, the same is also true for suspect *B*.

The outcome of the prisoner's dilemma is no different even if both suspects agree before being separated not to confess. Although the suspects know that they will receive only six months jail time if they both remain silent, each has a powerful incentive to double-cross the other to get a better deal. This is especially true since each suspect will begin to question the fidelity of the other. A large number of experiments have confirmed this prediction. In experiments conducted by Cooper, DeJong, Forsythe and Ross (1996), test subjects were asked to play the prisoner's dilemma twenty times against different anonymous rivals. As the players gained experience, the strategy profiles of the players converged toward the predicted Nash equilibrium. In the first five rounds, for example, rivals cooperated 43 percent of the time. This cooperation fell to just 20 percent in the last five rounds. The strictly dominant strategy for both suspects is to confess. The **strictly dominant strategy equilibrium** for this game is the strategy profile {*Confess, Confess*}. The payoffs for the equilibrium depicted in Figure 1.1 are highlighted in boldface. In this case, both suspects will receive five years in jail.

> **Strictly dominant strategy equilibrium** A Nash equilibrium that results when each player has, and adopts, a strictly dominant strategy.

The foregoing strategy profile is called a **Nash equilibrium** in honor of John Forbes Nash, Jr., who along with John Harsanyi and Reinhard Selten were awarded the 1994 Nobel Prize in economics for their pioneering work in game theory. A noncooperative game has a Nash equilibrium when all players adopt a strategy that is the best response to the strategies adopted by the rival players. When both players have a strictly dominant strategy, neither player can improve his or her payoff by changing strategies.

Nash equilibrium When each player adopts a strategy that is the best response to the strategies adopted by the rivals. A strategy profile is a Nash equilibrium when no player can improve his or her payoff by switching strategies.

Nash created quite a stir in the economics profession when he proposed the idea of a "fixed-point equilibrium" in 1950. This concept seemed to contradict Adam Smith's famous metaphor of the "invisible hand," which asserted that the welfare of society as a whole is maximized when each individual pursues his or her own private interests. According to the situation depicted in Figure 1.1, it is clearly in the best interest of both suspects to adopt the joint strategy of not confessing. This would result in a strategy profile in which the suspects spend only six months in jail.

The concept of a Nash equilibrium is powerful because it exists for all noncooperative games, although this may not immediately appear to be the case. Games involving pure (nonrandom) strategies may not always have a Nash equilibrium, but when transformed into games involving mixed (probabilistic) strategies, they always do. Unfortunately, while all noncooperative games involving pure (nonrandom) strategies have a Nash equilibrium, these outcomes may not be unique. Many such games have multiple Nash equilibria, in which case the problem becomes choosing the most plausible outcome. Another shortcoming is that the definition is silent about how players go about choosing an optimal strategy. In spite of these shortcomings, the concept of a Nash equilibrium is virtually unchallenged as a solution concept to situations involving noncooperative strategic behavior.

The prisoner's dilemma provides important insights into the strategic behavior of oligopolists. To see this, consider an industry consisting of just two firms (duopoly). Suppose that firms A and B are confronted with the decision of charging a high or low price for their product. Each firm recognizes that its payoff depends on the simultaneous decision of its rival. If firm A charges a high price and firm B charges a low price, for example, firm B will increase its market share at firm A's expense, and vice versa. On the other hand, if the two firms cooperate, they can behave as a profit-maximizing monopolist and both would benefit. It is possible to model the strategic behavior of these two firms as a two-player, noncooperative, simultaneous-move, one-time game. Suppose that the alternatives facing each firm are summarized in the pricing game depicted in Figure 1.2. The entries in each cell represent the expected profits from each combination of *high price* and *low price* strategies.

We may begin by asking whether either firm has a strictly dominant strategy. To answer this question, consider the problem from the perspective of firm B. If firm A charges a high price, it will be in firm B's best interest to charge a low price. Why? If firm B adopts a *high price* strategy then it will earn a profit of $1 million, compared with a profit of $5 million by adopting a *low price* strategy. On the other hand, if firm A charges a low price, firm B will earn a profit of $100,000 if it charges a high price and $250,000 if it charges a low price. Regardless of the strategy adopted by firm A, it will be in firm B's best interest to charge a low price. Thus, firm B's strictly dominant strategy is to charge a low price. What about firm A? Since the entries in the payoff matrix are symmetrical, we would expect a similar outcome. If firm B charges a high price, it will be in firm A's best interest to adopt a *low price* strategy since it will earn a profit of $5 million, compared with a profit of only $1 million by adopting a *high price* strategy. If firm B charges a low price, it will again be in firm A's best interest to charge a low price and earn a profit of $250,000, compared with a profit of only $100,000 by charging a high price. Thus, firm A's strictly dominant strategy is to charge a low price as well. Thus, in this noncooperative game where the pricing decision of one firm is independent of the pricing decision of the other firm, it pays for both firms to charge a low price, with each firm earning a profit of $250,000. The strategy profile {*Low price, Low price*} is a strictly dominant strategy equilibrium. In Figure 1.2 the payoffs for this strategy profile are highlighted in boldface.

FIGURE 1.2
Pricing Game

Firm B

		High price	Low price
Firm A	High price	($1 million, $1 million)	($100,000, $5 million)
	Low price	($5 million, $100,000)	**($250,000, $250,000)**

Payoffs: (Firm *A*, Firm *B*)

The reader should note that the strategy profile {*Low price, Low price*} in Figure 1.2 is a Nash equilibrium because each firm adopted a strategy that it believed was the best response to the strategy adopted by the other firm. Moreover, in this strictly dominant strategy equilibrium neither firm can improve its payoff by switching strategies. The interesting thing here is that the pursuit of individual self-interest results in a less-than-optimal outcome for both players. If both firms cooperate and charge a high price, each firm will earn profits of $1 million. But, cooperation will not work precisely because the strategy profile {*High price, High price*} is not a Nash equilibrium. Both firms can improve their payoffs by switching strategies. That is, firm *A* can earn profits of $5 million by charging a low price, provided that firm *B* continues to charge a high price. The same is true for firm *B*. The implications for such problems as environmental pollution, traffic congestion, and so on, suggest that society as a whole may be made worse off if the individual economic agents do not cooperate to achieve a mutually advantageous outcome.

Historically, formal collusive arrangements among firms, also known as cartels, have proven to be highly unstable. In such arrangements, the incentive for cartel members to cheat is strong. Even if such an arrangement is legal, distrust of each member' motives and intentions will compel each to violate the agreement. Economic history is replete with examples of cartels that have disintegrated because of the lure of ill-gotten gains at the expense of the other cartel members. For such cartel arrangements to endure, it must be possible to enforce the agreement by effectively penalizing cheaters. The conditions under which cheating is likely to occur will be discussed in Chapters 4 and 5.

———————————————— **Demonstration Problem 1.1** ————————————————

Why do fast-food restaurants tend to cluster in the immediate vicinity of each other? Consider the following situation involving the owners of two fast-food franchises, Burger Queen and Wally's, which are kitty-corner from each other in downtown Hashbrowntown. Route 795 was recently extended from Baconsville to Hashbrowntown. The Hashbrowntown exit ramp off Route 795 is five miles from the center of town. Although both restaurants are operating profitably, the owners of these franchises are considering relocating near the exit ramp. The restaurant owners have calculated that by relocating they will continue to receive some in-town business, but will also receive highway customers who use the exit as a rest stop. The payoff matrix for this game is illustrated in Figure D1.1. The first entry in each cell of the payoff matrix refers to the payoff to Wally's and the second entry refers to the payoff to Burger Queen.

a. Do the franchise owners have a strictly dominant strategy?
b. Is this strategy profile a Nash equilibrium?

FIGURE D1.1
Normal Form of the Static Game in Demonstration Problem 1.1

Burger Queen

Wally's		Exit ramp	Downtown
	Exit ramp	($150,000, $150,000)	($1 million, $100,000)
	Downtown	($100,000, $1 million)	($250,000, $250,000)

Payoffs: (Wally's, Burger Queen)

Solution

a. Both franchise owners have a strictly dominant strategy to relocate near the exit ramp. Consider the problem from Burger Queen's perspective. If Wally's relocates near the exit ramp, it will be in Burger Queen's best interest to relocate there as well since the payoff of $150,000 is greater than the alternative of $100,000 by remaining downtown. If Wally's decides to remain downtown, it will again be in Burger Queen's best interest to relocate near the exit ramp since the payoff of $1 million is greater than $250,000 by remaining downtown. Thus, Burger Queen's strictly dominant strategy is to locate near the exit ramp. Since the entries in the payoff matrix are symmetrical, the same must also be true for Wally's. Thus, the strictly dominant strategy equilibrium for this game is {*Exit ramp, Exit ramp*}.

b. Note that the optimal strategy for both franchise owners is to agree to remain downtown since the payoff to both fast-food restaurants will be greater. But, this would require cooperation between Burger Queen and Wally's. If collusive behavior is ruled out, the strictly dominant strategy equilibrium {*Exit ramp, Exit ramp*} is a Nash equilibrium since neither franchise can improve its payoff by choosing a different strategy.

SHORTCUT FOR FINDING PURE-STRATEGY NASH EQUILIBRIA

At this point the student may be somewhat unsettled by the rather tortuous approach used for finding Nash equilibria for prisoner's dilemma–type games. In the preceding section our search involved the systematic examination of the payoffs for each strategy profile one by one. In fact, there is an easier and more efficient way to facilitate our search for a Nash equilibrium for static games. This approach is illustrated in Figures 1.3 to 1.6, which replicates the pricing game depicted in Figure 1.2. As before, we begin by putting ourselves into the shoes of firm *A* and asking what is the best response to the strategy adopted by firm *B*. If firm *B* charges a high price, the best response is to charge a *low price* because it results in the higher payoff of $5 million. The student should underline or circle this payoff, which is shown as the highlighted underlined payoff in Figure 1.3.

The next step is to identify the best response if firm *B* adopts a low price strategy. Once again, the best response is to charge a low price because of the higher payoff, which is $250,000. The student should underline or circle this payoff. This is illustrated by the highlighted underlined payoff in Figure 1.4.

Now examine the game from the perspective of firm *B*. If firm *A* adopts a *high price* strategy, the best response by firm *B* is a *low price* strategy because of the higher payoff of $5 million. Underline or circle this payoff. This is illustrated by the highlighted underlined payoff in Figure 1.5. Finally, if firm *A* adopts a *low price* strategy the best response by firm *B* is to adopt a *low price* strategy with a payoff of $250,000. This choice is illustrated by the highlighted underlined

FIGURE 1.3
First Step in Finding a Nash Equilibrium for the Pricing Game

Firm B

		High price	Low price
Firm A	High price	($1 million, $1 million)	($100,000, $5 million)
	Low price	(**$5 million**, $100,000)	($250,000, $250,000)

Payoffs: (Firm *A*, Firm *B*)

FIGURE 1.4
Second Step in Finding a Nash Equilibrium for the Pricing Game

Firm B

		High price	Low price
Firm A	High price	($1 million, $1 million)	($100,000, $5 million)
	Low price	($5 million, $100,000)	(**$250,000**, $250,000)

Payoffs: (Firm *A*, Firm *B*)

FIGURE 1.5
Third Step in Finding a Nash Equilibrium for the Pricing Game

Firm B

		High price	Low price
Firm A	High price	($1 million, $1 million)	($100,000, **$5 million**)
	Low price	($5 million, $100,000)	($250,000, $250,000)

Payoffs: (Firm *A*, Firm *B*)

FIGURE 1.6
Final Step in Finding a Nash Equilibrium for the Pricing Game

Firm B

		High price	Low price
Firm A	High price	($1 million, $1 million)	($100,000, **$5 million**)
	Low price	($5 million, $100,000)	($250,000, **$250,000**)

Payoffs: (Firm *A*, Firm *B*)

payoff in Figure 1.6. To determine the pure-strategy profile that constitutes a Nash equilibrium, simply look for the cell in the payoff matrix in which both payoffs are underlined. Thus, the "solution" to the game depicted in Figure 1.2 is the strategy profile {*Low price, Low price*}. In the remainder of this text, we will dispense with the underlines and denote the players' choice of moves by highlighting the payoffs.

CHAPTER REVIEW

Game theory is the study of how rivals make decisions in situations involving strategic behavior. Strategic behavior is concerned with how individuals and groups make decisions when they recognize that their actions affect, and are affected by, the actions of other individuals or groups. The rules of the game dictate the manner in which the players move.

Players make moves that are based on strategies. A *strategy* is a game plan that defines a player's moves in situations involving choices. In a one-time game, a player's strategy and *move* are synonymous. In games involving multiple moves, a strategy is a complete description of a player's moves. A *pure strategy* is a complete and nonrandom game plan. A strategy may involve randomly mixing different strategies. Such strategies are referred to as *mixed strategies.*

In *simultaneous-move games,* also referred to as *static games,* players move at the same time. In a *sequential-move game,* the players take turns. Sequential-move games are also called *multistage* or *dynamic games.*

The prisoner's dilemma is an example of a *two-player, noncooperative, simultaneous-move, one-time game* in which both players have a *strictly dominant strategy.* Dominating all other strategies, a strictly dominant strategy results in the best payoff regardless of the strategies adopted by other players. A *Nash equilibrium* occurs in a noncooperative game when each player adopts a strategy that is the best response to the strategies adopted by other players. When a game has a Nash equilibrium, neither player can improve his or her payoff by changing strategies.

It is sometimes possible for players to improve their combined payoffs by coordinating their strategies. Such coordinated behavior among firms is called *collusion.* Historically, formal collusive arrangements, which are called *cartels,* have proven to be highly unstable since there is a strong incentive for members to cheat. For a cartel to endure, it must be possible to enforce the cooperative agreement among members by effectively penalizing cheaters.

CHAPTER QUESTIONS

1.1   What is meant by strategic behavior?

1.2   What is the difference between a player's strategy and a player's move?

1.3   In the card game "war," a standard deck of cards is shuffled and equally divided between two players. The cards are valued from lowest to highest 2–10, jack, queen, king, ace. Suits (clubs, diamonds, hearts, and spades) in this game are irrelevant. The players then recite the phrase "w-a-r spells war." As each letter in the word "war" is recited, each player places a card face down on the table. When the word "war" is recited, each player turns a card face up on the table. The player who shows the highest valued card wins the other players' cards. If two players show the same card, the process is repeated until a player wins. The game ends when both players have gone through the deck. The player with the greatest number of cards at the end of the game wins.

   An alternative way to play this game is to isolate the players in separate rooms, prohibiting communication between them. A third individual, the referee, goes to each room and asks each player to reveal his or her "war" card. After examining each player's "war" card the referee declares a winner.

   Both versions of this game may be called simultaneous-move games. Do you agree? If not, why not?

1.4   What is a strictly dominant strategy?

1.5   What is a strictly dominant strategy equilibrium?

1.6   A strictly dominant strategy equilibrium is not the same thing as a Nash equilibrium. Do you agree?

1.7   What is a Nash equilibrium?

1.8   The prisoner's dilemma is an example of a one-time, two-player, simultaneous-move, non-cooperative game. If the players are allowed to cooperate, a Nash equilibrium is no longer possible. Do you agree with this statement? If not, why not?

CHAPTER EXERCISES

1.1   In the country of Arcadia two equally sized companies, Auburn Motorcar Company and Cord Automobile Corporation, dominate the domestic automobile market. Each company can produce 500 or 750 midsized automobiles a month. This static game is depicted in Figure E1.1. The first entry in each cell of the payoff matrix represents the payoff to Auburn and the second entry represents the payoff to Cord. All payoffs are in millions of dollars.

    a.   Does either firm have a dominant strategy?
    b.   What is the Nash equilibrium strategy profile for this game?

1.2   Consider the game depicted in Figure E1.2 in which players Fred and Ethel have a choice of a Lucy or Desi strategy.

    a.   Does Fred or Ethel have a strictly dominant strategy?
    b.   What is the Nash equilibrium strategy profile for this game?

1.3   Suppose that Tsunami Corporation and Cyclone Company are considering a change in their pricing policies. The payoffs for this static game are depicted in Figure E1.3. The first entry in each cell of the payoff matrix refers to the payoff to Tsunami and the second entry refers to the payoff to Cyclone. All payoffs are in millions of dollars.

FIGURE E1.1
Normal Form of the Static Game in Chapter Exercise 1.1

		Cord	
		750 cars a month	*500 cars a month*
Auburn	*750 cars a month*	(5, 5)	(3, 6)
	500 cars a month	(6, 3)	(4, 4)

Payoffs: (Auburn, Cord)

FIGURE E1.2
Normal Form of the Static Game in Chapter Exercise 1.2

		Ethel	
		Lucy	*Desi*
Fred	*Lucy*	(4, 4)	(6, 1)
	Desi	(3, 3)	(5, 5)

Payoffs: (Fred, Ethel)

FIGURE E1.3

Normal Form of the Static Game in Chapter Exercise 1.3

		Cyclone	
		High price	*Low price*
Tsunami	*High price*	(6, 6)	(4, 8)
	Low price	(8, 4)	(5.5, 5.5)

Payoffs: (Tsunami, Cyclone)

FIGURE E1.4

Normal Form of the Static Game in Chapter Exercise 1.4

		Player B	
		Heads	*Tails*
Player A	*Yes*	(5, 10)	(7, 8)
	No	(−3, 3)	(4, −4)

Payoffs: (Player A, Player B)

FIGURE E1.5

Normal Form of the Static Game in Chapter Exercise 1.5

		Newsweek	
		CIA	*Terrorism*
Time	*CIA*	(10, 10)	(20, 80)
	Terrorism	(80, 20)	(40, 40)

Payoffs: (Time, Newsweek)

a. Does either firm have a strictly dominant strategy?
b. What strategy will be adopted by each firm?
c. Does this game have a Nash equilibrium strategy profile?

1.4 Consider the two-player, noncooperative, simultaneous-move, one-time game depicted in Figure E1.4. In this game, larger payoffs are preferred.

a. Does either player have a strictly dominant strategy?
b. Does this game have a Nash equilibrium strategy profile?

1.5 Suppose that the editors of *Time* and *Newsweek* are trying to decide upon their cover story for the week. There were two major news developments: The leaking of a clandestine CIA officer's identity and a terrorist suicide bombing in Amman, Jordan. It is important that the news weeklies not choose the same cover story since some readers will buy only one magazine when they might otherwise buy both. The choice of cover story depends on newsstand sales. The editors believe that 20 percent of newsstand buyers are interested in the CIA story, while 80 percent are interested in the terrorist bombing. If both news magazines run the same cover story, they will evenly share the market, but if they run different cover stories, each will get that share of the market. This static game is depicted in Figure E1.5.

    a. Does either news weekly have a strictly dominant strategy?

    b. What is the Nash equilibrium strategy profile for this game?

1.6 Determine the pure-strategy Nash equilibrium for the static game in Figure E1.6.

FIGURE E1.6

Normal Form of the Static Game in Chapter Exercise 1.6

Guildenstern

		Left	*Center*	*Right*
Rosencrantz	*Left*	(5, 4)	(6, 1)	(7, 3)
	Center	(3, 2)	(8, 5)	(4, 7)
	Right	(4, 1)	(10, 7)	(1, 9)

Payoffs: (Rosencrantz, Guildenstern)

APPENDIX: MATHEMATICS REVIEW

INTRODUCTION

Most of the concepts presented in this book will be illustrated using simple diagrams and arithmetic examples. In some instances, however, a somewhat higher level of mathematical proficiency is required. The purpose of this appendix is to review these mathematical techniques. This review, however, is not exhaustive. For example, our review of simple calculus will only consider the power-function, sums-and-differences, and product rules of differentiation. We will not consider the quotient and chain rules since they will not be needed in the text.

FUNCTIONAL RELATIONSHIPS

In mathematics, a functional relationship of the form

$$y = f(x) \tag{1A.1}$$

is read "y is a function of x." This relationship indicates that the value of y depends in some systematic way on the value of x. The variable y is referred to as the *dependent variable*. The variable x is referred to as the *independent variable*.

    The functional notation f in Equation (1A.1) can be regarded as a specific rule that defines the relationship between the values of x and y. When we assert, for example, that $y = f(x) = 3x$ the actual rule is made explicit. This rule, for example, asserts that when $x = 2$, $y = f(2) = 6$, and so on.

    The value of y may also be expressed as a function of more than one independent variable, that is:

$$y = f(x_1, ..., x_n) \tag{1A.2}$$

Suppose, for example, that $y = f(x_1, x_2) = 5x_1 + 2x_2$. This rule asserts that when $x_1 = 2$ and $x_2 = 3$, $y = f(2, 3) = 3(2) + 2(3) = 12$.

TABLE 1A.1
Total Revenue for Selected Output Levels

Q	TR
0	0
1	5
2	10
3	15
4	20
5	25
6	30

METHODS OF EXPRESSING ECONOMIC AND BUSINESS RELATIONSHIPS

Economic and business relationships may be represented in a variety of ways, including tables, charts, graphs, and algebraic expressions. Consider, for example, Equation (1A.3) which summarizes total revenue (*TR*) for a firm operating in a perfectly competitive industry.

$$TR = f(Q) = P_0 Q \qquad (1A.3)$$

In Equation (1A.3), P_0 represents the constant market determined selling price of commodity Q produced and sold during a given period of time. Suppose that the selling price is $5. The firm's total revenue equation is:

$$TR = 5Q \qquad (1A.4)$$

Total revenue for selected output levels may be expressed in tabular form, such as in Table 1A.1, or diagrammatically as in Figure 1A.1.

The total revenue (*TR*) in Figure 1A.3 illustrates the general class of mathematical relationships called linear functions, which may be written in the general form:

$$y = f(x) = a + bx \qquad (1A.5)$$

where a and b are known constants. The value a is called the *intercept* and the value b is called the *slope.* Mathematically, the intercept is the value of y when $x = 0$. Equation (1A.5) is said to be linear in x and y because the corresponding graph is a straight line. From Equation (1A.4), $a = 0$ and $b = 18$.

THE SLOPE OF A LINEAR FUNCTION

In Equation (1A.5), the slope b is defined as the change in the value of the dependent variable (the "rise") given a change in the value of the independent variable (the "run"). The value of the slope may be calculated using the equation:

$$\text{Slope} = \frac{\Delta y}{\Delta x} = \frac{y_2 - y_1}{x_2 - x_1} = \frac{f(x_2) - f(x_1)}{x_2 - x_1} = b \qquad (1A.6)$$

FIGURE 1A.1
Constant Selling Price and Linear Total Revenue

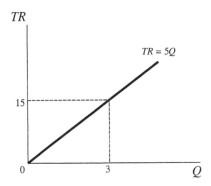

The symbol Δ denotes "change." In terms of the total revenue example above, consider the quantity-total revenue combinations $(Q_1, TR_1) = (1, 5)$ and $(Q_2, TR_2) = (3, 15)$. Since Equation (1A.4) is a straight line, any two coordinate points on the corresponding graph may be used to calculate the slope. In this case, the slope of Equation (1A.4) is:

$$b = \frac{\Delta TR}{\Delta Q} = \frac{TR_2 - TR_1}{Q_2 - Q_1} = \frac{f(Q_2) - f(Q_1)}{Q_2 - Q_1} \tag{1A.7}$$

Substituting the above coordinate values into Equation (1A.6) we obtain:

$$b = \frac{15 - 5}{3 - 1} = \frac{10}{2} = 5 = P_0 \tag{1A.8}$$

In this case, the slope of Equation (1A.4) is equal to the product price. Rearranging Equation (1A.6) we get:

$$b(x_2 - x_1) = (y_2 - y_1) \tag{1A.9}$$

Rearranging Equation (1A.9) this becomes:

$$y_2 = (y_1 - bx_1) + bx_2 \tag{1A.10}$$

Equation (1A.10) says that y_1 and y_2 are solutions to Equation (1A.5) and $x_1 \neq x_2$. If we are given specific values for y_1, x_1, and b, Equation (1A.10) reduces to Equation (1A.5) where $y = y_2$, $x = x_2$, and $a = y - bx$. Equation (1A.10) is referred to as the *slope-intercept form* of the linear equation.

AN APPLICATION OF LINEAR FUNCTIONS IN ECONOMICS

In economics, tables, graphs, and equations are often used to explain business and economic relationships. Such models are abstractions, and as such often appear unrealistic. Nevertheless, the models are useful in studying business and economic relationships. Managerial economic decisions should not be made without having first analyzed their possible implications. Consider, for example, the concept of the market from introductory economics, which is illustrated in Figure 1A.2.

The demand curve D slopes downward to the right, illustrating the inverse relationship between the quantity of output Q that consumers are willing and able to purchase at each price P. The

FIGURE 1A.2
Demand, Supply, and Market Equilibrium

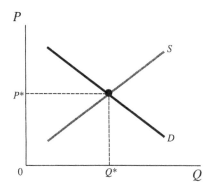

supply curve S illustrates the positive relationship between the quantity of output that suppliers are willing to bring to market at each price. Equilibrium in the market occurs at a price P^*, where the quantity demanded equals the quantity supplied Q^*.

The above simple model may also be expressed algebraically as:

$$Q_D = f(P) \tag{1A.11}$$

$$Q_S = g(P) \tag{1A.12}$$

where Q_D represents the market quantity demanded and Q_S represents the market quantity supplied. Market equilibrium occurs at the selling price where:

$$Q_D = Q_S \tag{1A.13}$$

If Q_D and Q_S are linearly related to price, Equations (1A.11) and (1A.12) may be written:

$$Q_D = a - bP \tag{1A.14}$$

$$Q_S = c + dP \tag{1A.15}$$

where a, b, c, and d are positive constants. By equating the quantity demanded with quantity supplied and solving, the equilibrium price P^* becomes:

$$P^* = \frac{a-c}{b+d} \tag{1A.16}$$

Substituting Equation (1A.16) into Equation (1A.14) or Equation (1A.15) we get the equilibrium quantity:

$$Q^* = a + b\left(\frac{a-c}{b+d}\right) = c + d\left(\frac{a-c}{b+d}\right) \tag{1A.17}$$

Example

The market demand and supply equations for a product are:

$$Q_D = 25 - 3P \tag{1A.18}$$

$$Q_S = 10 + 2P \qquad (1A.19)$$

To determine the equilibrium price and quantity, substitute Equations (1A.18) and (1A.19) into Equation (1A.13) and solve for the equilibrium price of $P^* = \$3$. The equilibrium quantity $Q^* = 16$ may be obtained by substituting the equilibrium price into either Equation (1A.18) or (1A.19).

INVERSE FUNCTIONS

Consider, again, Equation (1A.1). This function is said to be *one-to-one* if there exists a unique value of y for each value of x. This function is also said to be *onto* if for each value of y there exists a unique value of x. If the function in Equation (1A.1) is both one-to-one and onto, there exists a *one-to-one correspondence* between x and y. A nonnumerical example (Chiang 1984) of a relationship that is one-to-one, but not onto, is the relationship between fathers and sons. Each son has one, and only one, father, but each father may have more than one son. By contrast, the relationship between husbands and wives in a monogamous society is both one-to-one and onto. A one-to-one correspondence exists because each husband has one, and only one wife, and each wife has one, and only one, husband.

If Equation (1A.1) is a one-to-one correspondence, the function $x = g(y)$ is said to be the inverse function of Equation (1A.1), and *vice versa*. For example, the function $y = f(x) = 5 - 3x$ has the property that different values of x yield unique values of y. There exists the inverse function $x = g(x) = (5/3) - (1/3)y$ for which different values of y will yield unique values of x.

For the function $y = f(x)$ to have an inverse, it must be *monotonic*. Diagrammatically, it is easy to determine whether a function is monotonic by examining its slope. If the slope of the function is positive for all values of x, the function $y = f(x)$ is *monotonically increasing*. If the slope of the function is negative for all values of x, the function $y = f(x)$ is *monotonically decreasing*. Equation (1A.5) is monotonically increasing because its positive slope is constant. A negative value for b would indicate that the function is monotonically decreasing. Thus, the inverse of Equation (1A.5) is:

$$x = g(y) = -\frac{a}{b} + \frac{1}{b}y \qquad (1A.20)$$

which is also a monotonically increasing function.

Since all linear functions are monotonic, there is a corresponding inverse function for each, and *vice versa*.

Example

Consider the monotonically decreasing function:

$$y = f(x) = 2 - 3x \qquad (1A.21)$$

This function is one-to-one since for every value of x there is one, and only one, value for y. Solving for x we obtain the monotonically decreasing function:

$$x = g(x) = \frac{2}{3} - \left(\frac{1}{3}\right)y \qquad (1A.22)$$

This function is also one-to-one since for each value of y there is one, and only one, value for x. Since Equation (1A.20) is both one-to-one and onto, there is a one-to-one correspondence between x and y. Conversely, since Equation (1A.22) is also one-to-one and onto, there is a one-to-one correspondence between y and x.

NONLINEAR FUNCTIONS OF ONE INDEPENDENT VARIABLE

A distinguishing feature of a *linear function* is that it has a constant slope. In other words, the ratio of the change in the value of the dependent variable with respect to a change in the value of the independent variable is constant. In the case of *nonlinear functions,* however, the slope is variable. Graphs of nonlinear functions appear to be "curves" rather than straight lines.

Polynomials represent a class of functions that contain at lease one independent variable that is raised to a power whose absolute value is greater than unity. Two of the most common polynomial functions encountered in economics and business are *quadratic* and the *cubic equations*. The general forms of quadratic and cubic equations in the two-variable case are given by Equations (1A.23) and (1A.24), respectively.

$$y = a + bx + cx^2 \tag{1A.23}$$

$$y = a + bx + cx^2 + dx^3 \tag{1A.24}$$

For example, a firm's total cost (TC) of producing Q units of output might be:

$$TC = 10 - 5Q + Q^2 \tag{1A.25}$$

ECONOMIC OPTIMIZATION

Many problems in economics involve determining optimal solutions. For example, a decision maker might want to determine the level of output that will maximize profit subject to, say, a fixed operating budget. Economic optimization typically involves maximizing or minimizing some objective function, which may or may not be subject to one or more binding side constraints.

The process of economic optimization may be illustrated by considering the firm's profit (π) function, which is defined as:

$$\pi = TR - TC \tag{1A.26}$$

where TR represents total revenue and TC represents total economic cost. Substituting Equations (1A.4) and (1A.23) into Equation (1A.26) we get the firm's total profit equation:

$$\pi = 5Q - (10 - 5Q + Q^2) = -10 + 10Q - Q^2 \tag{1A.27}$$

From Equation (1A.27), Table 1A.2 summarizes the firm's total economic profit for selected levels of output.

From the information provided in Table 1A.2, the firm's profit reaches a maximum value of $15 at an output level of $Q = 5$. While this optimal value can be read directly from Table 1A.2, it may be determined directly from Equation (1A.27). A clue to how this might be accomplished may be gleaned from Figure 1A.3 in which a smooth curve had been fitted to the data in Table 1A.2. The reader should note that the slope (steepness) of the curve at the output level that maximizes profit is zero, that is, the curve is neither upward nor downward sloping. If it is possible to derive an equation for slopes corresponding to each value of the independent variable, it should be a relatively straightforward process for finding optimal values for the dependent variable without the use of tables or graphs. One such method is differential calculus.

Taking the *first derivative* of an equation results in another, usually different, equation that gives the values of the slope for each value of the independent variable. In the case of Equation

TABLE 1A.2
Total Profit for Selected Output Levels

Q	π
0	−10
1	−1
2	6
3	11
4	14
5	15
6	14
7	11
8	6
9	−1
10	−10

FIGURE 1A.3
Total Profit Curve from Table 1A.2

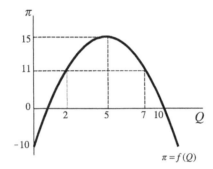

$\pi = f(Q)$

(1A.27), total profit is maximized or minimized at the output level where the slope of the profit function is zero. This is accomplished by taking the first derivative of the profit function, setting this equation equal to zero, and then solving for corresponding level of output. Before proceeding, however, we must first review some of the rules for taking first derivatives.

DERIVATIVE OF A FUNCTION

Recall that the slope of a function may be expressed by Equation (1A.6). This equation may be used to calculate discrete rates of change, such as the value of the slope of the cord AB in Figure 1A.4. As point B is brought arbitrarily closer to point A, the value of the slope of AB approaches the value of the slope of the function at the point A. At point A, the slope of a tangent is identical to the slope of the function. The slope of a function at a single point is an instantaneous rate of change, which is impossible to calculate using Equation (1A.6) (why?). It is possible, however, to determine the value of the slope at a single point by finding the first derivative of the function, and then calculating its value at x_1.

FIGURE 1A.4
Discrete Versus Instantaneous Rates of Change

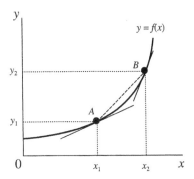

The first derivative of a function (dy/dx) is simply the slope of the function when the interval between x_1 and x_2 is infinitesimally small. Technically, the derivative is the *limit* of the ratio $(\Delta y)/(\Delta x)$ as Δx approaches zero, that is:

$$\frac{dy}{dx} = \lim_{\Delta x \to 0} \left(\frac{\Delta y}{\Delta x} \right) \tag{1A.28}$$

When the limit of a function as x approaches x_0 equals the value of the function at x_0, then the function is said to be *continuous* at x_0. Calculus provides us with a set of rules for finding the derivatives (slopes) of continuous functions. These derivatives can be used for finding local maxima and minima.

RULES OF DIFFERENTIATION

Having established that the derivative of a function is the limit of the ratio of the change in the dependent variable given a change in the value of the independent variable, we will now discuss some general rules for finding first derivatives and their applications. It should be emphasized that for a function to be differentiable at a given point it must be *well defined*, that is, it must be continuous or "smooth." It is not possible to find the derivative of a function if it is discontinuous (i.e., has a "corner") at that point. The interested student is referred to any standard calculus text for formal proofs of these propositions.

Power-Function Rule

A power function is of the form:

$$y = f(x) = ax^b \tag{1A.29}$$

where a and b are constant real numbers. The rule for finding the derivative of a power-function is:

$$\frac{dy}{dx} = f'(x) = bax^{b-1} \tag{1A.30}$$

Example

$$y = 4x^2 \tag{1A.31}$$

$$\frac{dy}{dx} = f'(x) = 2(4)x^{2-1} = 8x \tag{1A.32}$$

A special case of the power-function rule is the *identity rule*:

$$y = f(x) = x \tag{1A.33}$$

$$\frac{dy}{dx} = f'(x) = 1(1)x^{1-1} = 1x^0 = 1 \tag{1A.34}$$

Another special case of the power-function rule is the *constant-function rule*. Since $x^0 = 1$:

$$y = f(x) = ax^0 = a \tag{1A.35}$$

Thus,

$$\frac{dy}{dx} = f'(x) = 0\left(ax^{0-1}\right) = 0 \tag{1A.36}$$

that is, the derivative of a constant is zero.

Example

$$y = 5 = 5x^0 = 5 \times 1 \tag{1A.37}$$

$$\frac{dy}{dx} = f'(x) = 0\left(5x^{0-1}\right) = 0 \tag{1A.38}$$

Sums and Differences Rule

There are a number of economic and business relationships that are formed by combining one or more separate, but related, functions. A firm's total profit function, for example, is equal to total revenue minus total economic cost. Let $u = g(x)$ and $v = h(x)$. Suppose that:

$$y = f(x) = u \pm v = g(x) \pm h(x) \tag{1A.39}$$

$$\frac{dy}{dx} = f'(x) = \frac{du}{dx} \pm \frac{dv}{dx} = u'(x) \pm v'(x) \tag{1A.40}$$

Example

$$u = g(x) = 2x; \qquad v = h(x) = x^2 \tag{1A.41}$$

$$y = f(x) = g(x) + h(x) = 2x + x^2 \tag{1A.42}$$

$$\frac{dy}{dx} = f'(x) = 2 + 2x \tag{1A.43}$$

Example

$$y = 0.04x^3 - 0.9x^2 + 10x + 5 \tag{1A.44}$$

$$\frac{dy}{dx} = f'(x) = 0.12x^2 - 1.8x + 10 \tag{1A.45}$$

Example

From Equation (1A.27):

$$\frac{d\pi}{dQ} = 10 - 2Q \tag{1A.46}$$

Product Rule

Similarly, many business and economics relationships are the product of two or more separate, but related, functions. The total revenue of a monopolist, for example, is equal to price times output, where price is a function of output. Suppose that:

$$y = f(x) = uv = g(x) \cdot h(x) \tag{1A.47}$$

$$\frac{dy}{dx} = f'(x) = u\left(\frac{dv}{dx}\right) + v\left(\frac{du}{dx}\right) = uh'(x) + vg'(x) \tag{1A.48}$$

Example

$$y = 2x^2(3 - 2x) \tag{1A.49}$$

$$u = g(x) = 2x^2 \tag{1A.50}$$

$$\frac{du}{dx} = g'(x) = 4x \tag{1A.51}$$

$$v = (3 - 2x) \tag{1A.52}$$

$$\frac{dv}{dx} = h'(x) = -2 \tag{1A.53}$$

Substituting Equations (1A.50) to (1A.53) into Equation (1A.48) we get:

$$\frac{dy}{dx} = f'(x) = 2x^2(-2) + (3 - 2x)(4x) = -12x^2 + 12x \tag{1A.54}$$

Example

Suppose the inverse of the demand equation for the output of a monopolist is:

$$P = 100 - 2Q \tag{1A.55}$$

The monopolist's total revenue equation is:

$$TR = PQ = (100 - 2Q)Q = 100Q - 2Q^2 \qquad (1A.56)$$

$$\frac{dTR}{dQ} = 100 - 4Q \qquad (1A.57)$$

PROFIT MAXIMIZATION: THE FIRST-ORDER (NECESSARY) CONDITION

In this section, we will demonstrate how the ability to take a first derivative can be used to efficiently determine the level of output (Q) that maximizes profit (π) depicted in Figure 1A.3. We know by inspection that the value of the slope of the profit function is zero at $Q = 5$, that is, $d\pi/dQ = 0$. This is the first-order (necessary) condition for profit maximization. Taking the first derivative of Equation (1A.27) we get Equation (1A.46), which gives us the value of the slope for any value of Q. We are only interested, however, in the value of Q when the slope is equal to zero. Thus, we set Equation (1A.46) equal to zero and solve for $Q^* = 0$, which is precisely what we are looking for.

Unfortunately, simply setting the first derivative of the function equal to zero is not sufficient to ensure that we will achieve a maximum. The reason for this is that zero slopes occur for both local maxima *and* minima. To determine which, we will have to examine the second-order (sufficient) conditions as well.

PROFIT MAXIMIZATION: THE SECOND-ORDER (SUFFICIENT) CONDITION

A second-order (sufficient) condition for $y = f(x)$ to have a maximum at some value $x = x_0$ is for the second derivative (i.e., the derivative of the derivative) to be negative:

$$\frac{d\left(\frac{dy}{dx}\right)}{dx} = \frac{d^2y}{dx^2} = f''(x) < 0 \qquad (1A.58)$$

Inequality (1A.58) expresses the notion that for a maximum, as we "walk" over the top of the "hill," the slope of the function is initially positive, becomes increasingly less positive until we get to zero, and then becomes increasingly negative. In other words, the rate of change of the slope is negative. Inequality (1A.58) expresses this idea by noting that functions with a local maximum are said to be "concave downward" in the neighborhood of $x = x_0$. Similarly, the second-order condition for $f(x)$ to have a minimum at some value $x = x_0$ is:

$$\frac{d\left(\frac{dy}{dx}\right)}{dy} = \frac{d^2y}{dx^2} = f''(x) > 0 \qquad (1A.59)$$

Inequality (1A.59) says that as we walk through the lowest point in the "valley," our feet initially point downward, become increasingly less so until they are level, and then point increasingly upward. In other words, the rate of change of the slope is positive.

The first- and second-order conditions for a function with one independent variable to have a maximum or minimum are summarized in Table 1A.3. Consider, again Equation (1A.27). We have already demonstrated that this equation reaches a local maximum at $Q^* = 5$. To verify that

TABLE 1A.3
First- and Second-Order Conditions for a Local Optimum for Functions of
One Independent Variable

	Maximum	Minimum
First-order condition	$\dfrac{dy}{dx} = 0$	$\dfrac{dy}{dx} = 0$
Second-order condition	$\dfrac{d^2y}{dx^2} < 0$	$\dfrac{d^2y}{dx^2} > 0$

this is a local maximum, take the second derivative of this equation, which is the same thing as taking the first derivative of Equation (1A.46).

$$\frac{d^2\pi}{dQ^2} = -2 < 0 \qquad (1A.60)$$

Example

A firm's total revenue and total cost equations are:

$$TR(Q) = PQ = 20Q - 3Q^2 \qquad (1A.61)$$

$$TC(Q) = 2Q^2 \qquad (1A.62)$$

Combining Equations (1A.61) and (1A.62), the firm's total profit equation is:

$$\pi = TR - TC = 20Q - 5Q^2 \qquad (1A.63)$$

To determine the output level that maximizes total profit, the first-order condition for a local maximum to exist is:

$$\frac{d\pi}{dQ} = 20 - 10Q = 0 \qquad (1A.64)$$

Solving Equation (1A.64), the profit-maximizing output level is $Q^* = 2$ units. To verify that profit is maximized, take the derivative of Equation (1A.64):

$$\frac{d^2\pi}{dQ^2} = -10 < 0 \qquad (1A.65)$$

Since the value of the second derivative in Equation (1A.65) is negative, the sufficient condition for profit maximization is satisfied. To determine the firm's maximum profit, substitute $Q^* = 2$ into Equation (1A.63) and solve for $\pi^* = \$20$.

Example

A firm's *TR* and *TC* equations are:

$$TR(Q) = 45Q - 0.5Q^2 \qquad (1A.66)$$

$$TC(Q) = 2 + 57Q - 8Q^2 + Q^3 \qquad (1A.67)$$

The firm's total profit equation is:

$$\pi = TR - TC = -2 - 12Q + 7.5Q^2 - Q^3 \qquad (1A.68)$$

The first-order condition for a local optimum is:

$$\frac{d\pi}{dQ} = -12 + 15Q - 3Q^2 = 0 \qquad (1A.69)$$

By factoring Equation (1A.69) it may be demonstrated that Equation (1A.68) reaches local optima at $Q_1{}^* = 4$ and $Q_2{}^* = 1$. To determine if these values correspond to a local minimum or maximum, we could substitute these values into Equation (1A.68) and evaluate, or we can examine the values of the second derivatives:

$$\frac{d^2\pi}{dQ^2} = -6Q + 15 \qquad (1A.70)$$

At $Q_1{}^* = 4$:

$$\frac{d^2\pi}{dQ^2} = -6(4) + 15 = -24 + 15 = -9 < 0 \qquad (1A.71)$$

Inequality (1A.71) is the second-order condition for a local maximum. Substituting $Q_2{}^* = 1$ into Equation (1A.70) we get:

$$\frac{d^2\pi}{dQ^2} = -6(1) + 15 = -6 + 15 = 9 > 0 \qquad (1A.72)$$

Inequality (1A.72) is the second-order condition for a local minimum. Substituting $Q_1{}^* = 4$ into Equation (1A.68), we find that the firm's maximum economic profit is $6. At $Q_2{}^* = 1$, the firm's minimum economic profit (maximum economic loss) is –$7.5.

PARTIAL DERIVATIVES AND MULTIVARIATE OPTIMIZATION: THE FIRST-ORDER CONDITION

Most economic and business relationships involve more than one independent variable. For example, suppose that the demand for a firm's output is a function of price and the level of advertising expenditures (A) in thousands of dollars:

$$Q = f(P, A) \qquad (1A.73)$$

To determine the marginal effect of each independent variable on sales, we take the first derivative of the function with respect to each independent variable separately, treating the remaining variables as constants. This process, which is known as partial differentiation, is denoted by replacing d with ∂.

Example

Consider the following equation:

$$Q = f(P, A) = 80P - 2P^2 - PA - 3A^2 + 100A \qquad (1A.74)$$

Taking first partial derivatives with respect to P and A yields:

$$\frac{\partial Q}{\partial P} = f_P = 80 - 4P - A \tag{1A.75}$$

$$\frac{\partial Q}{\partial A} = f_A = -P - 6A + 100 \tag{1A.76}$$

To determine the values of the independent variables that maximize the objective function we set the first partial derivatives equal to zero. Solving the resulting equations simultaneously gives the sales maximizing values $P^* = \$16.52$ and $A^* = 13.92$. Substituting these values into Equation (1A.74) results in maximum sales of 1,356.52 units of output.

PARTIAL DERIVATIVES AND MULTIVARIATE OPTIMIZATION: THE SECOND-ORDER CONDITION

Unfortunately, a general discussion of the second-order conditions for multivariate optimization is beyond the scope of the present discussion.[1] We will, however, briefly consider the second-order (sufficient) conditions for a maximum and minimum in the case of two independent variables. Consider the following functional relationship:

$$y = f(x_1, x_2) \tag{1A.77}$$

The first-order conditions for a local maximum or minimum are:

$$\frac{\partial y}{\partial x_1} = f_1 = 0 \tag{1A.78a}$$

$$\frac{\partial y}{\partial x_2} = f_2 = 0 \tag{1A.78b}$$

The second-order conditions for a local maximum are:

$$\frac{\partial^2 y}{\partial x_1^2} = f_{11} < 0 \tag{1A.79a}$$

$$\frac{\partial^2 y}{\partial x_2^2} = f_{22} < 0 \tag{1A.79b}$$

$$\left(\frac{\partial^2 y}{\partial x_1^2}\right)\left(\frac{\partial^2 y}{\partial x_2^2}\right) - \left(\frac{\partial^2 y}{\partial x_1 \partial x_2}\right)^2 = f_{11}f_{22} - f_{12}^2 > 0 \tag{1A.79c}$$

The second-order conditions for a local minimum are:

$$\frac{\partial^2 y}{\partial x_1^2} = f_{11} > 0 \tag{1A.80a}$$

$$\frac{\partial^2 y}{\partial x_2^2} = f_{22} > 0 \tag{1A.80b}$$

$$\left(\frac{\partial^2 y}{\partial x_1^2}\right)\left(\frac{\partial^2 y}{\partial x_2^2}\right) - \left(\frac{\partial^2 y}{\partial x_1 \partial x_2}\right)^2 = f_{11}f_{22} - f_{12}^2 > 0 \tag{1A.80c}$$

Example

Consider, again, Equation (1A.74). Taking the second partial derivatives with respect to P and A we get:

$$f_{PP} = -4 < 0 \tag{1A.81}$$

$$f_{AA} = -6 < 0 \tag{1A.82}$$

Equations (1A.81a) and (1A.81b) satisfy Inequalities (1A.79a) and (1A.79b). What about Inequality (1A.79c)? Taking the cross partial derivatives we get:

$$f_{PA} = f_{AP} = -1 \tag{1A.83}$$

Substituting these results into Equation (1A.79c) we get:

$$f_{PP}f_{AA} - f_{PA}^2 = (-4)(-6) - (-1)^2 = 23 > 0 \tag{1A.84}$$

Thus, the solution values satisfy the conditions for a local maximum.

INDEFINITE INTEGRATION

Thus far we have been concerned with *differential* calculus in which we took the functional relationship $y = f(x)$ to derive another functional relationship $dy/dx = f'(x) = g(x)$. This second function was useful because it made it possible to identify relative maxima and minima. Suppose, however, that we know $dy/dx = f'(x) = g(x)$ and wish to recover the function $y = f(x)$. Suppose, for example, that we are given the equation:

$$\frac{dy}{dx} = 3x^2 \tag{1A.85}$$

From what functional relationship was Equation (1A.85) derived? We know from experience that this equation could have been derived from $y = x^3$, or $y = 100 + x^3$, or $y = -10,000 + x^3$, or $y = c + x^3$, where c is an arbitrary constant. In general, given Equation (1A.1), *differentiation* is the process of finding:

$$\frac{dy}{dx} = \frac{df(x)}{dx} = f'(x) = g(x) \tag{1A.86}$$

By contrast, the process of recovering Equation (1A.1) from Equation (1A.85) is *integration*. In fact, $f(x)$ is called the integral of $g(x)$. If we are given $g(x)$ and wish to recover $f(x)$, then the general solution is:

$$y = f(x) + c \tag{1A.87}$$

The term c is referred to as an arbitrary *constant of integration,* which may be unknown. Since $dy/dx = g(x)$, then:

$$dy = g(x)dx \tag{1A.88}$$

Integrating both sides of Equation (1A.88) we obtain:

$$\int dy = \int g(x)\,dx \tag{1A.89}$$

By definition, the left-hand side of Equation (1A.89) is equal to y. The integral of $g(x)dx$ is $f(x) + c$. Thus, Equation (1A.89) may be rewritten as:

$$y = \int g(x)\,dx + c = f(x) + c \tag{1A.90}$$

The second term in Equation (1A.90) is called an *indefinite* integral because c cannot be recovered from the integration procedure. For simple power functions, the process of integration is often straightforward. For example, if $g(x) = dy/dx = x^m$, its integral is:

$$y = \int x^m\,dx = \left(\frac{x^{m+1}}{m+1}\right) + c \tag{1A.91}$$

To verify this, simply take the derivative of Equation (1A.91). In fact, Equation (1A.91) is an integration rule that may be applied to a wide class of problems in business and economics. On the other hand, finding the integral:

$$y = \int x^2 e^x\,dx = x^2 e^x - 2x e^x + c \tag{1A.92}$$

may be quite challenging.

Examples

$$\int (2x + 100)\,dx = x^2 + 100x + c \tag{1A.93}$$

$$\int \left(4x^2 + 3x + 50\right) = \left(\frac{4}{3}\right)x^3 + \left(\frac{3}{2}\right)x^2 + 50x + c \tag{1A.94}$$

INTEGRALS AS THE AREA UNDER A CURVE

An important application of integrals stems from its interpretation as the area under a curve. Consider, for example, the marginal cost function $MC(Q)$ in Figure 1A.5.

Marginal cost is the change in total cost given a change in total output, Q. The process of adding up (or integrating) the cost of each additional unit of Q will result in the total cost of producing Q units less any costs not directly related to the production process, such as insurance premiums, fixed rental payments, nonrecoverable fees, etc. We shall refer to these "indirect" (to the actual production of Q) costs collectively as total fixed cost (*TFC*). Costs that vary directly with the level of output are referred to as total variable cost (*TVC*). Total cost is defined as:

$$TC(Q) = TFC + TVC(Q) \tag{1A.95}$$

Note that in Equation (1A.95), *TC* is a function of total output because *TVC* is a function of output. Marginal cost, which is illustrated in Figure 1A.5, is simply the first derivative of Equation (1A.95):

$$\frac{dTC(Q)}{dQ} = \frac{dTVC(Q)}{dQ} = MC(Q) \tag{1A.96}$$

FIGURE 1A.5
Integration as the Area Under a Curve

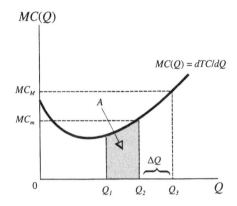

Integrating Equation (1A.96) we obtain:

$$\int MC(Q)dQ = TVC(Q) + c \tag{1A.97}$$

By integrating the marginal cost function we can recover total variable cost, but the constant of integration, which is equal to *TFC,* will remain elusive. This process is illustrated in Figure 1A.5 as the area beneath the curve $MC(Q)$ between Q_1 and Q_2. Let us denote the value of this area as A. Suppose that we wish to consider the effect of an increase in the value of the area under the curve resulting from an increase in output from Q_2 to Q_3, where $Q_3 = Q_2 + \Delta Q$.

In the interval Q_2 to Q_3 there is a minimum and maximum value of $MC(Q)$, which is designated MC_m, and MC_M, respectively. It must be the case that

$$MC_m \Delta Q \leq \Delta A \leq MC_M \Delta Q \tag{1A.98}$$

This is illustrated in Figure 1A.6 as the area of the shaded rectangle. Thus, estimating the value of ΔA using discrete changes in the value of Q results in an approximation of the increase in the value of the area under the curve.

How can we improve upon this estimate of ΔA? One way is to divide ΔQ into smaller intervals. This is illustrated in Figure 1A.7. Taking the limit as $\Delta Q \to 0$ "squeezes" the difference between MC_m and MC_M to its limiting value $MC(Q)$. Thus,

$$\lim_{\Delta Q \to 0} \left(\frac{\Delta A}{\Delta Q} \right) = \frac{dA}{dQ} = MC(Q) \tag{1A.99}$$

In Figure 1A.7, we estimated total cost of production from marginal cost over the output interval Q_1 to Q_3. This procedure may be summarized as:

$$TC = \int_{Q_1}^{Q_3} MC(Q)dQ = TVC(Q) + TFC \tag{1A.100}$$

As noted above, A and $TC(Q)$ differ only by the value of some arbitrary constant c, which in this case is *TFC.* In most cases, however, we are interested in estimating total cost over the entire range of output, which is summarized as:

$$TC = \int_0^{Q_3} MC(Q)dQ = TVC(Q) + TFC \tag{1A.101}$$

FIGURE 1A.6
Approximating the Increase in the Area Under a Curve

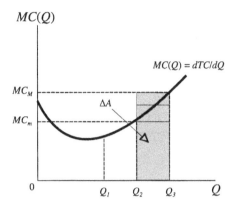

FIGURE 1A.7
**Improving the Estimated Value of the Area Under the
Marginal Cost Curve by "Squeezing" Q**

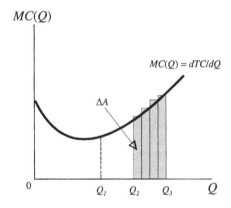

Example

Suppose that a firm's total fixed cost is $4,000 and marginal cost equation is:

$$MC(x) = 50x + 600 \qquad (1A.102)$$

The firm's total cost of production is:

$$TC(Q) = \int MC(Q)\,dQ = \int (50Q + 600)\,dQ = 25Q^2 + 600Q + 4{,}000 \qquad (1A.103)$$

ENDNOTE

1. For an excellent discussion of the second-order conditions for the multivariate case, see Eugene Silberberg, *The Structure of Economics: A Mathematical Analysis*, 2nd edition, McGraw-Hill, 1990, Chapter 4.

CHAPTER

2 NONCOOPERATIVE, ONE-TIME, STATIC GAMES WITH COMPLETE INFORMATION

In this chapter we will:

- *Analyze pure-strategy, noncooperative, static games with complete information;*
- *Deepen our understanding of strictly dominant strategies;*
- *Distinguish between zero-sum and non-zero-sum games;*
- *Expand on our discussion of a Nash equilibrium;*
- *Discuss weakly dominant, iterated strictly dominant, and nondominant strategies;*
- *Develop decision rules for selecting strategies in games with zero or multiple Nash equilibria.*

INTRODUCTION

In this chapter we will take a closer look at two-player, noncooperative, one-time, static games. In a static game, the players are ignorant of their rivals' decisions until after all moves have been made. In business, for example, a firm's profits may depend not only on the decisions made by its managers, but also on the decisions made by managers of rival firms as well. In noncooperative, one-time, static games, the players do not take possible future moves into consideration; they are interested only in the immediate consequences of their decisions. Understanding one-time static games is important because it is the first step to an understanding of a wider range of strategic situations involving static games that are played repeatedly. Competition among rival firms, for example, involves strategic interaction that occurs repeatedly over time. If firms cooperate to achieve mutually beneficial outcomes, the behavior of business rivals in one period will most certainly affect how the players interact in the future.

Although the description of this class of games sounds rather daunting, two-player, noncooperative, simultaneous-move, one-time games are the most basic of all games. Consider once again the game depicted in Figure 1.2. As we saw, the equilibrium strategy profile for this game was {*Low price, Low price*}, with a payoff to both players of $250,000. If both players were able to agree and bind themselves to a *high price* strategy, the payoff to both players would have been $1 million. Strategic situations in which such binding and enforceable agreements are possible are referred to as **cooperative games**. On the other hand, strategic situations in which there is no agreement between players are referred to as **noncooperative games**. Noncooperative games not only refer to situations in which players do not explicitly collude to "rig" the final outcome, but also rule out the possibility of implicit collusion. As before, we will continue to assume that the players are rational, that is, each player will behave in such a way as to achieve the best possible outcome.

> **Cooperative game** A strategic situation in which players agree to coordinate strategies, usually to achieve a mutually beneficial outcome.

FIGURE 2.1
Normal Form of a Two-Player, Static Game

Player B

		B1	B2
Player A	A1	**(100, 200)**	(**150**, 75)
	A2	(50, 50)	(100, **100**)

Payoffs: (Player A, Player B)

> **Noncooperative game** A strategic situation in which players do not agree to coordinate their strategies.

In a one-time game each player has one, and only one, move, and the players move simultaneously. The significance of these assumptions is that the players do not learn anything about a rival's intentions or strategies until the end of the game. For example, suppose two firms that are contemplating advertising campaigns to sell a product that will become obsolete at the end of the year. One firm is considering advertising on radio or in magazines, while the other firm is contemplating advertising on television or in newspapers. Both firms understand that their sales and profits depend on their ability to appeal to a common customer base. The challenge confronting the managers of both firms is to identify the advertising strategy that results in the largest payoff. Because this game is noncooperative, neither firm will know the advertising strategy of its rival until the game is over and sales figures tallied. Finally, we will assume that games are characterized by **complete information**, which means that the players, strategies, and payoffs are common knowledge. While information is complete, it is imperfect. By **imperfect information** we mean that the players do not know a rival's strategy choice until the end of the game. This assumption rules out many important, real-world situations such as auctions, contract negotiations, and so on. Games involving incomplete information will be deferred until later in the text.

> **Complete information** When all players' strategies and payoffs are common knowledge.

> **Imperfect information** When players move without knowledge of the moves of the other players.

Games may have either zero-sum or non-zero-sum payoffs. In a **zero-sum game**, one player's gain is exactly another player's loss. Examples of zero-sum games include poker, a bet on the toss of a fair coin, and so on. In a **non-zero-sum game**, both players may be made better or worse off. Most games are non-zero-sum games. The game depicted in Figure 2.1 is an example of a non-zero-sum game.

> **Zero-sum game** When one player's gain is another player's loss.

> **Non-zero-sum game** When one player's gain does not equal another player's loss. In a non-zero-sum game, both players may be made better or worse off.

STRICTLY DOMINANT STRATEGIES

In the game depicted in Figure 2.1, consider the strategy options available to player A. If player B adopts strategy B1, the payoffs will be 100 if player A adopts strategy A1 and 50 if player A adopts strategy A2. On the other hand, if player B adopts strategy B2, the payoffs will be 150 if player

A adopts strategy *A1* and 100 if player *A* adopts strategy *A2*. There is no question that player *A* should adopt strategy *A1*. The reason for this is that strategy *A1* will result in the largest payoff to player *A* regardless of the strategy adopted by player *B*. These best payoffs for player *A* are highlighted in boldface. In this case, player *A* is said to have a strictly dominant strategy because this strategy always results in the best payoff.

Application 2.1   Run with the Pitch

Strictly dominant strategies can be found almost anywhere you look. Dixit and Nalebuff (1991, Chapter 3) provide two entertaining examples. The first example involves a baseball game. Suppose that the New York Yankees are playing the Boston Red Sox. It is the bottom half of the sixth inning, although the top or bottom half of any inning will do. Alex Rodriguez is up at bat and Derek Jeter is on first base. There are two outs and the pitch count is two strikes and three balls. In this situation, Jeter must decide whether to adopt a wait-and-see or run-with-the-pitch strategy. With a wait-and-see strategy, he remains at first base until after the next pitch before deciding whether to advance. With a run-with-the-pitch strategy, he will run as the pitcher begins his delivery. Which strategy should Jeter adopt?

Running with the pitch is a strictly dominant strategy. Why? Because the payoff associated with a run-with-the-pitch strategy is no worse than the payoff from a wait-and-see strategy, and may be better. To see this, suppose that Rodriguez does not swing at the next pitch. If the pitch is a ball, Jeter will advance to second base regardless of the strategy selected. If the pitch is a strike, the inning is over. Either way, the Yankees are no worse regardless of the strategy adopted by Jeter. Suppose, on the other hand, that Rodriguez swings at the next pitch. If he misses, it is strike three and the side is out. If Rodriguez hits a foul ball that is caught, again the side is out. If the foul ball goes out of play, Jeter remains at first base and waits for the next pitch. In each case, the outcome is the same, regardless of the strategy adopted by Jeter. On the other hand, if Rodriguez hits a fair ball and makes it safely to first base, a wait-and-see strategy will only advance Jeter to second base. By contrast, a run-with-the pitch strategy might allow Jeter to advance to third base or home plate. Clearly, run-with-the-pitch strictly dominates wait-and-see.

What is the optimal strategy for player *B?* An examination of the normal-form game depicted in Figure 2.1 reveals that player *B*'s optimal strategy depends on the strategy adopted by player *A*. To see this, suppose that player *A* adopts strategy *A1*. In this case, player *B* should adopt strategy *B1* because it results in a payoff of 200, compared with a payoff of only 75 by adopting strategy *B2*. The better payoff of 200 is highlighted in boldface. On the other hand, if player *A* adopts strategy *A2*, player *B* should adopt strategy *B2* because it results in a payoff of 100, compared with a payoff of only 50 by adopting strategy *B1*. The better payoff of 100 is also highlighted in boldface. Although player *B* does not have a strictly dominant strategy, player *B* knows that player *A* will adopt his or her strictly dominant strategy. Thus, player *B* will adopt strategy *B1,* which results in a payoff of 200. The equilibrium strategy profile for this two-person, noncooperative, simultaneous-move, one-time game is {*A1, B1*}. Note that this is the only strategy profile for which *both* payoffs are highlighted in boldface. This strategy profile is a Nash equilibrium.

Application 2.2   Indiana Jones and the Last Crusade

The second example deals with the movie blockbuster *Indiana Jones and the Last Crusade*. In the climactic scene of the movie, Dr. Henry "Indiana" Jones, Jr. (portrayed by Harrison Ford), and his father, Professor Henry Jones, Sr. (Sean Connery), arrive at the site of Christendom's most sacred relic, the Holy Grail, at the same time as their evil Nazi nemesis, Walter Donovan (Julian Glover). To reach the chamber containing the Holy Grail, they need to avoid a series of lethal booby traps. Donovan lacks the knowledge to safely negotiate a passage to the Grail chamber. Believing that the "Jones boys" know the secret, Donovan shoots and mortally wounds Indiana's father. Realizing that only water from the Holy Grail will save his father, Indiana accepts the challenge.

After making it past the booby traps and upon reaching the chamber, Indiana is confronted with one last challenge. He must choose from among dozens of chalices, only one of which is the actual cup of Christ. While drinking from the correct cup grants eternal life (although you will have to spend eternity in the Grail chamber), drinking from the wrong cup will result in rapid aging and certain death. By this time, the Nazis have followed Indiana's lead and have made their way to the Grail chamber. Rather than waiting for Indiana to choose, Donovan asks his co-conspirator, Dr. Elsa Schneider (Alison Doody), to select from among the many cups. Upon drinking from the selected gold chalice, Donovan dies.* In the words of the Grail knight guarding the chamber, Donovan "chose poorly." Indiana Jones, by contrast, selects the plain wooden cup of a carpenter. When Jones drinks from the cup, the Grail knight declares: "You have chosen wisely." Indiana then uses the Grail to heal his father's mortal wound.

Did Indiana Jones play his strictly dominant strategy? Alas, no. Instead of first drinking from the Grail, Indiana Jones should have taken the Grail directly to his father. Why? If Indiana had "chosen poorly," drinking from the cup would have resulted in death for both the Joneses. By first letting his father drink, only his father would have died if he had picked the wrong cup. In this game, not drinking is a strictly dominant strategy. Alternatively, drinking is a strictly dominated strategy.

*In telling the story, Dixit and Nalebuff analyze the strategies open to Indiana Jones, but overlook the fact that the overeager Donovan chose first. It is left as an exercise for the viewer to determine whether Donovan had a strictly dominant strategy, and if so, what it was.

Application 2.3   The Good, the Bad and the Ugly

This example deals with the movie classic, *The Good, the Bad and the Ugly*.* In this legendary spaghetti western, Tuco (Eli Wallach) is a wanted outlaw whom Blondie (Clint Eastwood) hands over to the authorities in exchange for the reward, only to save Tuco from hanging by shooting the rope above his head. Tuco and Blondie split the bounty and repeat the scam in the next town. However, Blondie grows tired of Tuco and leaves him to fend for himself in the desert without water. Tuco survives the ordeal, makes his way to a town, steals a gun, captures Blondie, and takes him into the desert to return the favor. But, before Tuco can finish Blondie off, a runaway stagecoach full of dead and dying Confederate soldiers happens along. One of the dying soldiers, Bill Carson, reveals to Tuco the name of a graveyard in which $200,000 in stolen gold is buried. But, Carson will only disclose the name of the grave in exchange for water. As Tuco runs to his horse to retrieve his canteen, a nearly dead Blondie makes his way to the stagecoach. Before dying, Carson reveals the name of the grave to Blondie, but not the name of the cemetery. Tuco has no choice but to keep Blondie alive, since each has a piece of the puzzle.

Before Tuco and Blondie can get on with the treasure hunt, they stumble into a battle between Union and Confederate soldiers, who are fighting for control of a useless bridge. Tuco and Blondie decide to blow up the bridge, not only to get on with their quest, but also as a favor to a fatally wounded Union captain. While setting up the dynamite, Blondie reveals to Tuco that the gold is buried in the grave of Arch Stanton.

Unfortunately for Tuco and Blondie, "Angel Eyes" Sentenza (Lee Van Cleef) also knows the name of the cemetery, but not the grave. After blowing up the bridge, Tuco makes a mad dash for the graveyard, with Blondie and Angel Eyes in pursuit.

The film's climactic sequence is known as the "Ecstasy of Gold." Tuco locates Arch Stanton's grave, but is held at gunpoint by Blondie before he can start digging. Blondie is also held at gunpoint by Angel Eyes. We now learn that Blondie lied to Tuco: Arch Stanton's grave contains nothing more than a rotting corpse. To resolve the ensuing Mexican standoff, Blondie writes the name of the real grave on a stone, which he places name down on the ground. The trio takes triangulating positions around the stone and the shooting begins. Unbeknownst to Tuco, Blondie has emptied Tuco's gun, and he wins the shootout by killing Angel Eyes. Blondie tells Tuco that the treasure is buried in the grave marked "Unknown," which is right next to Arch Stanton's grave. As it turns out, Blondie did not write anything on the stone because nobody's name is on the grave. Tuco digs up the gold, only to find himself once again staring down the barrel of Blondie's gun. Blondie threatens to shoot Tuco if he does not hang himself from a nearby tree. After taking half the gold, Blondie rides away, leaving Tuco swinging in the breeze. In a dramatic twist, Blondie turns and shoots the rope above Tuco's head, thereby freeing him one last time.

Who should Blondie have killed in the final showdown? The answer, of course, is Angel Eyes because he knew Tuco's gun was unloaded. Blondie set up an interesting game in which only he had the information necessary to accurately predict the outcome. But, what if Tuco's gun had been loaded? Would this have affected the outcome of the game? As before, it is clearly in Angel Eyes' best interest to keep Blondie alive since he knows the location of the grave containing the treasure. For the same reason, it is also in Tuco's best interest to keep Blondie alive. Thus, keeping Blondie alive is a strictly dominant strategy for both Tuco and Angel Eyes. The shootout will involve only Tuco and Angel Eyes. But, what about Blondie? Since he needs neither Tuco nor Angel Eyes, his optimal strategy is to wait for the outcome of the duel between them, and then shoot the survivor.

* The author would like to express his appreciation to Professor John Teall of Rensselaer Polytechnic Institute for this interesting application.

NASH EQUILIBRIUM

The reader should note that for the game depicted in Figure 2.1 to have a Nash equilibrium it was not necessary for both players to have a strictly dominant strategy. The reader will recall that a two-person, noncooperative, simultaneous-move, one-time game has a **Nash equilibrium** when each player adopts a strategy that is the best response to the strategy adopted by the other players. More formally, a Nash equilibrium exists for any strategy profile $\{a, b\}$ whenever strategy a is player A's best response to strategy b adopted by player B, and strategy b is player B's best response to strategy a adopted by player A. When these games have a Nash equilibrium, neither player can improve upon his or her payoff by adopting a different strategy. For a Nash equilibrium to exist, it is necessary for only one of the players to have a strictly dominant strategy.

> **Nash equilibrium** Any strategy profile $\{a^*, b^*\}$ such that strategy a^* is player A's best response to strategy b^* adopted by player B, strategy b^* is player B's best response to strategy a^* adopted by player A.

Another interesting aspect of the equilibrium strategy profile depicted in Figure 2.1 is that player A did not receive the largest possible payoff by adopting a strictly dominant strategy. Having a strictly dominant strategy does not guarantee that a player will receive the best possible payoff. In the game depicted in Figure 2.1, the largest payoff is received by player *B* whose choice of strategy depended upon the strategy adopted by player *A*. Nash equilibria are appealing precisely because they are self-fulfilling outcomes to game theoretic problems. If either player expects the other to adopt a Nash equilibrium strategy, both parties will, if fact, choose a Nash equilibrium strategy. For Nash equilibria, actual and anticipated behavior are one and the same.

Example: Oil-Drilling Game

The concepts of a strictly dominant strategy and Nash equilibrium may be illustrated in the oil-drilling game.[1] The game begins when the Global Mining and Recovery (GLOMAR) Corporation purchases a two-year lease on land that lies directly above a four-million-barrel crude oil deposit with an estimated market value of $200 million, or $50 per barrel. The price per barrel of crude oil is not expected to change in the foreseeable future. In order to extract the oil, GLOMAR has the option of drilling a wide well or a narrow well. The accounting data for this project, which are summarized in Table 2.1, suggests that GLOMAR should drill a narrow well to earn the greatest profits. If GLOMAR drills a wide well, the entire deposit can be extracted in one year at a profit of $104 million. If GLOMAR drills a narrow well, it will take two years to extract the oil, but the profit will be $132 million. To avoid computational clutter, we will ignore future value of profits earned and invested at the end of the first year. The difference in profits is due to the fact that it is more expensive to drill a wide well than it is to drill a narrow well.

Enter the Petroleum Exploration (PETROX) Company. PETROX has purchased a two-year lease on land adjacent to the land leased by GLOMAR. The land leased by PETROX also lies above the same crude oil deposit. If both companies sink wells of the same size at the same time, each company will extract half of the total crude oil reserve. This accounting data is summarized in Table 2.2. If both GLOMAR and PETROX sink wide wells, each company will extract two million barrels in six months, with each earning profits of $24 million. On the other hand, if each company sinks a narrow well, it will take one year for GLOMAR and PETROX to extract their respective shares, with each earning profits of $52 million.

Finally, if one company drills a wide well, while the other company drills a narrow well, the first company will extract three million barrels and the second company will extract only one million barrels. This accounting data is summarized in Table 2.3. In this case, the company drilling the wide well will earn profits of $70 million, while the company drilling the narrow well will earn only $12 million. The payoffs in millions of dollars for different combinations of drilling strategies are summarized in the normal-form game depicted in Figure 2.2.

In the oil-drilling game, the strategy profiles are {*Narrow, Narrow*}, {*Narrow, Wide*}, {*Wide, Narrow*}, and {*Wide, Wide*}. The respective payoffs from each strategy profile are (52, 52), (12, 70), (70, 12), and (24, 24). Unlike the game depicted in Figure 2.1, the payoffs depicted in Figure 2.2 are symmetrical. Which strategy should each company adopt?

First consider the options confronting GLOMAR. The size of the well drilled by GLOMAR depends on the size of the well drilled by PETROX. If PETROX sinks a narrow well, GLOMAR's best strategy is to drill a wide well because of the greater payoff. If PETROX sinks a wide well, once again GLOMAR's best strategy is to drill a wide well. In other words, regardless of the strategy adopted by PETROX, GLOMAR's best strategy is to sink a wide well. Because the payoffs are symmetrical, the same is also true for PETROX. The oil-drilling game depicted in Figure 2.2 may look familiar because it is a variation of the prisoner's dilemma discussed in Chapter 1. The distinguishing characteristic of both games is that each player has a strictly dominant strategy.

TABLE 2.1
GLOMAR's Profits from Extracting and Selling Oil Reserves

	Narrow well	Wide well
Drilling cost	$28 million	$56 million
Pumping cost	$40 million	$40 million
Total production cost	$68 million	$96 million
Total revenues	$200 million	$200 million
Total net profit	$132 million	$104 million

TABLE 2.2
GLOMAR's and PETROX's Profits from Drilling the Same-Size Well

	Narrow well	Wide well
Drilling cost	$28 million	$56 million
Pumping cost	$20 million	$20 million
Total production cost	$48 million	$76 million
Total revenues	$100 million	$100 million
Total net profit	$52 million	$24 million

TABLE 2.3
GLOMAR's and PETROX's Profits from Drilling Different-Size Wells

	Narrow well	Wide well
Drilling cost	$28 million	$56 million
Pumping cost	$10 million	$30 million
Total production cost	$38 million	$86 million
Total revenues	$50 million	$150 million
Total net profit	$12 million	$70 million

FIGURE 2.2
Normal Form of the Two-Player Oil-Drilling Game

		GLOMAR	
		Narrow	Wide
PETROX	Narrow	(52, 52)	(12, **70**)
	Wide	(**70**, 12)	(**24, 24**)

Payoffs: (PETROX, GLOMAR)

STRICTLY DOMINANT STRATEGY EQUILIBRIUM

Since a *wide* strategy will be chosen by both companies regardless of the strategy adopted by their rivals, a *wide* strategy strictly dominates a *narrow* strategy. Stated differently, a *narrow* strategy is strictly dominated by a *wide* strategy. The Nash equilibrium strategy profile is {*Wide, Wide*} with payoffs to both companies of $24 million. Since both companies have the same strictly dominant strategy, the Nash equilibrium for this game is called a strictly dominant strategy equilibrium.

――――――――――― **Demonstration Problem 2.1** ―――――――――――

Pie Eye Brewery and Red Nose Lager are considering their upcoming advertising campaigns. Pie Eye is considering advertising on radio or in newspapers, while Red Nose is thinking about television or magazine advertising. Figure D2.1 summarizes the expected increase in profits to each beer company from alternative combinations of advertising strategies.

    a. Does either firm have a strictly dominant strategy?
    b. What is the Nash equilibrium for this game?

FIGURE D2.1
Normal Form of the Static Game in Demonstration Problem 2.1

Red Nose

Pie Eye		Magazines	Television
	Radio	(**$5 million, $6 million**)	(**$10 million**, $1 million)
	Newspapers	($2 million, **$8 million**)	($2 million, $3 million)

Payoffs: (Pie Eye, Red Nose)

Solution

    a. Both companies have a strictly dominant strategy. Suppose that Red Nose decides to advertise in magazines. Pie Eye's best response is to advertise on radio because it will yield a $5 million increase in profits, compared with $2 million by advertising in newspapers. If Red Nose chooses to advertise on television, once again it will be in Pie Eye's best interest to advertise on radio because it will generate $10 million in increased profits, compared with $2 million by advertising in newspapers. Regardless of the advertising strategy adopted by Red Nose, Pie Eye will advertise on radio. Thus, Pie Eye's strictly dominant strategy is to advertise on radio.

        What about Red Nose? If Pie Eye advertises on radio, it will be in Red Nose's best interest to advertise in magazines because doing so will generate $6 million in increased profits, compared with only $1 million by advertising on television. If Pie Eye advertises in newspapers, once again it will be in Red Nose's best interest to advertise in magazines because doing so will result in an $8 million increase in profits, compared with $3 million by advertising on television. Thus, advertising in magazines is a strictly dominant strategy for Red Nose.

    b. Since both Pie Eye Brewery and Red Nose Lager have strictly dominant strategies, the strictly dominant strategy equilibrium for this game is {*Radio, Magazines*}. This strategy profile is a Nash equilibrium because the strategy adopted by each beer company is the best response to the strategy adopted by its rival. Moreover, neither company can improve its payoff by switching strategies.

FIGURE 2.3
Static Game with a Strictly Dominant Strategy Equilibrium

GLOMAR

		Narrow	Wide
PETROX	*Narrow*	(52, 52)	(12, **70**)
	Wide	**(70, 25)**	(**24**, 24)

Payoffs: (PETROX, GLOMAR)

A variation on the oil-drilling game is depicted in Figure 2.3. In this game, the payoff to GLOMAR by adopting a *narrow* strategy in response to a *wide* strategy by PETROX has been increased from $12 million to $25 million. How does this modification alter the equilibrium to this game? The reader should verify that while PETROX still has a strictly dominant *wide* strategy, GLOMAR no longer does. GLOMAR's best strategy depends on the strategy adopted by PETROX. As before, if PETROX adopts a *narrow* strategy, it will be in GLOMAR's best interest to adopt a *wide* strategy. If PETROX adopts a *wide* strategy, GLOMAR's best response is to adopt a *narrow* strategy

The Nash equilibrium strategy profile for the static game depicted in Figure 2.3 is {*Wide, Narrow*}. The reason for this is that GLOMAR knows that PETROX is rational, and will adopt its strictly dominant *wide* strategy. GLOMAR's best response is to adopt a *narrow* strategy because of the larger payoff. Thus, the Nash equilibrium strategy profile for this game is {*Wide, Narrow*}. The payoffs are highlighted in boldface. This equilibrium strategy profile is a Nash equilibrium because both companies adopted a strategy that is the best response to the strategy adopted by its rival. Moreover, neither company can improve its payoff by adopting a different strategy. As we saw in our discussion of the static game depicted in Figure 2.1, it is not necessary for both companies to have a strictly dominant strategy for a unique Nash equilibrium to exist. It is only necessary for one of the companies, in this case PETROX, to have a strictly dominant strategy for the game to have a unique Nash equilibrium.

———————————— **Demonstration Problem 2.2** ————————————

The two leading firms in the highly competitive running shoe industry, Treebark and Adios, are considering increases in their advertising expenditures. Both companies are considering buying advertising space in *Joggers' World,* the leading national magazine about recreational, long-distance running, or buying air time on *KNUT,* an all-talk, all-sports, all-the-time radio station. Figure D2.2 summarizes the payoffs in millions of dollars associated with each combination of advertising strategies.

   a. Does either company have a strictly dominant strategy?
   b. Based on your answer to question *a*, what is the Nash equilibrium for this game?

Solution

   a. Treebark has a strictly dominant strategy, which is to advertise in *Joggers' World.* If Adios advertises in *Joggers' World,* Treebark's best response is to advertise in *Joggers' World* as well. If Adios advertises on *KNUT,* once again Treebark's best response is to advertise in *Joggers' World.*

FIGURE D2.2
Static Game in Demonstration Problem 2.2

Treebark

		Joggers' World	KNUT
Adios	Joggers' World	**(1, 2)**	(0.3, 0.35)
	KNUT	(0.75, **0.75**)	(**2.5**, 0.5)

Payoffs: (Adios, Treebark)

b. Adios does not have a strictly dominant strategy. If Treebark advertises in *Joggers' World,* Adios's best response is also to advertise in *Joggers' World.* On the other hand, if Treebark advertises on *KNUT,* Adios's best strategy is to advertise on *KNUT.* Adios's best strategy depends on what Treebark does. Since we know that Treebark's strictly dominant strategy is to advertise in *Joggers' World,* it will be in the best interest of Adios to advertise in *Joggers' World* as well. Thus, the Nash equilibrium strategy profile for this game is {*Joggers' World, Joggers' World*} with payoffs to Adios and Treebark of $1,000,000 and $2,000,000, respectively. Neither company can improve its payoff by adopting a *KNUT* strategy.

Example: Internet Service Game

Mr. Smith is considering signing a contract with Blue Streak, an Internet service provider. Blue Streak must decide between providing high- or low-speed service. Some of the cost to Blue Streak is independent of whether or not Mr. Smith signs a service contract. Mr. Smith is only interested in signing a high-speed service contract with Blue Streak. Although the quality of the service provided by Blue Streak can be written into the contract, Mr. Smith knows that he cannot credibly verify this. He must decide, therefore, whether or not to sign the service contract. The normal form of this game is depicted in Figure 2.4. The payoff matrix for this game summarizes the level of satisfaction (utility) enjoyed by each player from alternative combinations of strategies.

The reader should verify that Mr. Smith does not have a strictly dominant strategy. By contrast, Blue Streak's strictly dominant strategy is to provide low-speed service. If Blue Streak is rational and adopts its strictly dominant strategy, it will be in Mr. Smith's best interest not to sign a service contract. Thus, the Nash equilibrium strategy profile for this game is {*Low speed, Don't sign*}. The payoffs to Blue Streak and Mr. Smith are highlighted in boldface. Even though only Blue Streak has a strictly dominant strategy, the strategy profile {*Low speed, Don't sign*} is a Nash equilibrium because neither player can obtain a better payoff by adopting a different strategy.

The important thing about the Internet service game is that the best outcome for both players is for Blue Streak to provide high-speed service and for Mr. Smith to sign a service contract. Unfortunately, this strategy profile is not a Nash equilibrium. Recall that Mr. Smith is not able to verify the speed of service provided by Blue Streak. Thus, Blue Streak will be tempted to violate the terms of its own contract by switching to low-speed service. With a {*High speed, Sign*} strategy profile, both players will have an incentive to switch strategies to obtain an improved payoff.

WEAKLY DOMINANT STRATEGIES

Consider a new variation of the oil-drilling game depicted in Figure 2.5. In this game, PETROX does not have a strictly dominant strategy. To see why, suppose that GLOMAR adopts a *narrow*

FIGURE 2.4
Internet Service Game

Mr. Smith

		Sign	Don't sign
Blue Streak	High speed	(2, **2**)	(0, 1)
	Low speed	(**3**, 0)	(**1, 1**)

Payoffs: (Blue Streak, Mr. Smith)

FIGURE 2.5
Static Game in Which Both Players Have a Weakly Dominant Strategy

GLOMAR

		Narrow	Wide
PETROX	Narrow	(52, 52)	(**12, 70**)
	Wide	(**70, 12**)	(**12, 12**)

Payoffs: (PETROX, GLOMAR)

strategy. As before, PETROX should choose a *wide* strategy because it results in the largest payoff. On the other hand, if GLOMAR chooses a *wide* strategy, PETROX will be indifferent to choosing a *narrow* or a *wide* strategy since both will result in a profit of $12 million. Thus, *wide* is no longer a strictly dominant strategy for PETROX. Analogously, if PETROX adopts a *narrow* strategy, GLOMAR should also adopt a *wide* strategy because it results in the largest payoff. On the other hand, if PETROX adopts a *wide* strategy, GLOMAR will also be indifferent to choosing a *narrow* or a *wide* strategy. For both PETROX and GLOMAR, *wide* is a **weakly dominant strategy**. A strategy is weakly dominant if no other strategy results in a better outcome regardless of the strategies adopted by the other players. Stated differently, a *wide* strategy is preferred to a *narrow* strategy because it will never result in a lower payoff and may result in a larger payoff. Strictly dominant and weakly dominant strategies are collectively referred to as **dominant strategies**.

> **Weakly dominant strategy** A strategy that results in a payoff that is no lower than any other payoff regardless of the strategy adopted by the other player.
>
> **Dominant strategy** A strictly or weakly dominant strategy.

A rational player will always adopt a weakly dominant strategy. The reason for this is straightforward. To see why, suppose that PETROX opts to drill a narrow well. If GLOMAR drills a narrow well its payoff is $52 million. If GLOMAR drills a wide well its payoff is $70 million. In this case, it will be in GLOMAR's best interest to drill a wide well. On the other hand, if PETROX drills a wide well, GLOMAR is indifferent to drilling a narrow or a wide well since both will result in a payoff of $12 million. Which strategy will GLOMAR adopt? Clearly, GLOMAR should still drill a wide well. Why? Because by drilling a narrow well, the best that GLOMAR can do is a profit of $52 million. On the other hand, if GLOMAR drills a wide well then it could earn a profit of $70 million. Since the game is symmetrical, the same line of reasoning also applies to PETROX. Thus, each company should adopt its weakly dominant strategy.

In the static game depicted in Figure 2.5, the **weakly dominant strategy equilibrium** is for both players to drill a wide well. The weakly dominant strategy profile for this game is {*Wide, Wide*}. The payoffs are highlighted in boldface. The strategy {*Wide, Wide*} is a Nash equilibrium because both companies adopted a strategy that is the best response to the strategy adopted by the other company, and because neither player can improve its payoff by switching strategies. It should be noted that while neither company can improve its payoff by switching strategies at the strategy profiles {*Wide, Narrow*} and {*Narrow, Wide*}, they do not constitute Nash equilibria. The reason for this is that they do not represent best-response strategies.

> **Weakly dominant strategy equilibrium** A Nash equilibrium that results when both players adopt a weakly dominant strategy.

—————————————— **Demonstration Problem 2.3** ——————————————

Firm *X* is considering entering a market in which firm *Y* is a monopolist. Firm *Y* must decide how to respond to this challenge. If firm *X* decides to *enter* the market, firm *Y* can adopt an *aggressive* pricing strategy or adopt a *passive* pricing strategy. If firm *Y* adopts an aggressive pricing strategy, firm *X* will lose $2.5 million, although firm *Y* will only earn $1 million. If firm *Y* adopts a *passive* pricing strategy, firm *X* will earn $3 million and firm *B* will earn $4 million. Finally, if firm *X* decides to stay out of the market, firm *Y* will earn $7 million.

a. Does either player in this game have a strictly dominant strategy? Explain.
b. What is the Nash equilibrium for this game?

Solution

a. The normal form for this game is depicted in Figure D2.3. Neither firm has a strictly dominant strategy. If firm *Y* adopts an *aggressive* pricing strategy, it will be in firm *X*'s best interest to stay out since entering the market will result in a loss of $2.5 million. If firm *Y* adopts a *passive* pricing strategy, firm *X* will enter the market and earn profits of $3 million. Since firm *X*'s strategy depends upon the strategy adopted by firm *Y*, firm *X* does not have a strictly dominant strategy. Now, consider the game from firm *Y*'s perspective. If firm *X* enters the market, firm *Y* will respond with a *passive* pricing strategy because it will result in the highest payoff of $4 million. If firm *X* stays out of the market, firm *Y* will be indifferent between an *aggressive* and a *passive* pricing strategy. Firm *Y* should adopt a *passive* pricing strategy since the worst it can do is earn profits of $4 million, and could earn $7 million. Thus, firm *Y* has a weakly dominant *passive* strategy.

FIGURE D2.3
Static Game in Demonstration Problem 2.3

		Firm *Y*	
		Aggressive	*Passive*
Firm *X*	*Enter*	(−2.5, 1)	**(3, 4)**
	Stay out	**(0, 7)**	(0, 7)

Payoffs: (Firm *X*, Firm *Y*)

b. Since firm Y will play its weakly dominant *passive* strategy, firm X will enter the market and earn $3 million. Thus, the strategy profile {*Enter, Passive*} is a Nash equilibrium because both firms adopted a strategy that it believes is the best response to the strategy adopted by its rival. Moreover, neither firm can improve its profits by adopting a different strategy. While this is also true for the strategy profile {*Stay out, Aggressive*}, this is not a Nash equilibrium because it does not represent best-response strategies.

In the preceding paragraphs we saw that in noncooperative, simultaneous-move, one-time games, a rational player should always adopt his or her dominant strategy. A player with a dominant strategy should not be concerned with a rival's behavior. The only thing that matters is selecting a strategy that always results in the best payoff. This leads us to our first general principle for playing noncooperative games:

> ***Principle:*** A rational player should adopt a dominant strategy whenever possible. Alternatively, a rational player should not adopt a dominated strategy.

We also observed that for a game to have a unique Nash equilibrium it is only necessary for one player to have a dominant strategy. The reason for this is that the player without a dominant strategy will predict that a rational rival will adopt a dominant strategy and adopt the best strategy in response. This brings us to our second general principle:

> ***Principle:*** A rational player believes that a rational rival will always adopt a dominant strategy and avoid a dominated strategy whenever possible, and will act on that belief. Moreover, a player believes that rivals think the same way, and will also act on their beliefs.

ITERATED ELIMINATION OF STRICTLY DOMINATED STRATEGIES

So far, we have considered noncooperative, simultaneous-move, one-time games involving two players with just two strategies to choose from. Suppose, however, that two players in a game are required to select from among three or more strategies. In games where both players have a strictly dominant strategy, finding the Nash equilibrium is straightforward. On the other hand, suppose that neither player has a strictly dominant strategy. Is it still possible for a two-player, multistrategy game to have a unique Nash equilibrium? To answer this question, consider the version of the oil-drilling game depicted in Figure 2.6.

The payoff matrix for this static game introduces a third strategy—*don't drill*. The reader should verify that neither PETROX nor GLOMAR has a strictly dominant strategy. Which strategy should both companies adopt?

Let us consider the problem from GLOMAR's perspective. Suppose that PETROX adopts a *don't drill* strategy. In this case, GLOMAR should adopt a *narrow* strategy because it results in the largest payoff of $132 million. On the other hand, if PETROX chooses a *narrow* or *wide* strategy, GLOMAR should adopt a *wide* strategy, which results in payoffs of $70 million and $24 million, respectively. Since *narrow* and *wide* dominate *don't drill, don't drill* is said to be a **strictly dominated strategy**. Playing a strictly dominated strategy will always result in a worse payoff regardless of the strategies adopted by the other players.

Strictly dominated strategy A strategy that is dominated by every other strategy.

FIGURE 2.6
Static Game with an Iterated Strictly Dominant Strategy Equilibrium

		GLOMAR		
		Don't drill	Narrow	Wide
	Don't drill	(0, 0)	(0, **132**)	(0, 104)
PETROX	Narrow	(**132**, 0)	(52, 52)	(12, **70**)
	Wide	(104, 0)	(**70**, 12)	(**24, 24**)

Payoffs: (PETROX, GLOMAR)

Since the payoff matrix in Figure 2.6 is symmetrical, *don't drill* is a strictly dominated strategy for PETROX as well. Since PETROX and GLOMAR will never adopt a *don't drill* strategy, this strategy may be eliminated from further consideration. The resulting payoff matrix reduces to the two-strategy game in Figure 2.2. The reader should verify that the resulting normal-form game does, indeed, have a strictly dominant strategy equilibrium, which is the strategy profile {*Wide, Wide*}. Thus, the strategy profile {*Wide, Wide*} is a Nash equilibrium for the three-strategy game summarized in Figure 2.6. The payoffs for this game are highlighted in boldface. The equilibrium strategy profile {*Wide, Wide*} is called an **iterated strictly dominant strategy equilibrium** because it was obtained after eliminating the strictly dominated *don't drill* strategy.

> **Iterated strictly dominant strategy equilibrium** A strictly dominant strategy equilibrium that is obtained after all strictly dominated strategies have been eliminated.

The rationale behind this approach to finding a Nash equilibrium strategy profile for games involving multiple strategies can be found in Sherlock Holmes's famous assertion: "When you have eliminated the impossible, whatever remains, *however improbable,* must be the truth."[2] Actually, this assertion by Doyle's famous super sleuth is an overstatement. Noncooperative, one-time, static games with a large number of strategies may require several iterations before an equilibrium is found. Moreover, removing all strictly dominated strategies does not necessarily guarantee that the game will have a unique Nash equilibrium. On the other hand, if a strictly dominant strategy emerges after strictly dominated strategies are eliminated, the Nash equilibrium strategy profile is an iterated strictly dominant strategy equilibrium. As long as both players have a strictly dominant strategy, the order in which strictly dominated strategies are eliminated is irrelevant. We will still end up with a strictly dominant strategy equilibrium. On the other hand, if strictly dominant strategies are replaced by weakly dominant strategies, the order in which weakly dominated strategies are removed could affect the outcome of the game.

———————————————— **Demonstration Problem 2.4** ————————————————

Consider the noncooperative, simultaneous-move, one-time game depicted in Figure D2.4.1 in which larger payoffs are preferred. Does this game have a Nash equilibrium? Explain.

Solution

The reader should verify that neither player has a strictly dominant strategy. The reader should also verify that for player *A,* strategy *A3* is strictly dominated by every other strategy. Thus, this strategy should be eliminated from further consideration. Likewise, for player *B,* strategy *B3* is strictly dominated by every other strategy. Thus, this strategy will also

FIGURE D2.4.1

Static Game in Demonstration Problem 2.4

Player B

		B1	B2	B3	B4
	A1	(1, 0)	(2, 3)	(4, 0)	(1, **4**)
Player A	A2	(2, 1)	(**4, 5**)	(3, 1)	(3, 4)
	A3	(2, **4**)	(1, 3)	(2, 0)	(0, 1)
	A4	(**3**, 2)	(3, **4**)	(2, 1)	(**5**, 3)

Payoffs: (Player A, Player B)

FIGURE D2.4.2

First Iterated Payoff Matrix for Demonstration Problem 2.4

Player B

		B1	B2	B4
	A1	(1, 0)	(2, 3)	(1, **4**)
Player A	A2	(2, 1)	(**4, 5**)	(3, 4)
	A4	(**3**, 2)	(3, **4**)	(**5**, 3)

Payoffs: (Player A, Player B)

FIGURE D2.4.3

Second Iterated Payoff Matrix for Demonstration Problem 2.4

Player B

		B2	B4
Player A	A2	(**4, 5**)	(3, 4)
	A4	(3, **4**)	(**5**, 3)

Payoffs: (Player A, Player B)

be eliminated from further consideration. The resulting normal-form game when strategies A3 and B3 are eliminated is depicted in Figure D2.4.2.

The reader should now verify that neither player has a strictly dominant strategy in the normal-form game depicted in Figure D2.4.2. For player A, strategy A1 is strictly dominated by strategies A2 and A4. Thus, strategy A1 should be eliminated from further consideration. Likewise, for player B, strategy B1 is strictly dominated by strategies B2 and B4. Thus, strategy B1 should be eliminated from further consideration. The resulting normal-form game is depicted in Figure D2.4.3.

Finally, the reader should verify that although player A does not have a strictly dominant strategy, player B's strictly dominant strategy is B2. Thus, player A will adopt strategy A2, which results in the largest payoff. Since player B expects player A to adopt strategy A2, it will be in player B's best interest to adopt strategy B2 since this results in a payoff of 5, compared with a payoff of 4 by selecting strategy B4. The strategy profile for this game is {*A2, B2*} is a Nash equilibrium (why?), with payoffs to player *A* and player *B* of 4 and 5, respectively.

In this section we found that in noncooperative, simultaneous-move games with multiple strategies that a player may not have a dominant strategy. On the other hand, such games may have one or more strictly dominated strategies. A strictly dominated strategy will always result in a payoff that is worse than some other strategy. Just as a player will always adopt a strictly dominant strategy, a player should always avoid a strictly dominated strategy. By eliminating strictly dominated strategies, a dominant strategy may emerge in the process. The leads us to our next general principle for playing noncooperative games.

> *Principle:* A player should successively eliminate all strictly dominated strategies. If a dominant strategy emerges, it should be adopted.

THREE-PLAYER GAMES

In the previous section we analyzed static, normal-form games involving two players and multiple pure strategies. Is it possible to analyze static, pure-strategy games involving multiple players in the same manner? Unfortunately, analyzing static games in pure strategies becomes exponentially more difficult as the number of players increases. In fact, analyzing games involving more than three players requires the use of more advanced mathematical techniques. In this section we will illustrate the procedure for finding a Nash equilibrium for pure-strategy games involving three players.

Figure 2.7 summarizes another variation of the oil-drilling game. In this version there are now three oil companies that have purchased leases on land lying above the same crude oil deposit. The third player in this game is the Continental Exploration Company (COEXCO). As in the original version of this game, each company must decide whether to drill a *wide* or a *narrow* well. As before, drilling cost depends on the size of the well and pumping costs are proportional to the amount of oil extracted by each company. Since there are now three players, the share to each company of the four-million-barrel crude oil deposit, with an estimated market value of $200 million, will be proportionately less.

To find the pure-strategy Nash equilibrium for this game we proceed as follows. First, treat each payoff matrix involving PETROX and GLOMAR as a separate game. For example, for the payoff matrix on the left, in which COEXCO adopts a *narrow* strategy, PETROX will also adopt a *narrow* strategy if GLOMAR adopts a narrow or wide strategy. The payoffs are highlighted in boldface.

FIGURE 2.7
Three-Player Static Oil-Drilling Game

COEXCO

		Narrow		*Wide*	
		GLOMAR		**GLOMAR**	
		Narrow	*Wide*	*Narrow*	*Wide*
PETROX	*Narrow*	**(25, 25, 25)**	**(12, 24, 12)**	**(12, 12, 24)**	(4, 8, 8)
	Wide	(24, **12, 12**)	(8, 8, **4**)	(8, **4**, 8)	(3, 3, 3)

Payoffs: (PETROX, GLOMAR, COEXCO)

Thus, for the payoff matrix on the left, PETROX has a strictly dominant *narrow* strategy. Similarly, regardless of the strategy adopted by PETROX for the payoff matrix on the left, GLOMAR also has a strictly dominant strategy. These payoffs are also highlighted in boldface.

Next, consider the payoff matrix on the right, in which COEXCO adopts a *wide* strategy. It is left as an exercise for the reader to verify that both players again have a strictly dominant *narrow* strategy. These payoffs are also highlighted in boldface.

The final step in our search for a pure-strategy Nash equilibrium is to find COEXCO's best strategy by comparing the payoffs in both games. If PETROX and GLOMAR adopt a *narrow* strategy, the highest payoff for COEXCO is $25 million by also adopting a *narrow* strategy. If PETROX adopts a *narrow* strategy and GLOMAR adopts a *wide* strategy, the highest payoff for COEXCO is $12 billion by adopting a *narrow* strategy. Similarly, if PETROX adopts a *wide* strategy and GLOMAR adopts a *narrow* strategy, the highest payoff for COEXCO is $12 million by adopting a *narrow* strategy, and so on. The only cell of the payoff matrix in Figure 2.7 in which all of the payoffs are highlighted in boldface is the strategy profile {*Narrow, Narrow, Narrow*}, which is a unique Nash equilibrium for this game.

NONDOMINANT STRATEGIES

In each of the games thus far considered, either or both players had a dominant strategy, which resulted in a unique Nash equilibrium. But, what if neither player has a dominant strategy? Is it still possible for a game to have a unique Nash equilibrium? We will explore this question by considering yet another variation of the oil-drilling game. The revised payoff matrix is depicted in Figure 2.8.

The reader should verify that in this game neither company has a dominant strategy. Both strategies, *narrow* and *wide,* are nondominant because neither strategy dominates the other. A **nondominant strategy** is neither strictly dominant nor weakly dominant. Since neither PETROX nor GLOMAR has a dominant strategy, the optimal strategy for either company will depend on what each believes will be the strategy adopted by its rival. To see this, suppose the GLOMAR believes that PETROX plans to adopt a *narrow* strategy. In this case, it will be in GLOMAR's best interest to adopt a *wide* strategy. On the other hand, if GLOMAR believes that PETROX plans to adopt a *wide* strategy, it will be in GLOMAR's best interest to adopt a *narrow* strategy. Similarly, if PETROX believes that GLOMAR is going to adopt a *narrow* strategy, it will be in PETROX's best interest to adopt a *wide* strategy. The equilibrium strategy profile will be {*Wide, Narrow*}. But, if PETROX believes that GLOMAR is going to adopt a *wide* strategy, the best response by PETROX is to adopt a *narrow* strategy. The important thing is that each company's optimal strategy depends upon what it *believes* will be the strategy adopted by its rival. These beliefs result in two possible Nash equilibria: {*Wide, Narrow*} and {*Narrow, Wide*}. These strategy profiles are

FIGURE 2.8

Static Oil-Drilling Game with Nondominant Strategies

		GLOMAR	
		Narrow	Wide
PETROX	Narrow	(52, 52)	**(26, 70)**
	Wide	**(70, 26)**	(24, 24)

Payoffs: (PETROX, GLOMAR)

said to be *self-fulfilling* in the sense that if both companies believe that the outcome of a game is a particular strategy profile, it will be in their best interest to adopt that strategy.

> **Nondominant strategy** A strategy that is neither strictly nor weakly dominant. A player's best strategy depends on what he or she believes is the strategy adopted by a rival.

The concept of a Nash equilibrium generated considerable excitement when it was first proposed in the 1950s. This does not mean, of course, that every noncooperative game has a unique Nash equilibrium. Simple solutions to problems involving complex human behavior are the exception, rather than the rule. Most games involving multiple players and pure strategies have multiple Nash equilibria, or none at all. In the game depicted in Figure 2.8, {*Narrow, Wide*} and {*Wide, Narrow*} are Nash equilibria because each represents a company's best response to the strategy adopted by its rival, and because neither company can earn a better payoff by adopting a different strategy. In games involving multiple Nash equilibria, it will be difficult to predict the strategies of the other players without more information concerning the nature of the relationship between the players, the context in which the game is being played, and whether cooperation between the players is possible. In cases where cooperation is not possible, a single strategy profile may stand out because the players share a common understanding of the "environment" in which the game is being played. In the literature on game theory, this shared understanding is sometimes referred to as "conventional wisdom." The resulting strategy profile is called a "focal-point equilibrium," which will be discussed at greater length in the next chapter.

Example: Internet Service Game Again

The problems associated with adopting an appropriate strategy when neither player has a strictly dominant strategy may be further illustrated by revisiting the Internet service game discussed earlier in this chapter. Recall from Figure 2.4 that Mr. Smith did not have a strictly dominant strategy, but Blue Streak's strictly dominant *low speed* strategy resulted in the unique Nash equilibrium strategy profile {*Low speed, Don't sign*}. In that game, increasing Mr. Smith's satisfaction with high-speed service would not have affected the outcome unless Blue Streak had an incentive to provide high-speed service, which it did not. Now, let us change the terms of the game by assuming that Blue Streak offers Mr. Smith the option of cancelling his subscription if he is dissatisfied with the speed of its Internet service. The normal form for this new game is summarized in Figure 2.9. The only alteration in this game from the one depicted in Figure 2.4 is the lower payoff to Blue Streak from the strategy profile {*Low speed, Sign*}. The opt-out clause in the contract results in the same low payoff to Blue Streak as would occur if Mr. Smith chose not to sign the contract in the first place. In fact, Mr. Smith still prefers not to sign an Internet service contract if Blue Streak provides low-speed service.

In the game depicted in Figure 2.9, neither player has a dominant strategy. The optimal strategy for Blue Streak and Mr. Smith depends upon the strategy adopted by the other player. The reader should verify that this game has two Nash equilibria: {*High speed, Sign*} and {*Low speed, Don't sign*}. These strategy profiles are self-fulfilling in the sense that if both players believe that the outcome of a game is a particular strategy profile, it will be in the best interest of each to adopt that strategy. The payoffs for the two Nash equilibria for the game depicted in Figure 2.9 are highlighted in boldface.

In games involving multiple Nash equilibria, it is often difficult to predict the outcome without additional information. If cooperation is possible, {*High speed, Sign*} is the most likely strategy profile since it results in the best outcome for both players. Even when cooperation is not possible, a unique Nash equilibrium may present itself because the players share a common understanding of the problem. The strategy profile {*High speed, Sign*} for the game depicted in Figure 2.9 is an

FIGURE 2.9
Revised Internet Service Game

Mr. Smith

		Sign	Don't sign
Blue Streak	*High speed*	**(2, 2)**	(0, 1)
	Low speed	(1, 0)	**(1, 1)**

Payoffs: (Blue Streak, Mr. Smith)

example of a focal-point equilibrium. The reason for this is that both players believe it will be in the best interest of each to adopt that particular strategy because of the bigger payoff. In other games involving multiple Nash equilibria, determining the equilibrium strategy profile may not be as straightforward, in which case it may be necessary for players to choose from among competing strategies on the basis of some other selection criterion. Two commonly used criteria for selecting a strategy in games with zero or multiple Nash equilibria in which a focal-point equilibrium does not exist are the *maximin* and *minimax regret* decision rules.

Application 2.4   A Beautiful Mind

The 2001 Academy Award for Best Picture was presented to *A Beautiful Mind,* which was loosely based on the best-selling biography of the same name by Sylvia Nasar (1998). The book recounts the life of John Forbes Nash, Jr., who, along with John Harsanyi and Reinhard Selten, was awarded the 1994 Nobel Prize in economics for his pioneering work in game theory. The movie purportedly relates the moment when Professor Nash, who as a doctoral candidate in mathematics at Princeton University was desperately in search of a dissertation topic, experienced an epiphany that ultimately leads to his now famous equilibrium. According to the movie, this momentous event occurred, of all places, in a tavern. In the scene, John Nash (portrayed by Russell Crowe) is enjoying a few adult beverages with his classmates when into the tavern walk several young women, one of whom is a striking blonde and the rest are brunettes. The juvenile banter that ensues clearly suggests that the blonde is the target of choice. It is at this point in the movie that John Nash receives his most famous insight, which seemingly takes him a full year to put down on paper. Nash reasoned that if everyone goes for the blonde, not only do they block one another, they will also fail to attract any of the brunettes since they will be insulted at being second choice. From this, Nash concludes that the optimal strategy, that is, the strategy profile that results in the best outcome for the group, is for everyone to ignore the blonde and go for the brunettes. While this strategy will result in no one getting the blonde, at least each will have someone to go home with.

The proposed outcome in which Nash and his classmates ignore the blonde and go for the brunettes is presumably a Nash equilibrium to the game described in the above scene. Unfortunately, it is not. Recall that in a Nash equilibrium every player adopts a strategy that is the best response to the strategies adopted by rivals, and no player can obtain a better payoff by switching strategies. This is clearly not the case in the above scene. The reason for this is that if everyone goes for a brunette, there will be no competition for the blonde, in which case it will be in the best interest of at least one of the classmates to adopt a *blonde* strategy. In fact, there are several strategy profiles that constitute a Nash equilibrium, but ignoring the blonde is not included among them. To see why, consider Figure A2.4. To keep things simple, assume that there are only two players (student A and student B) and two strategies (*blonde* and *brunette*). The payoffs in parentheses represent utility indices of happiness—the larger the number, the greater the overall level of satisfaction.

FIGURE A2.4
Beautiful Mind Game

Student B

		Blonde	Brunette
Student A	Blonde	(0, 0)	**(2, 1)**
	Brunette	**(1, 2)**	(1, 1)

Payoffs: (Student A, Student B)

There are four strategy profiles for this game: {*Blonde, Blonde*}, {*Blonde, Brunette*}, {*Brunette, Blonde*}, and {*Brunette, Brunette*}. The payoffs for a {*Blonde, Blonde*} strategy profile is (0, 0) since if both players go for the blonde they will block each other and neither will end up with a date. The payoffs for the {*Blonde, Brunette*} strategy profile is (2, 1), which suggests that the blonde is preferred to the brunette. If both students adopt a brunette strategy, the payoffs are (1, 1) indicating that while neither student will get the best payoff, each will have a date.

The first thing to note is that neither student has a dominant strategy. If student B adopts a *blonde* strategy, it will be in the best interest of student A to adopt a *brunette* strategy, and vice versa. It turns out that this game has two Nash equilibria, and {*Brunette, Brunette*} is not one of them. Why? Because if both players adopt a *brunette* strategy, it will be in both players' best interest to switch to a blond strategy since to do so promises a larger payoff. Of course, this leads to circular reasoning since if both adopt a *blonde* strategy it will be in the best interest of both to switch to a *brunette* strategy. The Nash equilibrium strategy profiles for this game are {*Blonde, Brunette*} and {*Brunette, Blonde*}.

MAXIMIN DECISION RULE

We saw in the version of the oil-drilling game depicted in Figure 2.8 and the Internet service game depicted in Figure 2.9 that neither player had a dominant strategy. Both of these games had multiple Nash equilibria. If a focal-point equilibrium does not exist for games involving multiple Nash equilibria, how should a player choose which strategy to adopt, and will the resulting equilibrium strategy profile be a Nash equilibrium? One possible approach is for a player to select a strategy based upon an arbitrary decision rule. One such decision rule is for a player to adopt a **maximin strategy**, which is also referred to as a **secure strategy**.

> **Maximin (secure) strategy** A strategy that selects the best payoff from among the worst payoffs. By adopting a maximin strategy, a player avoids the worst possible outcome.

A maximin strategy is one of several arbitrary decision rules that may be adopted by players in games involving multiple Nash equilibria. Each of these decision rules has an underlying rationale that reflects the personality and objectives of the players who adopt them. A player who adopts a maximin strategy has a very conservative personality. A player adopts a maximin strategy for essentially one of two reasons. Either the player is very risk averse, or the payoff associated with incorrectly predicting a rival's strategy is too onerous to contemplate, as might occur, for example, in times of war when the lives of the soldiers are at stake, or in business when adopting the wrong strategy could lead to bankruptcy. The objective of a player who adopts a maximin strategy is to maximize his or her sense of security. A maximin strategy guarantees that a player will receive the best of the worst possible payoffs.

A player who adopts a maximin strategy begins by examining the worst payoffs associated with every strategy profile. That player then adopts the strategy associated with the best of these worst payoffs. Somewhat surprisingly, while a player who adopts a maximin strategy may not actually obtain the best of the worst payoffs, that player is assured of avoiding the worst of all possible payoffs.

Consider, again, the game depicted in Figure 2.8. What is PETROX's maximin (secure) strategy? To determine this, suppose that GLOMAR adopts a *narrow* strategy. If PETROX adopts a *narrow* strategy, its payoff will be $52 million. If PETROX adopts a *wide* strategy, its payoff will be $70 million. The strategy that results in the lowest payoff is clearly *narrow*. Now, suppose that GLOMAR instead adopts a *wide* strategy. If PETROX adopts a *narrow* strategy, its payoff will be $26 million. If PETROX adopts a *wide* strategy, its payoff will be $24 million. In this case, the strategy with the lowest payoff is *wide*. PETROX's maximin strategy is *narrow* because it results in the best of the worst payoffs, which is $52 million. Stated differently, regardless of the strategy adopted by GLOMAR, PETROX will avoid the worst payoff of $24 million. In this case, regardless of the strategy adopted by GLOMAR, the worst that PETROX can do is to earn a payoff of $26 million. Since the payoff matrix in this game is symmetrical, GLOMAR's secure strategy is also *narrow*. Thus, the maximin strategy profile for this game is {*Narrow, Narrow*}, in which case both companies earn a profit of $52 million.

The interesting thing about the maximin strategy profile {*Narrow, Narrow*} is that it is not a Nash equilibrium. To begin with, since the maximin decision rule is arbitrary, neither company's strategy is the best response to the strategy adopted by its rival. Moreover, if either company believes that its rival will adopt its maximin strategy, that company can improve its payoff by adopting a *wide* strategy. If both companies believe that its rival will adopt its maximin strategy, the resulting strategy profile {*Wide, Wide*} is also not a Nash equilibrium. Continuing along this line of reasoning, both companies will decide to adopt a *narrow* strategy, and so on and so forth. Although this back and forth reasoning is enough to give you a headache, in the end both companies may just decide to throw up their hands, play their maximin strategy, and let the chips fall where they may.

As the reader may have guessed, adopting a maximin strategy in non-zero-sum games suffers from an important shortcoming. As an arbitrary decision rule, using a maximin decision rule fails to consider the optimal decisions of rivals, which could result in significant opportunity costs. To see this, consider again Figure 2.9. If both players adopt a maximin strategy, the maximin strategy profile for this game is {*Low speed, Don't sign*} with payoffs of (1, 1). As we saw earlier, however, it is entirely rational for both players to focus on the strategy profile that results in the best possible outcome for both players, which in this case would be for Blue Streak to adopt a high-speed strategy and for Mr. Smith to sign an Internet service contract. The resulting payoff for the strategy profile {*High speed, Sign*} is double the payoff that results when both players employ a maximin decision rule.[3]

_____ **Demonstration Problem 2.5** _____

Consider the noncooperative, simultaneous-move, one-time game depicted in Figure D2.5 in which larger payoffs are preferred.

    a. Does either player in this game have a strictly dominant strategy? Explain.
    b. What is the strategy profile for this game if both players adopt a secure strategy?
    c. Is this strategy profile a Nash equilibrium?

Static Game in Demonstration Problem 2.5

Player B

		B1	B2
Player A	A1	**(100, 100)**	(75, 75)
	A2	(−100, 90)	**(200, 200)**

Payoffs: (Player A, Player B)

Solution

a. Neither player has a strictly dominant strategy. If player A adopts strategy A1, player B will adopt strategy B1 because it will result in a payoff of 100, compared with a payoff of 75 if player B adopts strategy B2. If player A adopts strategy A2, player B will adopt strategy B2 because it will result in a payoff of 200, compared with a payoff of 90 by adopting strategy B1. Thus, player B does not have a strictly dominant strategy. What about player A? If player B adopts strategy B1, player A will adopt strategy A1 because it results in a payoff of 100, compared with a payoff of −100 by adopting strategy A2. If player B adopts strategy B2, player A will adopt strategy A2 because it results in a payoff of 200, compared with a payoff of 100 by adopting strategy B1. Thus, player A also does not have a strictly dominant strategy.

b. Using a maximin strategy, each player will select the largest payoff from among the worst possible payoffs. If player B plays strategy B1, the minimum payoff for player A is −100 by adopting strategy A2. If player B plays strategy B2, the minimum payoff for player A is 75 by adopting strategy A1. Player A's maximin strategy is A1 because it results in the best of the two worst payoffs. What about player B's secure strategy? If player A plays strategy A1, the minimum payoff for player B is 75 by choosing strategy B2. If player A plays strategy A2, the minimum payoff for player B is 90 by playing strategy B1. The secure strategy for player B is to play strategy B1 because it results in the best of the two worst payoffs. Thus, the secure strategy profile for this game is {A1, B1}.

c. The secure strategy profile for this game {A1, B1} is a Nash equilibrium, but it is not the only one. The strategy profile {A2, B2} is also a Nash equilibrium. If player A does not adopt a secure strategy, but instead adopts strategy A2, it will be in player B's best interest to adopt strategy B2, which results in the highest payoff of 200. Likewise, if player B does not adopt its secure strategy, but instead adopts strategy B2, it will be in player A's best interest to adopt strategy A2, which also results in a payoff of 200. The strategy profile for this combination of strategies is {A2, B2}. Which strategy should the players adopt? In games involving multiple Nash equilibria, it is difficult for a player to predict a rival's strategy without additional information about the players' relationship, the context in which the game is played, and whether communication between the players is possible. In some cases, a unique equilibrium may present itself because the players share a common understanding of the problem. Such a strategy profile is called a focal-point equilibrium.

Using a maximin strategy may even be used when a player has a dominant strategy. Consider the game depicted in Figure 2.10. The reader should verify that player B has a strictly dominant

FIGURE 2.10
Risk Aversion and a Maximin Strategy

Player B

		B1	B2
Player A	A1	(**100**, 0)	(100, **100**)
	A2	(−1,000, 0)	(**200, 100**)

Payoffs: (Player A, Player B)

strategy, while player A does not. Strategy B2 results in the largest payoff for player B regardless of the strategy adopted by player A. Of course, player A knows this and will probably adopt strategy A2, which will result in the largest payoff. Thus, the resulting Nash equilibrium strategy profile for this game is {A2, B2}. The payoffs for this outcome are highlighted in boldface.

Suppose, however, that in the game depicted in Figure 2.10 player A believes that player B might not, in fact, play his or her strictly dominant strategy. This possibility might arise if player B has a history of making mistakes. When risk and uncertainty are introduced, the game changes. Depending upon the level of risk aversion, it might be in player A's best interest to adopt a maximin strategy, especially if the expected loss by choosing the wrong strategy is very large. If player B adopts the strictly dominant strategy B2, player A can earn a payoff of 200 by adopting strategy A2. On the other hand, if player A believes that there is a possibility that player B could adopt strategy B1, the payoff for player A will be an unacceptably large loss of 1,000. Prudence dictates adopting a secure strategy A1 that guarantees a payoff that is no lower than 100, which may be preferable to an uncertain payoff of 200 by adopting the strictly dominant strategy A2. If, in fact, player B does adopt a strictly dominant strategy, the outcome is the strategy profile {A1, B2} with payoffs of (100, 100). It is left as an exercise for the reader to determine whether this strategy profile constitutes a Nash equilibrium.

Example: Touchdown Game

Suppose that the New York Giants and the Baltimore Ravens football teams are in the fourth quarter of a game with seconds remaining on the clock. There is enough time left for one last play. The score is Baltimore Ravens 13 and the New York Giants 17. The Baltimore Ravens have the ball on the New York Giants' 8-yard line. There are no time-outs for either side. A field goal for 3 points will not help Baltimore. To win the game, Baltimore must score a touchdown for 6 points. Both sides must decide on a strategy for the final play of the game. The objective of both teams is to maximize the probability of winning the game. Both head coaches are aware of the strengths and weaknesses of their team's rival. The probabilities of either team winning the game from alternative offensive and defensive strategies are summarized in Figure 2.11. The student should note that the sum of the probabilities in each cell is 100 percent. This is an example of a zero-sum game—if one team wins, the other team loses.

An examination of the payoff matrix in Figure 2.11 will verify that neither team has a dominant strategy. If the New York Giants, for example, adopt a pass defense, the best offensive play for the Baltimore Ravens is to run the ball because the probability of winning the game is 60 percent, compared with 50 percent by adopting a pass offense. If the Giants adopt a run defense, the best offensive play for the Ravens is to pass the ball since this will result in an 80 percent chance of winning the game, compared with a 20 percent chance of winning by running the ball. On the other hand, if the Ravens decide to pass the ball, the best strategy for the Giants is a pass defense, which results in a 50 percent chance of winning, as opposed to 20 percent by adopting a

FIGURE 2.11
Touchdown Game

Baltimore Ravens

		Pass offense	Run offense
New York Giants	Pass defense	(**50**, 50)	(40, **60**)
	Run defense	(20, **80**)	(**80**, 20)

Payoffs: (New York Giants, Baltimore Ravens)

run defense. Finally, if the Ravens decide to run the ball, the best strategy for the Giants is a run defense, which results in an 80 percent chance of winning, compared with only 40 percent by adopting a pass defense. The reader should verify that a Nash equilibrium does not exist for this game. There is no strategy profile for which a change in strategy will result in a higher probability of winning for either team.

Since neither team has a strictly dominant strategy, what strategy should each head coach adopt? If we assume that both head coaches are very risk averse, both teams may adopt a maximin strategy. If the head coach of the New York Giants decides to play a pass defense, the worst the team can do is a 40 percent probability of winning the game. If the Giants decide to play a run defense, the worst that the team can do is a 20 percent chance of winning the game. Since a 40 percent probability of winning the game is the largest of the two worst payoffs, playing a pass defense is the Giants secure strategy.

Now consider the secure strategy of the Baltimore Ravens. If the Ravens play a pass offense, the worst the team can do is a 50 percent probability of winning the game. If the Ravens decide to play a run offense, the worst the team can do is a 20 percent probability of winning the game. Since a 50 percent probability of winning the game is the largest of the two worst payoffs, playing a pass offense is the Ravens' secure strategy. When both teams play their maximin strategy, the maximin strategy profile for this version of the touchdown game is {*Pass defense, Pass offense*}.

─────────────────── **Demonstration Problem 2.6** ───────────────────

Suppose that in the touchdown game the probabilities of either team winning from alternative offensive and defensive strategy profiles are summarized in the game depicted in Figure D2.6.

    a. Does either team have a dominant strategy?
    b. Suppose that both coaches are extremely risk averse and adopt a maximin strategy. What strategy will each coach likely adopt for the last play of the game?
    c. Does this game have a Nash equilibrium?

Solution

    a. Neither team has a strictly dominant strategy. If the New York Giants adopts a *Pass defense*, the best that the Baltimore Ravens can do is to adopt a *run offense*. If the Giants adopt a *run defense*, the best offensive play for the Ravens is to adopt a *pass offense*. As for the New York Giants, if the Ravens adopt a *pass offense*, the best strategy for the Giants is a *pass defense*. If the Ravens decide upon a *run offense*, the best strategy for the Giants is a *run defense*. Since the best strategy for both teams

FIGURE D2.6
Static Game in Demonstration Problem 2.6

Baltimore Ravens

		Pass offense	Run offense
New York Giants	Pass defense	(**70**, 30)	(20, **80**)
	Run defense	(10, **90**)	(**55**, 45)

Payoffs: (New York Giants, Baltimore Ravens)

depends on the strategy adopted by the other team, neither team has a dominant strategy.

b. Assuming again that both head coaches are risk averse, both teams should adopt a maximin strategy. If the New York Giants adopt a *pass defense,* the worst the team can do is a 20 percent chance of winning the game. If the Giants decide to play a *run defense,* the worst that the team can do is a 10 percent chance of winning the game. Since a 20 percent probability of winning the game is the largest of the two worst payoffs, the Giants' secure strategy is to play a *pass defense.*

What is the maximin strategy of the Baltimore Ravens? If the Ravens play a *pass offense,* the worst the team can do is a 30 percent chance of winning the game. If the Ravens decide to play a *run offense,* the worst the team can do is a 45 percent chance of winning the game. Since a 45 percent chance of winning the game is the largest of the two worst payoffs, the Ravens' maximin strategy is to run the ball. Thus, the maximin strategy profile for this version of the touchdown game is {*Pass defense, Run offense*}.

c. This game does not have a Nash equilibrium. To see this, suppose that the Giants believe that the Ravens will adopt its secure *run offense* strategy. In this case, it will be in the Giants' best interest to adopt a *run defense* strategy, and not its *pass defense* secure strategy. But, if the Ravens believe that the Giants will adopt a *run defense* strategy, it will be in the Ravens' best interest to switch strategies and adopt a *pass offense* strategy. Of course, the Giants will reason that if the Ravens adopt a *pass offense* strategy, it will be in their best interest to switch strategies and adopt a *pass defense,* in which case the Ravens will switch to a *run offense* strategy, and so on.

MINIMAX REGRET DECISION RULE

Another decision rule for selecting a strategy that is frequently mentioned in the literature on game theory is the **minimax regret strategy**. According to this decision rule, a player should minimize the regret associated with adopting an incorrect strategy. Also known as a **Savage strategy**, this decision rule is based on the psychologically plausible premise that a player who adopts a strategy believing that a rival will adopt a given strategy will feel regret if the rival adopts some other strategy, which results in a lower payoff. The positive difference between a given payoff and the best possible payoff is referred to as **opportunity loss**. A player who adopts a minimax regret strategy attempts to minimize *ex post* feelings of regret associated with incorrectly predicting a rival's strategy. Unlike the maximin decision rule, which reflects a player's attitude toward risk, the minimax regret strategy emphasizes the after-the-fact consequences of making an incorrect decision.

Minimax regret (Savage) strategy A strategy that minimizes the opportunity loss of an incorrect decision.

FIGURE 2.12
PETROX's Opportunity Loss (Regret) Payoff Matrix

		GLOMAR's strategy	
		Narrow	*Wide*
PETROX's opportunity loss	*Narrow*	$70 – $52 = $18	$26 – $26 = $0
	Wide	$70 – $70 = $0	$26 – $24 = $2

Opportunity loss The difference between a given payoff and the best possible payoff.

To illustrate the rationale behind a minimax decision rule using a numerical example, consider Figure 2.12, which summarizes the opportunity losses to PETROX associated with each of the four strategy profiles in the oil-drilling game depicted in Figure 2.8. The cells in this matrix summarize the difference between the maximum possible payoff and the actual payoff to PETROX associated with whatever strategy is adopted by GLOMAR. These opportunity losses are either positive or zero since we are subtracting the payoff from a given strategy from the best possible payoff. For example, if GLOMAR adopts a *narrow* strategy and PETROX adopts a *narrow* strategy, the payoff to PETROX will be $52 million. Since PETROX could have adopted a *wide* strategy with a payoff of $70 million, the opportunity loss is $18 million. If PETROX adopts a *wide* strategy, the opportunity loss will be zero since this results in a payoff of $70 million, which is also the largest payoff. While PETROX will regret adopting an incorrect *narrow* strategy, it will not regret adopting a *wide* strategy. By contrast, suppose that GLOMAR adopts a *wide* strategy and PETROX adopts a *narrow* strategy. In this case, the payoff to PETROX is $26 million, which is the best possible payoff. Since the opportunity loss is zero, PETROX has no regrets about adopting a *narrow* strategy. On the other hand, if PETROX adopts a *wide* strategy, the payoff will be $24 million, with an opportunity loss of $2 million. Thus, PETROX will adopt a *wide* strategy because it will minimize the opportunity loss (regret) of adopting an incorrect strategy. Since the payoff matrix in Figure 2.8 is symmetric, a *wide* strategy will be GLOMAR's minimax regret strategy as well.

To quickly determine a player's minimax regret strategy, examine the payoff matrix itself. In Figure 2.8, if GLOMAR adopts a *narrow* strategy, the difference in payoffs for PETROX by adopting an "incorrect" *narrow* strategy is $18 million. If GLOMAR adopts a *wide* strategy, the difference in payoffs for PETROX by adopting an "incorrect" *wide* strategy is $2 million. Since the opportunity loss of $2 million is less than the opportunity loss of $18 million, PETROX will adopt a *wide* strategy since this minimizes the opportunity loss associated with making an incorrect decision. Similarly, if PETROX adopts a *narrow* strategy, the difference in payoffs for GLOMAR by adopting a *narrow* strategy is $18 million. If PETROX adopts a *wide* strategy, the difference in payoffs for GLOMAR by adopting a *wide* strategy is $2 million. Again, since the opportunity loss of $2 million is less than the opportunity loss of $18 million, GLOMAR will adopt the strategy that minimizes these opportunity losses, which is a *wide* strategy. If both companies adopt a minimax regret strategy, the minimax regret strategy profile for this game is {*Wide, Wide*}. Compare this with the strategy profile {*Narrow, Narrow*} that results when both companies adopt a maximin (secure) strategy. The reader should verify that just as in the case of the maximin strategy profile {*Narrow, Narrow*}, the minimax regret strategy profile {*Wide, Wide*} is not a Nash equilibrium.

Example: Umbrella Game

While getting ready for work, Moira glances out the window and observes that the sky is overcast. Fearing the worst, she tunes in to the Weather Channel and learns that there is a 50 percent chance

FIGURE 2.13
Moira's Payoffs in the Umbrella Game

Nature

Moira		Sunny	Rainy
Moira	Umbrella	2	5
	No umbrella	7	1

of rain. Moira has to decide whether or not to carry an umbrella. In the umbrella game there are two players, Moira and Nature. In this game, Nature randomly selects from between two strategies: *Sunny* or *rainy*. Moira's choice is between *umbrella* and *no umbrella*. If Moira adopts an *umbrella* strategy and it rains, then she has chosen correctly. If she adopts a *no umbrella* strategy and the skies clear, once again she has made the correct choice. But, if Moira brings an umbrella and the day is bright and sunny, she chooses incorrectly, in which case she will be inconvenienced by carrying an unneeded umbrella. Alternatively, if Moira decides to leave the umbrella home and it rains, she will suffer the discomfort of spending the day cold and wet, and might even come down with a cold or flu. Moira's options are summarized in Figure 2.13. The payoffs are utility indices representing Moira's preferences for each strategy profile. A larger utility index represents a greater level of happiness. Clearly, Moira most prefers the strategy profile {*No umbrella, Sunny*} and least prefers the strategy profile {*No umbrella, Rainy*}. Finally, she prefers the strategy profile {*Umbrella, Sunny*} only marginally more than the strategy profile {*No umbrella, Rainy*}.

The reader should verify that Moira does not have a dominant strategy. Moira's best strategy depends on the strategy "adopted" by Nature. What strategy should Moira adopt? The answer depends on which incorrect decision will cause Moira the greatest regret. Moira's decision will be based on a desire to minimize her *ex post* feelings of regret in the event of an incorrect choice. Using the minimax regret decision rule, the strategy profile {*No umbrella, Rainy*} results in an opportunity loss of 4, whereas the strategy profile {*Umbrella, Sunny*} results in an opportunity loss of 5. Since the objective is to minimize *ex post* feelings of regret from choosing an incorrect strategy, Moira will bring an umbrella.

The minimax regret decision rule also suffers from the same shortcoming of the maximin decision rule in that it fails to consider the strategies of rivals. We saw in the game depicted in Figure 2.8 that if both players adopt a minimax regret decision rule, the resulting strategy profile is {*Wide, Wide*} with payoffs of (24, 24), which *is not* a Nash equilibrium. If PETROX uses a maximin decision rule and GLOMAR uses the minimax regret decision rule, then the resulting strategy profile is {*Narrow, Wide*}, which *is* a Nash equilibrium. To make matters worse, suppose that both players focus on the strategy profile {*Narrow, Narrow*} because this promises the best combined outcome for both companies. This strategy profile also does not constitute a Nash equilibrium. As with the application of the minimax decision rule, the problem with the minimax regret decision rule is that it denies the rationality properties inherent in a Nash equilibrium.

───────────── **Demonstration Problem 2.7** ─────────────

Consider, again, the touchdown game depicted in Figure 2.11.

    a. Suppose that the New York Giants and Baltimore Ravens adopt a minimax regret decision rule. What strategy will each team adopt for the last play of the game?

    b. Is the minimax regret strategy profile for this game a Nash equilibrium?

Solution

a. If the Baltimore Ravens adopt a *pass offense* strategy, the difference in payoffs for the New York Giants from alternative strategies is 30 percent. If the Ravens adopt a *run offense* strategy, the difference in payoffs for the Giants from alternative strategies is 40 percent. Since the opportunity loss of 30 percent is less than the opportunity cost of 40 percent, the Giants will adopt a *run defense* minimax regret strategy. Likewise, if the Giants adopt a *pass defense* strategy, the difference in payoffs for the Ravens from alternative strategies is 10 percent. If, on the other hand, the Giants adopt a *run defense* strategy, the difference in payoffs from alternative strategies is 60 percent. Since the opportunity cost of 10 percent is less than the opportunity cost of 60 percent, the Ravens will adopt a *pass offense* minimax regret strategy. Thus, the minimax regret strategy profile for this game is {*Run defense, Pass offense*}, which is not the same thing as the maximin strategy profile {*Pass defense, Pass offense*}.

b. The minimax regret strategy profile {*Run defense, Pass offense*} is not a Nash equilibrium. The reason for this is that if the Giants believe that the Ravens will adopt its minimax regret *pass offense* strategy, it will be in the Giants' best interest not to adopt its minimax regret *run defense* strategy, but instead switch to a *pass defense* strategy because of its larger payoff. Of course, if the Ravens believe that the Giants plan to adopt a *pass defense* strategy, the Ravens will adopt a *run offense* strategy. Knowing this, the Giants will stick with its minimax regret *run defense* strategy, and so on. Thus, the minimax regret strategy profile {*Run defense, Pass offense*} is not a Nash equilibrium.

─────────────────── **Demonstration Problem 2.8** ───────────────────

Consider, again, the touchdown game depicted in Figure D2.6. Suppose that both coaches adopt a minimax regret strategy. What strategy will each coach adopt for the last play of the game?

Solution

If the Baltimore Ravens adopt a *pass offense* strategy, the difference in payoffs for the New York Giants from alternative strategies is 60 percent. If the Ravens adopt a *run offense* strategy, the difference in payoff for the Giants from alternative strategies is 35 percent. The Giants will adopt a *pass defense* strategy, which is the smaller of these opportunity costs. Similarly, if the Giants adopt a *pass defense* strategy, the difference in payoffs for the Ravens from alternative strategies is 50 percent. If the Giants adopt a *run defense* strategy, the difference in payoffs for the Ravens from adopting alternative strategies is 45 percent. Since the opportunity cost of 45 percent is less than the opportunity cost of 50 percent, the Ravens will adopt a *run offense* strategy. Thus the minimax regret strategy profile for this game is {*Pass defense, Run offense*}.

CHAPTER REVIEW

This chapter examined *two-player, noncooperative, static, one-time games.* In a *noncooperative game* the players do not engage in collusive behavior. Such games may have zero-sum or non-zero-sum outcomes. In a *zero-sum game*, one player's gain is another player's loss. In a *non-zero-sum game*, one player's gain does not equal another player's loss. In a non-zero-sum game, both players may be made better or worse off. In a *one-time game* each player has one, and only one, move. In static, one-time games, a rival's strategy is unknown until the end of the game. The prisoner's dilemma is an example of a two-player, noncooperative, non-zero-sum, static, one-time game.

A *strictly dominant strategy* is strictly preferred to every other strategy. A strictly dominant strategy has the best payoff regardless of the strategies adopted by the other players. A player with a strictly dominant strategy, however, does not necessarily receive the largest payoff associated with all possible strategy profiles.

Many, but not all, solutions to problems in game theory have a *Nash equilibrium.* A game has a *Nash equilibrium* when each player adopts a strategy that is believed to be the best response to the strategies adopted by the other players. When a game has a Nash equilibrium, the players' payoffs cannot be improved by adopting a different strategy. A *strictly dominant strategy equilibrium* is a Nash equilibrium that results when all players adopt a strictly dominant strategy.

Adopting a *weakly dominant strategy* results in a payoff that is no lower than any other payoff, regardless of the strategy adopted by a rival. A *weakly dominant strategy equilibrium* results when each player adopts his or her weakly dominant strategy.

Noncooperative, one-time, static games with a large number of strategies may require several iterations before an equilibrium is found. If the players have a strictly dominant strategy after all strictly dominated strategies have been removed, the equilibrium strategy profile is called an *iterated strictly dominant strategy equilibrium.* If each player has a strictly dominant strategy, the order in which strictly dominated strategies are eliminated is irrelevant. If strictly dominant strategies are replaced by weakly dominant strategies, the order in which weakly dominated strategies are removed could change the outcome of the game.

A *nondominant strategy* is a strategy that does not dominate any other strategy. An optimal strategy depends upon what each player believes will be the strategies adopted by the other players. Games in which no player has a dominant strategy often do not have a unique Nash equilibrium, or have multiple Nash equilibria.

In games involving multiple Nash equilibria, it is difficult for a player to predict his or her rival's strategy without additional information about the nature of the relationships among the players, the context in which the game is being played, and whether cooperation between the players is possible. In some cases, an equilibrium may stand out because the players share a common understanding of the problem. Such an outcome is referred to as a *focal-point equilibrium.*

If a focal-point equilibrium does not exist, one possible approach is for each player to adopt a *maximin decision rule.* Also referred to as a *secure strategy,* this approach selects the best payoff from among the worst possible payoffs. Maximin strategies are most appropriate for extremely risk-averse players, or in situations involving disastrous outcomes.

An alternative to the maximin decision rule is a *minimax regret decision rule.* According to this approach, a player should minimize feelings of regret associated with adopting an incorrect strategy. This strategy is based on the psychologically plausible premise that a player who adopts a strategy based on the belief that a rival will adopt a given strategy will feel regret if that rival adopts some other strategy with results in a lower payoff. This difference in payoffs is referred to as *opportunity loss.* A player who adopts a minimax regret strategy is attempting to minimize *ex post* feelings of regret associated with incorrectly guessing a rival's strategy. Unlike the maximin decision rule, which reflects a player's attitude toward risk, the minimax regret strategy emphasizes the after-the-fact consequences of making an incorrect decision.

CHAPTER QUESTIONS

2.1   Explain the difference between moves and strategies.

2.2   Suppose you and a group of your co-workers have decided to have lunch at a Japanese restaurant. It has been decided in advance that the lunch bill will be divided equally. Each person in the group is concerned about his or her share of the bill. Without explicitly agreeing to do so, each person will order from among the least expensive items on the menu. Do you agree? Explain.

2.3   In a two-player, one-time, static game, if one player has a dominant strategy, the second player will never adopt a maximin strategy. Do you agree? Explain.

2.4   In a two-player, static game with a strictly dominant strategy equilibrium, at least one of the players will adopt a secure strategy. Do you agree? If not, why not?

2.5   Explain the difference between a strictly dominant strategy and a weakly dominant strategy.

2.6   Under what circumstances will a strictly dominant strategy lead to a strictly dominant strategy equilibrium?

2.7   The existence of a Nash equilibrium confirms Adam Smith's famous metaphor of the invisible hand. Do you agree with this statement? If not, why not?

2.8   Explain the difference between a strictly dominant strategy and an iterated strictly dominant strategy.

2.9   Explain the difference between a strictly dominant strategy and a weakly dominated strategy.

CHAPTER EXERCISES

2.1   Argon Airlines and Boron Airways are two equally sized, commercial air carriers that compete for passengers along the lucrative Boston–Albany–Buffalo route. Both firms are considering offering discount air fares during the traditionally slow month of February. This static game is depicted in Figure E2.1. Payoffs are in millions of dollars.

   a.   If larger payoffs are preferred, does either firm have a strictly dominant strategy?
   b.   What is the Nash equilibrium strategy profile for this game? Explain.

2.2   Consider the static game depicted in Figure E2.2.

   a.   For what values of x is *strategy 2* a dominant for player A?
   b.   For what values of x is *strategy 3* a strictly dominant strategy for player B?
   c.   For what values of x is the strategy profile {*Strategy 2, Strategy 3*} a weakly dominant strategy Nash equilibrium?
   d.   For what values of x is the strategy profile {*Strategy 1, Strategy 4*} a weakly dominant strategy Nash equilibrium.

2.3   Suppose that an industry consists of two firms, Magna Company and Summa Corporation. Each firm produces an identical product, and each is trying to decide whether to expand (*Expand*) or not to expand (*Do not expand*) its production capacity for the next operating period. Assume that each firm is currently operating at full capacity. The trade-off facing each firm is that expansion will result in greater sales, but increased output will put downward pressure on prices and revenues. This one-time, static game is depicted in Figure E2.3. Payoffs are in millions of dollars.

   a.   If larger payoffs are preferred, does either firm have a dominant strategy?
   b.   What is the Nash equilibrium strategy profile for this game?

2.4   Consider the noncooperative, static, one-time game depicted in Figure E2.4.

FIGURE E2.1
Static Game in Chapter Exercise 2.1

		Boron	
		Discount	No discount
Argon	Discount	(2, 3)	(7.5, 1)
	No discount	(1.5, 6)	(3, 2)

Payoffs: (Argon, Boron)

FIGURE E2.2
Static Game in Chapter Exercise 2.2

		Player B	
		Strategy 3	Strategy 4
Player A	Strategy 1	$(3, 6 - x)$	(5, 4)
	Strategy 2	(4, 5)	$(6 - x, 3)$

Payoffs: (Player A, Player B)

FIGURE E2.3
Static Game in Chapter Exercise 2.3

		Summa	
		Do not expand	Expand
Magna	Do not expand	(25, 25)	(15, 30)
	Expand	(30, 15)	(20, 20)

Payoffs: (Magna, Summa)

FIGURE E2.4
Static Game in Chapter Exercise 2.4

		Player B	
		Strategy A	Strategy B
Player A	Strategy A	(20, 20)	(5, 25)
	Strategy B	(25, 5)	(2, 2)

Payoffs: (Player A, Player B)

a. If larger payoffs are preferred, does either player have a dominant strategy?
b. If player B believes that player A will adopt *strategy A,* what strategy should player B adopt?
c. If player B believes that player A will adopt *strategy B,* what strategy should player B adopt?
d. Does this game have a unique Nash equilibrium?

e. What is the strategy profile for this game if both players adopt a secure strategy?

2.5 Consider, again, the one-time, static game depicted in Figure E2.4.

a. Does either player have a dominant strategy?
b. If both players adopt a minimax regret strategy, what is the strategy profile for this game?
c. How does your answer to part b differ when both players adopt a maximin strategy? How do you explain your results?

2.6 Firm *A* is considering entering a market in which firm *B* is a monopolist. Firm *B* must decide how to respond to this challenge. If firm *A* decides to *enter,* firm *B* can either adopt an *aggressive* pricing strategy or a *passive* pricing strategy. If firm *B* adopts an *aggressive* pricing strategy, firm *A* will lose $5 million and firm *B* will earn $1 million. On the other hand, if firm *B* adopts a *passive* pricing strategy, firm *A* will earn $3 million and firm *B* will earn $7 million. Finally, if firm *A* decides to *stay out* of the market, firm *B* will earn $14 million regardless of the pricing strategy adopted.

a. Does either firm have a dominant strategy?
b. What is the Nash equilibrium strategy profile for this game?

2.7 Tom Teetotaler and Brandy Merrybuck are tobacconists specializing in three brands of pipe weed: Barnacle Bottom, Old Toby, and Southern Star. Both shops, Red Pony and Blue Dragon, are small and only have enough storage space to stock two brands. Teetotaler and Merrybuck are trying to determine the optimal combination of brands to carry. Expected earnings in this noncooperative, one-time, static game are depicted in Figure E2.7. What is the Nash equilibrium strategy profile for this game?

2.8 Suppose that the payoffs for the game in chapter exercise 2.7 are depicted in Figure E2.8.

a. Does either tobacconist have a strictly dominant strategy?
b. Does either tobacconist have a strictly dominated strategy?
c. What is the Nash equilibrium strategy profile for this game?

2.9 Consider the game depicted in Figure E2.9.

a. Does either firm have a dominant strategy?
b. What is the Nash equilibrium strategy profile, and why?

2.10 Consider the touchdown game depicted in Figure E2.10. The payoffs represent the probabilities of either team winning the game from alternative offensive and defensive strategies.

a. If larger payoffs are preferred, does either team have a dominant strategy?
b. If both teams are extremely risk averse, what decision rule is each team likely to adopt?
c. Does this game have a Nash equilibrium?

2.11 Consider the one-time, static game depicted in Figure E2.11 involving two petroleum-refining companies, Oxxon and Nonox. Each company is considering opening a gasoline station either at a busy *intersection* in the middle of town or near the exit ramp of an *interstate* highway. The payoffs in this game represent monthly profits in thousands of dollars.

a. If larger profits are preferred, does either company have a dominant strategy? Explain.
b. What is the Nash equilibrium strategy profile for this game, and why?

2.12 Consider the one-time, static game depicted in Figure E2.12 involving two competing firms that are considering alternative advertising strategies. Firm *A* is trying to decide whether to

FIGURE E2.7
Static Game in Chapter Exercise 2.7

		Blue Dragon		
		Barnacle Bottom	*Old Toby*	*Southern Star*
	Barnacle Bottom	(150, 150)	(100, 200)	(50, 250)
Red Pony	*Old Toby*	(200, 100)	(200, 200)	(150, 300)
	Southern Star	(250, 50)	(300, 150)	(300, 300)

Payoffs: (Red Pony, Blue Dragon)

FIGURE E2.8
Static Game in Chapter Exercise 2.8

		Blue Dragon		
		Barnacle Bottom	*Old Toby*	*Southern Star*
	Barnacle Bottom	(150, 150)	(300, 250)	(350, 200)
Red Pony	*Old Toby*	(300, 100)	(200, 200)	(150, 300)
	Southern Star	(250, 50)	(260, 150)	(300, 300)

Payoffs: (Red Pony, Blue Dragon)

FIGURE E2.9
Static Game in Chapter Exercise 2.9

		Blue Dragon		
		Barnacle Bottom	*Old Toby*	*Southern Star*
	Barnacle Bottom	(150, 150)	(200, 250)	(250, 350)
Red Pony	*Old Toby*	(250, 125)	(175, 200)	(270, 245)
	Southern Star	(350, 250)	(150, 275)	(200, 300)

Payoffs: (Red Pony, Blue Dragon)

FIGURE E2.10
Static Game in Chapter Exercise 2.10

		New England Patriots	
		Pass	*Run*
Carolina Panthers	*Pass defense*	(65, 35)	(27, 73)
	Run defense	(74, 26)	(30, 70)

Payoffs: (Carolina Panthers, New England Patriots)

FIGURE E2.11
Static Game in Chapter Exercise 2.11

Nonox

Oxxon		Intersection	Interstate
	Intersection	($150,000, $150,000)	($50,000, $200,000)
	Interstate	($200,000, $50,000)	($75,000, $75,000)

Payoffs: (Oxxon, Nonox)

FIGURE E2.12
Static Game in Chapter Exercise 2.12

Firm B

Firm A		Radio	Newspapers
	Television	(9, 7)	(14, 6)
	Magazines	(7, 11)	(8, 7)

Payoffs: (Firm A, Firm B)

advertise on *television* or in *magazines*. Firm *B* is trying to decide whether to advertise on *radio* or in *newspapers*. Payoffs are annual profits in millions of dollars.

a. If larger payoffs are preferred, does either firm have a dominant strategy? Explain.
b. What is the Nash equilibrium strategy profile for this game, and why?

2.13 Fly-by-night Airlines and Going-going-gone Airways are considering whether to switch from their current standard fare schedule or implement a frequent-flyer program. Figure E2.13 summarizes the monthly profits in thousands of dollars from alternative pricing strategies.

a. If larger payoffs are preferred, does either air carrier have a dominant strategy? Explain.
b. Does this game have a Nash equilibrium?
c. What is the strategy profile for this game if both air carriers adopt a maximin decision rule?
d. If both air carriers adopt their maximin strategy, is the resulting strategy profile a Nash equilibrium?

2.14 Two radio stations, KRZY and KRUD, are contemplating three possible broadcast formats: Rock, country, and talk radio. Surveys indicate that the market share for these three formats are 40 percent, 30 percent, and 20 percent of the listening audience, respectively. If both stations select the same format, they will split that portion of the market. If the stations choose different formats, they will get the total share of that portion of the market. The stations' revenues from the sale of advertising are proportional to their market shares. Figure E2.14 summarizes the payoffs to each station from alternative strategy profiles.

a. If larger market shares are preferred, does either radio station have a dominant strategy?
b. Does this game have a Nash equilibrium?
c. If both firms adopt a maximin decision rule, what is the strategy profile for this game?

FIGURE E2.13
Static Game in Chapter Exercise 2.13

Going-going-gone

		Standard	Frequent flyer
Fly-by-night	Standard	(250, 275)	(210, 350)
	Frequent flyer	(325, 190)	(200, 150)

Payoffs: (Fly-by-night, Going-going-gone)

FIGURE E2.14
Static Game in Chapter Exercise 2.14

KRUD

		Rock	Country	Talk
	Rock	(20, 20)	(40, 30)	(40, 20)
KRZY	Country	(30, 40)	(15, 15)	(30, 20)
	Talk	(20, 40)	(20, 30)	(10, 10)

Payoffs: (KRZY, KRUD)

FIGURE E2.15
Static Game in Chapter Exercise 2.15

Player B

		B1	B2	B3
	A1	(3, 3)	(4, 1)	(5, 4)
Player A	A2	(2, 2)	(3, 1)	(3, 3)
	A3	(2, 2)	(3, 2)	(2, 4)

Payoffs: (Player A, Player B)

2.15 Consider the noncooperative, one-time, static game depicted in Figure E2.15.

a. If larger payoffs are preferred, does either player have a dominant strategy?
b. What is the Nash equilibrium strategy profile for this game, and why?

ENDNOTES

1. The oil-drilling game presented in this and subsequent sections was inspired by a game of the same name found in Bierman and Fernandez (1998, Chapter 1).
2. See Sir Arthur Conan Doyle, *The Sign of Four* (1890), Chapter 6.
3. While adopting a maximin strategy may be a bad idea for zero-sum games, this is precisely the strategy that should be adopted in zero-sum games using mixed strategies. We will have more to say about this in Chapter 6.

3 FOCAL-POINT AND EVOLUTIONARY EQUILIBRIA

In this chapter we will:

- *Discuss possible solutions to pure-strategy, static games with multiple Nash equilibria;*
- *Introduce coordination games;*
- *Explore the concept, and the implications, of focal-point equilibria;*
- *Discuss the concept of framing and its usefulness in identifying focal-point equilibria;*
- *Introduce evolutionary game theory.*

INTRODUCTION

In the first two chapters we saw that when either player in a noncooperative, one-time, static game has a dominant strategy, a unique Nash equilibrium strategy profile will exist. If neither player has a dominant strategy, the game may have multiple Nash equilibria, or may not have a Nash equilibrium at all. When this happens, do players have any recourse other than using an arbitrary decision rule, such as adopting a maximin or minimax regret strategy? There is a substantial body of literature that attempts to identify the conditions whereby a single strategy profile emerges as being the most plausible. These conditions usually have something to do with the context within which the game is played, the nature of the relationships between and among the players, or some shared interest that leads players to "zero in" on a particular strategy profile. While formalizing these conditions has proven to be elusive, it may still be possible to identify a unique, pure-strategy Nash equilibrium.

MULTIPLE NASH EQUILIBRIA

To illustrate the problem, consider the oil-drilling game depicted in Figure 2.8. In this game, each player's optimal strategy depends on the strategy adopted by its rival. If GLOMAR believes that PETROX plans to drill a narrow well, GLOMAR's best strategy is to drill a wide well. If GLOMAR believes that PETROX intends to drill a wide well, GLOMAR's best strategy is to drill a narrow well. Because of the symmetry of the payoff matrix, the same reasoning also applies to PETROX. Thus, the oil-drilling game depicted in Figure 2.8 has two Nash equilibria: {*Narrow*, *Wide*} and {*Wide*, *Narrow*}. The optimal strategy for either company depends on the strategy adopted by its rival.

The existence of multiple equilibria sometimes leads to rather interesting results. If either GLOMAR or PETROX in Figure 2.8 believes the other is planning to drill a narrow well, both will respond by drilling wide wells, in which case the strategy profile is {*Wide*, *Wide*}. Conversely, if either company believes the other plans to drill a wide well, the strategy profile is {*Narrow*, *Narrow*}. Neither of these strategy profiles constitutes a Nash equilibrium since both companies can improve upon their payoffs by switching strategies. On the other hand, if PETROX adopts a

FIGURE 2.8
Static Oil-Drilling Game with Nondominant Strategies

GLOMAR

		Narrow	Wide
PETROX	Narrow	(52, 52)	**(26, 70)**
	Wide	**(70, 26)**	(24, 24)

Payoffs: (PETROX, GLOMAR)

FIGURE 3.1
Battle-of-the-Sexes Game

Rhett

		Tavern	Restaurant
Scarlett	Tavern	**(10, 20)**	(−59, −25)
	Restaurant	(−5, −4)	**(24, 5)**

Payoffs: (Scarlett, Rhett)

narrow strategy, and GLOMAR adopts a *wide* strategy, or if GLOMAR adopts a *narrow* strategy and PETROX adopts a *wide* strategy, the resulting strategy profiles {*Wide, Narrow*} and {*Narrow, Wide*}, both are Nash equilibria since neither firm can do better by switching strategies.

Example: Battle-of-the-Sexes Game

A well-known example of a strategic situation involving multiple Nash equilibria is the battle-of-the-sexes game. This game involves a couple, Rhett and Scarlett, who are independently planning how to spend the evening together. Scarlett would like to have dinner at an expensive restaurant, while Rhett would like to eat barbecue at a neighborhood tavern. This noncooperative, one-time, static game with complete information is depicted in Figure 3.1. The payoff matrix summarizes the couple's subjective evaluation of the satisfaction from any combination of dating strategies. The reader should confirm that the Nash equilibrium strategy profiles for this game are {*Tavern, Tavern*} and {*Restaurant, Restaurant*}. The payoffs for these strategy profiles are highlighted in boldface. As in the oil-drilling game depicted in Figure 2.8, the choice of dating strategies by Rhett and Scarlett depends on what each believes will be the most likely strategy adopted by the other.

The battle-of-the-sexes game is an example of a **coordination game**. Coordination games with multiple Nash equilibria involve social situations in which rational players attempt to coordinate their moves without explicitly cooperating to reach a mutually beneficial outcome. Players in coordination games should not be viewed as rivals. Rather, they should be considered partners whose objective is to maximize joint payoffs. A failure by these players to properly coordinate their moves will result in a strategy profile with less than optimal combined payoffs.

Thomas Schelling (1960) investigated the conditions under which a single Nash equilibrium strategy profile might stand out from among multiple Nash equilibria in coordination games.[1] Such a strategy profile, if it exists, was referred to by Schelling as a **focal-point (Schelling-point) equilibrium**. The reason why a single strategy profile emerges is because the players share a common understanding of the environment in which the game is being played. The conditions that define focal-point equilibria are often subjective and lack well-defined parameters. A focal-point equilibrium may not easily be put into words. In 1964, Supreme Court Justice Potter Steward tried to described "hard-core" pornography by observing: "I shall not today attempt further to define the kind of material I understand to be embraced . . . [b]ut I know it when I see it. . . ."[2] A focal-point equilibrium is something like that. Focal-point equilibria are also said to be **self-fulfilling** if the players believe it is in their best interest to adopt a particular strategy profile.

> **Coordination game** A game in which rational players attempt to coordinate their strategies to achieve a mutually beneficial outcome.
>
> **Focal-point (Schelling-point) equilibrium** When a single strategy profile stands out from among multiple Nash equilibria because the players share a common understanding of the environment in which the game is being played.
>
> **Self-fulfilling strategy profile** When the players believe it is in their collective best interest to adopt a particular strategy profile.

FOCAL-POINT EQUILIBRIUM

To illustrate the concept of a focal-point equilibrium, consider the situation in which a father and his young son become separated during a visit to an amusement park. Suppose that they had made no prior arrangement about where to meet if they were separated. How likely is it that both will think to look for each other at the same location? This coordination game has multiple Nash equilibrium strategy profiles, such as the amusement park's main gate or the office of park security. But will a particular strategy profile stand out from all the others? Undoubtedly, a shared cultural background, belief system, life experiences, and so on will contribute to a common understanding of the problem, and its most likely solution. Will this "conventional wisdom" lead to a unique Nash equilibrium? This is the idea underlying the concept of a focal-point equilibrium.

Dixit and Nalebuff (1991, Chapter 3) provide another example in which conventional wisdom may lead players to focus on a particular strategy profile. What is the answer to the question, "Which side of the road should I drive on?" There are two possible answers: Drive on the left or drive on the right. The problem is that a simple description of the game, including the players, the order of moves, strategies, and payoffs may not be sufficient to identify a unique Nash equilibrium strategy profile. Why? Because the answer depends on where the question is asked. If I ask the question in Beijing, Montreal, New York City, or Paris, the answer is: "On the right-hand side." If I ask the same question in Hong Kong, London, Sydney, or Tokyo, the answer is: "On the left-hand side." This question cannot be answered by looking for dominant strategies, or by eliminating dominated strategies. The players are drawn to a unique strategy profile because they share a common understanding of the context in which the question is being asked. In this case, knowledge of the local rules of the road provides the answer. Schelling (1960) illustrated the concept of a focal-point equilibrium by conducting experiments using the following "abstract puzzles."

1. A coin is flipped and two players are instructed to call "heads" or "tails." If both players call "heads," or both call "tails," both players win a prize. If one player calls "heads" and the other calls "tails," neither player wins a prize.
2. A player is asked to circle one of the following six numbers: 7, 100, 13, 261, 99, and 555. If all of the players circle the same number, each wins a prize. Otherwise, they receive nothing.
3. A player is asked to put a check mark in one of the sixteen squares in Figure 3.2. If all of the players check the same square, each wins a prize. Otherwise, they receive nothing.
4. Two players from New Haven, Connecticut, are instructed to meet somewhere in New York City. The players are not told where the meeting is to take place. The players in

FIGURE 3.2
Sixteen Squares Experiment

this game have never been placed in this situation before, nor are they permitted to communicate with each other. Each player must guess the most likely location. If they guess correctly, each wins a prize. Otherwise, they win nothing.

5. In the previous scenario, the players are told the date, but not the time of the meeting. Each player is asked to choose a time that the meeting should take place. If each chooses the same time, each player wins a prize. Otherwise, they win nothing.

6. Players are asked secretly to write down numbers. If all players write the same number, each player wins a prize. Otherwise, they receive nothing.

7. Players are asked to choose an amount of money. If all players choose identical amounts, each wins that amount. Otherwise, they receive nothing.

8. A player is asked secretly to divide $100 into two piles labeled *A* and *B*. Another player is asked to do the same. If the amounts in both piles coincide, each player receives $100. Otherwise, they receive nothing.

9. The results of the first ballot in an election are:

Smith	19 votes
Jones	28 votes
Brown	15 votes
Robinson	29 votes
White	9 votes

A second ballot is then taken. A group of players is asked to predict which candidate will receive a majority of votes on the second ballot. If a player makes a correct prediction, she or he wins a prize. If the player makes an incorrect prediction, she or he wins nothing.

Each of the above coordination games has multiple Nash equilibria. Schelling found in an "unscientific sample of respondents" that people tended to focus on just a few such equilibria (focal points). In problem 1, Schelling found that 86 percent of the respondents chose "heads." In problem 2, the first three numbers received 90 percent of the votes with the number 7 leading the number 100 by a slight margin. Number 13 came in third. In problem 3, 59 percent of the respondents chose the square in the upper left corner, while 93 percent of the respondents selected a square in the same diagonal. In problem 4, a majority of the respondents proposed meeting at the information booth in Grand Central Station, and nearly all of them agreed to meet at 12:00 noon. In problem 6, two-fifths of all respondents chose the number 1. In problem 7, 29 percent of the respondents chose $1 million, and only 7 percent chose cash amounts that were not multiples of 10. In problem 8, 88 percent of the respondents put $50 into each pile. Finally, in problem 9, 91 percent of the respondents chose Robinson.

Schelling found that the participants in these experiments tended to choose focal points, even when it was not in their individual best interest to do so. Consider, for example, the following variation of experiment 1 in which players A and B were asked to call "heads" or "tails." The players are not permitted to communicate with each other. If both players call "heads," player A gets $3 and player B gets $2. If both players call "tails," player A gets $2 and player B gets $3. As before, if one player calls "heads" and the other calls "tails," neither player wins a prize. In this scenario Schelling found that 73 percent of respondents chose "heads" when given the role of player A. More surprisingly, 68 percent of respondents in the role of player B still chose "heads" despite the lower payoff. Recall that if both players attempt to win $3, each will win nothing.

The final example of a game involving multiple Nash equilibria (Kreps 1987, 1990) involves an experiment involving pairs of college students. In each trial, the college students were provided with a list of eleven cities (Atlanta, Boston, Chicago, Dallas, Denver, Kansas City, Los Angeles, New York, Philadelphia, Phoenix, and San Francisco). Each city was assigned an arbitrary ranking index number in terms of its importance to commerce, science, the arts, and so on. New York was assigned the highest ranking (100) and Kansas City the lowest (1). Each pair of students was asked independently to assign each city to one of two subsets. Cooperation between the students was not permitted. A requirement of the selection process was that one subset of cities must include Boston, while the other subset must include San Francisco. The students were told that if a city appeared on one list but not the other, the student listing the city would be rewarded with a dollar amount equal to the city's ranking index. If a city appeared on both lists, each would lose a dollar amount that is double the city's ranking index. Finally, if the students partitioned all eleven cities without duplication, their total winnings would be tripled.

This coordination game involves 512 Nash equilibria. In nearly every experiment, however, the Boston list included New York and Philadelphia, and, somewhat less frequently, Chicago. The San Francisco list invariably included Los Angeles, Phoenix, and Denver, and less frequently Dallas and Kansas City. The reason for these assignments appears to have been based on geography, with the Mississippi River being the dividing line. The students' common, albeit imprecise, knowledge of the geography of the United States seems to have helped them identify focal-point equilibria. In a few experiments, the geographic distribution appeared to be based on the distinction between Snowbelt and Sunbelt states, although Atlanta's presence on both lists casts suspicion on this selection criterion. In this coordination game, the selection criterion seemed to be based on geography, but could have been based on almost anything, such as cultural norms, attitudes or perceptions, or whether a city's baseball team was in the American or National League.

A more straightforward example of a focal-point equilibrium is depicted in Figure 3.3. The reader should verify that this oil-drilling game has two pure-strategy Nash equilibrium strategy profiles: {Narrow, Narrow} and {Wide, Wide}. In the absence of a clear-cut strategy profile, each company might adopt an arbitrary decision rule, such as a maximin or minimax regret strategy,

FIGURE 3.3
Oil-Drilling Game with Two, Pure-Strategy Nash Equilibria

		GLOMAR	
		Narrow	Wide
PETROX	Narrow	**(52, 52)**	(23, 50)
	Wide	(50, 23)	**(24, 24)**

Payoffs: (PETROX, GLOMAR)

which were discussed in the previous chapter. But, wait a minute! The correct strategy choice would be obvious to both companies. The strategy profile {*Narrow, Narrow*} "stands out" because it results in the best payoff for both players. This strategy profile is a focal-point equilibrium.

─────────────── **Demonstration Problem 3.1** ───────────────

O'Sullivan's and Dimaggio's are sports bars kitty-corner to each other across Division Street on the north side of Chicago. The owners of these establishments are contemplating promotional strategies to increase their business during the NCAA college basketball tournament, otherwise known as "March Madness." Both owners are considering purchasing advertising space in the *Chicago Tribune* and *Chicago Sun-Times*, or sponsoring an Irish social group, such as the Ancient Order of Hibernians (AOH), in the St. Patrick's Day Parade on March 17. In exchange for financial support, the AOH has promised to display the sponsor's name prominently on its parade banner. The expected profits in thousands of dollars from alternative combinations of promotional strategies are summarized in the payoff matrix of the normal-form game depicted in Figure D3.1.

FIGURE D3.1
Static Game in Demonstration Problem 3.1

		Dimaggio's	
		Hibernians	*Newspapers*
O'Sullivan's	*Hibernians*	(12, 12)	**(15, 25)**
	Newspapers	**(25, 15)**	(12, 12)

Payoffs: (O'Sullivan's, Dimaggio's)

a. Does either owner have a dominant strategy? Explain.
b. Does this game have a unique Nash equilibrium?
c. Do you believe that a focal-point equilibrium exists for this game?

Solution

a. In the game depicted in Figure D3.1, neither owner has a dominant strategy. If O'Sullivan's adopts a *Hibernians* strategy, Dimaggio's will adopt a *newspapers* strategy because of the larger payoff. If O'Sullivan's adopts a *newspapers* strategy, Dimaggio's will adopt a *Hibernians* strategy. Both players' strategy depends on the strategy adopted by his or her rival.
b. This game has two Nash equilibria: {*Hibernians, Newspapers*} and {*Newspapers, Hibernians*}. For each strategy profile, neither player can improve its payoff by switching strategies.
c. The strategy profile {*Hibernians, Newspapers*} appears to constitute a focal-point equilibrium. The reason for this is that the owner of Dimaggio's may justly believe that O'Sullivan's is culturally obligated to sponsor AOH.

Example: Cold War Game

Between the end of World War II and 1991, the United States and its allies, also known as the Western bloc, were engaged in an open and restricted rivalry with the former Union of Soviet

Socialist Republics (Soviet Union) and its allies, which were collectively referred to as the Eastern bloc. The genesis of the rivalry between the United States and the Soviet Union stemmed from differences in economic and political ideologies, as well as mutual distrust following the defeat of German National Socialism in 1945. This rivalry came to be known as the Cold War because it did not actually lead to a direct military confrontation. The threat of a "hot war" between the United States and the Soviet Union, however, was all too real, and the stakes were very high. Both sides possessed, and threatened to use, massive stockpiles of nuclear weapons if their "national interests" were jeopardized. Yet, in spite of the deep distrust and suspicion that dominated international relations during this period, the world managed to avoid a nuclear holocaust. Why? Game theory provides an insight into the dynamics of geopolitical relations and helps explain why a nuclear confrontation was avoided.

Although the United States and the Soviet Union each had the capability to obliterate its rival, some military strategists believed that whoever launched their nuclear weapons first would be able to survive the conflict with "acceptable losses." Of course, both sides would be better off by not attacking, leaving the diplomatic door open to a peaceful resolution of national differences. As it turned out, this is precisely what happened. Although the Cold War game is more appropriately modeled as a sequential-move game, it is possible to analyze the challenges confronting the antagonists by presenting this game in normal format. This two-player, noncooperative, static game is depicted in Figure 3.4.

In the Cold War game, each superpower must choose between a *first strike* or *second strike* strategy. With a *second strike* strategy, a superpower will retaliate if attacked. In this game, a *first strike* by either superpower results in total annihilation of its rival, or a payoff of $-\infty$. A *first strike* by one superpower will lead to retaliation by the other superpower, resulting in a loss to the aggressor of $-L$, where $-L > -\infty$. While the loss to the aggressor is substantial, these losses are deemed "acceptable." If neither superpower strikes first, each enjoys a positive gain. The gain to the United States is denoted G_{US}, while the gain to the Soviet Union is denoted G_{SU}. Does the game depicted in Figure 3.4 have a unique, pure-strategy Nash equilibrium?

In this coordination game, the strategy profiles {*First strike*, *First strike*} and {*Second strike*, *Second strike*} constitute Nash equilibria since neither superpower can improve its payoff by unilaterally switching strategies. Take, for example, the Nash equilibrium strategy profile {*First strike*, *First strike*} in which both superpowers strike first, which leads to total annihilation. The reader might well question whether this strategy profile is consistent with the rationality assumption. If cooperation is possible, it would be in both sides' best interest to negotiate a mutually beneficial nonaggression treaty, which is what, in fact, happened. Even if overt cooperation is not possible, a single Nash equilibrium strategy profile stands out because rational superpowers share a common understanding of the problem. This is an important observation, and may be the best explanation for why a nuclear Armageddon never occurred. In fact, rationality can be an important selection criterion when identifying a focal-point equilibrium. If the rationality assumption is correct, the strategy profile {*Second strike*, *Second strike*} is a focal-point equilibrium.

Another way to analyze the Cold War game is to recognize that *second strike* is a weakly dominant strategy for both superpowers. To see this, suppose that the Soviet Union adopts a *first strike* strategy. Regardless of whether the United States strikes first or retaliates, the outcome for the United States is total annihilation. If revenge and inflicting greater death and destruction is ruled out as a rational response to a *first strike* by the Soviet Union, the United States will be indifferent between a *first strike* and a *second strike* strategy. On the other hand, if the Soviet Union adopts a *second strike* strategy, it is clearly in the best interest of the United States to adopt a *second strike* strategy as well. Thus, *second strike* is a weakly dominant strategy for the United States since no other strategy can result in a better outcome, regardless of the strategy adopted by the Soviet Union. Since the payoff matrix is symmetrical, *second strike* is also a weakly dominant strategy

FIGURE 3.4
Cold War Game

Soviet Union

		First strike	Second strike
United States	First strike	$(-\infty, -\infty)$	$(-L, -\infty)$
	Second strike	$(-\infty, -L)$	(G_{US}, G_{SU})

Payoffs: (United States, Soviet Union)

for the Soviet Union. Of course, this analysis depends critically on the assumption of rationality. Would the solution to this game have been different if the leader of one superpower had been rational while the leader of the other superpower had been a psychopath? Would the events of the first half of the twentieth century have unfolded differently if policymakers in London and other European capitals had not assumed rationality on the part of the leadership of the Third Reich in the years leading up to World War II?

DEVELOPING A THEORY OF FOCAL-POINT EQUILIBRIA

Schelling himself did not offer a formal theory to explain the existence of focal-point equilibria. Indeed, game theorists have not even been able to agree on a formal definition of a "focal point." Be that as it may, there appear to be a number of qualitative factors that help to explain their existence. Symmetry and uniqueness appear to be important selection criteria. In games involving multiple Nash equilibria, the distribution of payoffs, symmetric notions of equity, efficiency, and fairness appear to be important identifying principles. Moreover, when one identifying principle suggests a unique equilibrium, and another suggests multiple equilibria, the first principle tends to dominate. Roth and Schoumaker (1988) discovered that implicit agreements may also emerge from a process of adaptive expectations. Observing how players have played the game in the past affects how players play the game in the future.

In the preceding chapter we introduced two general principles to guide a player in his or her search for an optimal strategy in noncooperative, static games. The first principle asserts that a rational player should always adopt a dominant strategy. If a player does not have a dominant strategy, proceed to the second principle of strategic behavior: Successively eliminate all dominated strategies. If a dominant strategy emerges in a simplified game, the player should adopt it. When either or both players have a dominant strategy, a unique Nash equilibrium will exist. But, if neither player has a dominant strategy, the result may be multiple Nash equilibria, or none. When this happens, we are left with our third general principle for finding an optimal strategy. If the first two principles fail to produce a Nash equilibrium, look for the outcome in which each player's strategy is the best response to the strategy adopted by rivals. This strategy profile is a focal-point equilibrium.

> *Principle:* If looking for a dominant strategy or eliminating dominated strategies fails to produce a unique Nash equilibrium, search for an outcome in which each player's strategy is the best response to the strategy adopted by rivals. This is a focal-point equilibrium.

The economic significance of focal-point equilibria becomes readily apparent when we consider cooperative, non-zero-sum, repeated, static games, which will be discussed at length in the next two chapters. In those instances where explicit collusive agreements are prohibited, the existence of focal-point equilibria suggests that tacit collusion, coupled with a trigger strategy and an effective policing mechanism, may be possible.

FRAMING

In the previous section we noted that in games involving multiple Nash equilibria, notions of equity and fairness appear to be important guiding principles when attempting to identify focal points. Somewhat surprisingly, concepts of fairness and equity often appear to violate the assumption of rational behavior. Studies have shown that the context within which choices are presented is important to decision makers. **Framing** refers to the idea that people often make decisions based upon a familiar frame of reference. This frame of reference makes it possible for individuals to make comparisons and select from among different alternatives. One such frame of reference is the conventional notion of "fairness."

> **Framing** When players make decisions based on a familiar frame of reference.

The idea of framing can be illustrated with the following simple experiment, which is referred to as the *ultimatum game*. In this game, a test group of individuals is divided into two equally sized groups. The first group is referred to as givers, while the second group is referred to as takers. The givers are offered $10 under the following condition: The givers can keep, or give away, any portion of the $10 provided that a taker agrees with the distribution. The problem confronting the givers is how much of the $10 to keep and how much to give to the takers. Each member of the group of givers is asked to record his or her proposed allocation on a piece of paper. These proposals are then randomly distributed among the takers. If a taker rejects the proposed allocation, the giver and taker receive nothing.

According to the principle of rationality, receiving any amount is better than getting nothing. Thus, one would expect that the givers would choose to offer takers the smallest possible amount, which the takers would readily accept. Surprisingly, in repeated experimental trials, givers tended to offer takers $4 or $5, but never more than $5. Offers of $1, $2, or $3 were nearly always rejected, and offers of $4 were sometimes rejected. This reaction by takers appears irrational, if not downright ungrateful, since something is better than nothing. Yet, the frame of reference of both givers and takers is an amount that is conventionally considered to be fair, which in the ultimatum game is an even split.

People often reject offers they consider unfair, even though accepting the offer will make them marginally better off. Consider, for example, the pricing behavior of the retail giant Toys "R" Us. Over the years, crazes such as Pokémon trading cards, Tickle Me Elmo and Cabbage Patch Kids dolls, Power Rangers action figures, Tamagotchi virtual pets, and so on have resulted in very high resale prices. For example, at the height of its popularity, the retail price of a Tamagotchi was around $12, which was well below the market-clearing resale price of more than $50. At the retail price, retailers' profit margins were around $10 per unit. Could Toys "R" Us have taken advantage of the market shortage of Tamagotchis by charging the much higher resale price? Certainly they could have, but they did not. Why did the retailer continue to charge the wholesale price plus their usual markup? The reason is that the company was sensitive to customers' notions of fairness. Had the company originally charged the higher price, this would not have been as much of an issue. But, the perception that a company is engaging in "price gouging" behavior is considered taboo. To paraphrase a saying in the retail industry, if you price gouge at Christmas, they won't be back at Easter. This is why Home Depot does not charge higher prices for snowblowers during blizzards, and why Safeway does not raise prices for groceries and bottled water during hurricanes. They are reluctant to offend customer sensibilities.

The concept of framing can also be applied to situations in which players make decisions with less-than-perfect information. People often refer to things with which they are familiar when making decisions because it provides them with a certain comfort level, even though those decisions are irrational according to economic theory. To see this, consider your answer to the following multiple-choice question:

1. How many seven-letter words ending in "ing" would you expect to find in four pages of a novel (about 2,000 words)?

    (a) 0
    (b) 1–2
    (c) 3–4
    (d) 5–7
    (e) 8–10
    (f) 11–15
    (g) 16+

Now, consider your answer to the following multiple-choice question:

2. How many seven-letter words ending in "_n_" would you expect to find in four pages of a novel (about 2,000 words)?

    (a) 0
    (b) 1–2
    (c) 3–4
    (d) 5–7
    (e) 8–10
    (f) 11–15
    (g) 16+

Clearly, the answer to the question 2 must include the answer to question 1. Yet, because people are familiar with words ending in "ing" they tend to think that they occur more often. Choices based on familiarity help explain the common answer to the question: Do murders or suicides occur more often New York City? It has been demonstrated that people tend to reply "murder," even though suicides are much more prevalent. Why? In the absence of complete information, the more people are familiar with something, the more likely they are to select it.

CASE STUDY 3.1: AZT VERSUS VIDEX

Framing can be an important identifying principle when searching for a focal-point equilibrium. Framing refers to situations when players make strategic choices based on comparisons. An example of this occurred in 1987 when Burroughs-Wellcome (now GlaxoSmithKline) introduced its anti-AIDS medicine azidothymidine, which was marketed under the brand name AZT. This life-extending medical breakthrough for treating the HIV virus met with a decidedly negative public reaction. Gay-rights organizations lambasted the company for the drug's $10,000 to $12,000 per year price tag. The company was criticized for exploiting its monopoly on the drug to engage in price gouging practices. The company responded by mounting a largely unsuccessful advertising campaign to justify this price, which was due to high research and development costs. In spite of these efforts, the ongoing public hostility led Burroughs-Wellcome to lower the price of AZT in 1989 to about $6,500 to $8,000 per year.

On October 9, 1991, the U.S. Food and Drug Administration approved the anti-AIDS drug didanosine, which was developed by Bristol Myers Squibb and marketed under the brand name Videx. Although Videx was less effective than AZT, the public reaction was enthusiastic due to its lower annual price tag of $1,800. The dramatic difference in public reaction was due to the price difference between AZT and Videx. The public appreciated the introduction of Videx all the more because it had something with which to compare its price.

The existence of multiple Nash equilibria makes it difficult to predict a unique strategy profile in pure strategies. Game theorists have devoted considerable time and energy attempting to codify the existence of focal-point equilibria. Our discussion of focal-point equilibria is representative of some of the current research in this direction. For a unique strategy profile to stand out from among multiple Nash equilibria, players must share some common understanding of special circumstances influencing the game. Perhaps even more important than identifying a unique Nash equilibrium is figuring out why players are drawn to such outcomes in the first place. Recall that the concept of a Nash equilibrium critically depends on the assumption of rationality. This assumption says that players will always adopt a strategy that is the best response to the strategies adopted by rivals. For this to occur, however, all players must have both complete information and unbridled confidence in their rivals' rationality. Even the most ardent apostle of rationality must admit that this set of assumptions is not always satisfied in reality. It is possible, for instance, that players may be guided more by instinct than by reason? And if so, what is the genesis of this sort of behavior? Perhaps it is the result of an evolutionary process involving countless generations responding to external stimuli.

EVOLUTIONARY GAME THEORY

Beginning with Charles Darwin (1859), theories of evolution have been instrumental in providing insights into the physical and mental attributes of all living creatures. Might it not also be possible for a theory of evolution to explain animal behavior? John Maynard Smith's (1982) pioneering work in **evolutionary game theory** posits that animal behavior can be explained in terms of instinctual strategies that are genetically passed along from generation to generation. In a process analogous to how a species evolves physical and mental abilities, successful strategies will improve that species' survival rate and increase the chances that these genes will be passed along to future generations. New strategies arise from mutations. If these mutations are unsuccessful, survival rates will fall and the mutation will not be passed along to future generations. If these mutations are successful, they will come to define the behavior of future generations.

> **Evolutionary game theory** A branch of game theory that posits that animal behavior can be explained in terms of instinctual strategies that are genetically passed along from generation to generation.

Evolutionary game theory has attracted considerable attention from the economics profession. For one thing, evolutionary game theory may provide an avenue for identifying a unique strategy profile from among multiple Nash equilibria. Perhaps more importantly, evolutionary game theory suggests that equilibrium does not depend on the assumption of player rationality. It recognizes the possibility that instinct, and not reason, explains many strategy choices.

The population-dynamic aspect of evolutionary games is significant precisely because it does not depend on the assumption that players are sophisticated enough to adopt strategies based on the belief that rivals are rational. The assumption of rationality is replaced with the much weaker assumption of **reproductive success**, which asserts that successful strategies replace unsuccessful strategies over time. The notion of reproductive success can be traced to the application of game theory to problems in evolutionary biology (see, for example, Maynard Smith [1982], and Hofbauer and Sigmund [1998]). Animals in the wild are frequently involved in situations involving a choice of strategies. Pruett-Jones and Pruett-Jones (1994), for example, applied the principles of game theory to analyze the seemingly strange behavior of bowerbirds. Male bowerbirds construct elaborate nests, or bowers, to attract female bowerbirds. Once they have built their bowers, they attempt to destroy the bowers of neighboring males. If a male leaves his nest and marauds the nests of other males, his nest will be destroyed. The time spent destroying and repairing bowers could be used more profitably foraging or mating. What explains this behavior?

> **Reproductive success** A weaker assumption than rationality which asserts that successful strategies replace unsuccessful strategies over time.

In evolutionary game theory, a pure strategy is animal behavior that is "programmed" by its DNA. Genes that comprise strands of DNA "compete" with each other by being passed along to future generations in varying proportions. The genes of these animals are encoded with behavioral strategies. Animals that are successful in the battle for survival tend to have more offspring. Thus, the proportion of those animals that have genes encoded with the genetically superior survival strategy will increase until they dominate animals whose genes are genetically encoded with the inferior survival instincts. An evolutionarily stable strategy is one in which the population is resistant, at least initially, to the introduction of a mutant alternative strategy.

Pruett-Jones and Pruett-Jones attempted to explain the unusual behavior of male bowerbirds by assuming two strategies: *Marauding* and *guarding*. A male that adopts a *marauding* strategy spends all his time visiting and destroying the nests of rivals. A male that adopts a *guarding* strategy spends all his time guarding the nest. The reproductive success of each male bowerbird depends not only on his strategy, but on the strategy adopted by rival bowerbirds. Pruett-Jones and Pruett-Jones found that *marauding* results in a higher payoff than *guarding* regardless of the strategy adopted by neighboring bowerbirds, although *guarding* males with *guarding* neighbors do better than *marauding* males with *marauding* neighbors. Thus, *marauding* constitutes an **evolutionary equilibrium** in games involving *marauding* and *guarding* gene pools. More importantly, the Pruett-Joneses found that *marauding* constitutes an evolutionarily stable strategy.

> **Evolutionary equilibrium** A population-dynamic equilibrium strategy profile that results when successful strategies replace unsuccessful strategies. In the end, only successful strategies will be selected.

An interesting example of the emergence of an evolutionary equilibrium comes from the economics of networks, which revolves around the idea of **positive feedback effects** (Boyes 2004, p. 192), also known as **positive network externalities**. Positive feedback refers to the idea that the benefits received by a player joining a **network** are also shared by existing members of the same network. A network may be described as any situation in which groups of players sharing a common technology incur lower costs and/or receive greater benefits than if individual players used different technologies. The idea behind positive network externalities is that the value of network membership rises with an increase in the number of members. An interesting example of positive network externalities is instant messaging.

Positive feedback effects (positive network externalities) When the benefits received by a player joining a network are shared by existing members of the same network. The value of network membership rises with an increase in the number of members.

Network When the activities of groups of players sharing a common technology have lower costs and greater benefits than if individual players use different technologies.

Several Internet service providers (ISPs) offer their customers instant messaging services, such as America Online's Instant Messenger, Microsoft's MSN Messenger, and Yahoo! Messenger. **Instant messaging** refers to the real-time exchange of text messages by users of the Internet. How might the value of network membership increase with the number of members? Suppose that XYZ Instant Messenger (XIM) has 1,000 members who can communicate with each other, but cannot communicate with members of other instant messaging networks. If membership in this instant messaging network were to increase to 1 million users, the value of membership in this network would be much higher.

Instant messaging The exchange of real-time text messages between and among users of the Internet.

A numerical example can be used to highlight the benefits of positive network externalities that can lead to an evolutionary equilibrium. Assume that the value of membership in a network is proportional to the number of members according to the expression $n(n-1) = n^2 - n$. Suppose that the value of an individual member of a network is \$1. The value of this network with $n = 10$ members is $10^2 - \$10 = \$90$. If membership increases tenfold to 100 members, the value of the network increases to $100^2 - 100 = \$9,900$, or a 110-fold increase! The greater-than-proportional increase in a network's value serves a lure to new members. In the extreme, this pattern intensifies and accelerates until a single dominant network emerges.

When shopping for an instant messaging service, a player would be naturally drawn to the network with the greatest number of members. Does this mean that the evolutionary equilibria only exist when all players belong to the same network? Not necessarily. For one thing, different networks must be near-perfect substitutes for each other. Moreover, it must not be possible for members of one network to easily interface with members of another network. Positive feedback effects existed for members of early e-mail networks that could not communicate with each other. This situation persisted until software was developed that made communication across different e-mail networks possible. As a result, membership in any single e-mail network no longer constituted an evolutionary equilibrium. Changes in computer software technology resulted in multiple Nash equilibria.

Example: Hawk–Dove Game

The fundamental principles underlying evolutionary game theory are illustrated in the famous hawk–dove game, which is depicted in Figure 3.5. Animal A and animal B can adopt one of two possible strategies when competing over a food source. The animal can be either aggressive (*hawk*) or passive (*dove*). Consider the game from animal A's perspective. Suppose animal B is aggressive and fights for food. If animal A fights, they will equally share the benefits of food (b) less the cost (c) of injuries received in the confrontation. We will assume that $b > c$. If animal A does not fight, animal B will receive all of the benefits of the food, while animal A goes hungry. If animal B does not fight and animal A fights, animal A will receive all of the benefits and animal B will go hungry. Finally, if both animals do not fight, the benefits are divided equally with no costs associated with fighting.

FIGURE 3.5
Hawk–Dove Game

Animal *B*

		Hawk	Dove
Animal *A*	*Hawk*	*((b − c)/2, (b − c)/2)*	*(b*, 0)
	Dove	(0, *b*)	*(b/2, b/2)*

Payoffs: (Animal *A*, Animal *B*)

Clearly, the best outcome is for both animals not to fight, but this is not an evolutionary equilibrium. The reason for this is that animals from a large population are randomly combined. The genes of the animals adopting a survival strategy will be passed along to their descendants. A strategy is evolutionarily stable if it supersedes other strategies that are introduced into the population by mutation. The strategy profile {*Dove, Dove*} is not evolutionarily stable. To see this, suppose that the whole population initially adopts a *dove* strategy. Now suppose that a random mutation introduces a small number of hawks into this population. The hawks will survive and pass this behavior along to the next generation. Provided that c is not greater than b, the evolutionarily stable equilibrium for this game is the strategy profile {*Hawk, Hawk*}. Over time hawks will come to dominate the population until a new, more successful, mutation is introduced into the gene pool.

CASE STUDY 3.2: EVOLUTIONARY EQUILIBRIUM IN BROADBAND TECHNOLOGY?[3]

A steadily increasing percentage of American households and businesses are active Internet users. At present, there are two main broadband technologies in use. Digital subscriber line (DSL) utilizes a copper wire infrastructure originally developed to provide home and office telephone service. By contrast, cable uses a coaxial cable infrastructure that was originally installed to provide pay television service. Both DSL and cable can provide a variety of Internet services, including teleconferencing, interactive entertainment, and distance learning. While both technologies currently are widely used, will competition eventually lead to an evolutionary equilibrium in which one technology—such as DSL, cable, or satellite—dominates the broadband market? Or, will multiple evolutionary equilibria prevail, consisting of different broadband technologies?

For any single technology to dominate the market, certain conditions must be satisfied, such as the existence of economies of scale in production, positive feedback effects, or strong consumer preference for one technology. While economies of scale appear to exist in satellite and cable broadband distribution, it does not appear that either medium has a distinct competitive advantage. In addition, there does not appear to be a strong consumer preference for a particular broadband technology. Finally, there are no significant positive feedback effects associated with any of these broadband technologies.

The broadband market consists of several competing technologies, none of which appear to have a distinct competitive advantage. On the other hand, the demand for broadband services by Internet users is very sensitive to advances in computer hardware, software, and delivery technology. While an evolutionary equilibrium in broadband technology does not appear to exist at present, there is no guarantee that this situation might not change very rapidly.

CHAPTER REVIEW

The existence of a Nash equilibrium is a very powerful concept. Unfortunately, the existence of multiple Nash equilibria makes it difficult to predict player strategies. In games involving multiple Nash equilibria, the optimal strategy for either player depends upon what each player believes to be a rival's strategy. Thomas Schelling examined the conditions under which a single Nash equilibrium might "stand out" from the others. Schelling referred to this as a *focal-point equilibrium*.

Although there is no formal theory to explain the existence of a focal-point equilibrium, there appear to be several qualitative factors that help to explain their existence. Symmetry and uniqueness, distribution of payoffs, and notions of equity, efficiency, and fairness appear to be important identifying principles. Moreover, when one identifying principle suggests the existence of a unique Nash equilibrium, and another identifying principle suggests the existence of multiple Nash equilibria, the first principle tends to dominate. Implicit agreements may also come about through a process of adaptive expectations. Observing how other players have played the game in the past affects how other players play the game in the future.

The economic significance of focal-point equilibria is especially important for cooperative, non-zero-sum, repeated, static games. In those instances where explicit collusive agreements are prohibited, the existence of focal-point equilibria suggests that tacit collusion, coupled with a policing mechanism that involves trigger strategies, may be possible.

An *evolutionary equilibrium* is a population-dynamic equilibrium strategy profile that results when successful strategies replace unsuccessful strategies. In the end, only successful strategies will be selected. The population-dynamic aspect of evolutionary games is significant precisely because it does not depend on the assumption of player rationality, but on the much weaker assumption of *reproductive success*. This assumption asserts that successful strategies will be used more frequently over time than unsuccessful strategies, and only successful strategies will endure.

CHAPTER QUESTIONS

3.1   What is a focal point?

3.2   Focal-point equilibria are only possible when there is a unique Nash equilibrium. Do you agree? Explain.

3.3   Explain how focal points can be used to find solutions to games involving multiple Nash equilibria.

3.4   The use of focal points is only useful in games in which one or both players have a dominant strategy. Do you agree with this statement? Explain.

3.5   If neither player in a noncooperative, one-time, static game has a dominant strategy, multiple Nash equilibria are not possible, but a unique focal-point equilibrium is. Do you agree?

3.6   What is framing? In what way might the idea of framing contribute to the development of a theory of focal-point equilibrium?

3.7   Explain the concept of an evolutionary equilibrium.

3.8   The assumption of reproductive success for populations is essentially the same as the assumption of player rationality. Do you agree with this statement? Explain.

3.9   Game theorists have applied the concept of evolutionary equilibrium to explain social norms regarding sexual behavior such as monogamy, polygamy, and adultery. How would you construct such an argument?

3.10 Does evolutionary game theory provide any insights into changes in social attitudes about racism, age discrimination, sex discrimination, and homosexuality?

3.11 If nature plays less of a role in determining physical survival as a civilization progresses, what other factors might explain human evolution?

CHAPTER EXERCISES

3.1 We saw earlier in this chapter that an established rule provided the answer to the question: Which side of the road should I drive on? Now, consider the following situation in which no such formal rule exists. What should Phoebe and Chloe do if their telephone conversation gets cut off? Should Phoebe call Chloe, or should Chloe call Phoebe? Explain your reasoning. How many possible outcomes are there for this game? Illustrate your answer.

3.2 Suppose two individuals are hunting hare or stag. Each person could hunt hares alone, but because of its size and weight, the hunters can only be successful bringing down a stag by coordinating their efforts. Sharing a stag is preferred to sharing a hare. The hunters' strategy choices and utility payoffs are depicted in Figure E3.2.

a. Does either hunter have a dominant strategy?
b. Does this game have a unique Nash equilibrium?
c. Do you believe that this game has a focal-point equilibrium? Why?

3.3 Two motorists approach an intersection at the same time. On each corner is a stop sign. Driver A is driving north, while driver B is driving west. At each corner of the intersection is a stop sign. The intersection game is illustrated in the Figure E3.3. Which driver should enter the intersection first, or should they both enter the intersection at the same time? Does it matter whether a driver goes straight, turns right, or turns left? Is there a focal-point equilibrium for this game? Why?

3.4 Suppose that there are two landowners on the same side of a river. It is in each landowner's best interest to build and maintain a levee, but only if adjacent landowners build and maintain levees as well. The reason for this is that if an adjacent levee is not properly maintained, all landowners will suffer damages in the event of a flood. The landowners' strategy choices and payoffs in thousands of dollars are summarized in Figure E3.4.

a. Does either landowner have a dominant strategy?
b. Does this game have a unique Nash equilibrium?
c. Do you believe that this coordination game has a focal-point equilibrium? Why?

FIGURE E3.2
Static Game in Chapter Exercise 3.2

Hunter B

		Stag	Hare
Hunter A	Stag	(5, 5)	(0, 2)
	Hare	(2, 0)	(2, 2)

Payoffs: (Hunter A, Hunter B)

FIGURE E3.3
Intersection Game in Chapter Exercise 3.3

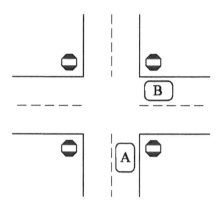

FIGURE E3.4
Levee Game in Chapter Exercise 3.4

		Landowner *B*	
		Maintain	*Don't maintain*
Landowner *A*	*Maintain*	(−5, −5)	(−10, −8)
	Don't maintain	(−8, −10)	(−8, −8)

Payoffs: (Landowner *A*, Landowner *B*)

ENDNOTES

1. In 2005, political scientist Thomas Schelling shared the Nobel Prize in economics with mathematician Robert Aumann for their work in game theory. The Nobel committee lauded their work for "having enhanced our understanding of conflict and cooperation." Schelling is probably best known for his innovation of the payoff matrix, which is used in analyzing static games. Schelling once quipped: "If I am ever asked whether I ever made a contribution to game theory, I shall answer yes. Asked what it was, I shall say the invention of staggered payoffs in a matrix . . . I did not suppose that the invention was patentable, so I made it freely available and hardly anybody except my students takes advantage. I offer it to you free of charge."
2. Jacobellis v. Ohio, 378 U.S. 184.197 (164).
3. This case study was inspired by "Winner-Takes-All" in Boyes (2004, pp. 188, 205–206).

CHAPTER

4 INFINITELY REPEATED, STATIC GAMES WITH COMPLETE INFORMATION

In this chapter we will:

- *Analyze pure strategy, cooperative, infinitely repeated, static games with complete information;*
- *Examine the long-term viability of cooperation in static games that are played an infinite number of times;*
- *Develop a general rule to identify when cheating in cooperative, infinitely repeated games is likely to occur;*
- *Discuss credible threats and trigger strategies;*
- *Review collusive behavior and enforcement mechanisms;*
- *Examine the determinants of collusive agreements among business rivals.*

INTRODUCTION

Proctor & Gamble (P&G) and Unilever are giants in the household-products industry. An American company based in Cincinnati, Ohio, P&G is credited with a number of ground-breaking marketing innovations that have become the industry standard. In 1880, for example, P&G became the first company to launch national and direct-to-consumer advertising campaigns. The company literally created the "soap opera" by sponsoring radio and television dramas that targeted homemakers. P&G invented the first fluoride-based toothpaste, the first synthetic detergent, and the first disposable diaper. More than twenty of P&G's brands are billion-dollar sellers, including Actonel, Always/Whisper, Bounty, Charmin, Crest, Downy, Folgers, Iams, Olay, Pampers, Pantene, Pringles, Tide, and Wella. Unilever, P&G's main rival, is a member of the Unilever Group, which is jointly owned by the Netherlands-based Unilever N.V. and the United Kingdom–based Unilever PLC. The company's brand names include Calvin Klein, Birds Eye, Dove, and Lipton.

For decades, P&G and Unilever were locked in intense global competition. In recent years, however, the two household-products giants have come to an understanding not to "blind side" each other by unexpectedly switching market strategies (Boyes 2004, Chapter 13). Neither P&G nor Unilever, for example, will slash prices nor introduce new product lines without sufficient prior notice of its intentions. Neither will either company launch an advertising campaign that attacks the other through product comparisons.

Suppose that senior management at P&G decided to violate this truce to increase its market share at Unilever's expense. What are the possible consequences for both companies of upsetting the status quo? Would Unilever retaliate, and if so, how? Would P&G's actions provoke a price, product, or advertising war that could ultimately leave both companies worse off? Is there anything that P&G could do to prevent Unilever from retaliating? Can Unilever punish P&G for breaking the peace?

Up to this point we have considered the strategic behavior of players involved in one-time, static games with complete information. We discovered that many of these games were characterized by the prisoner's dilemma. Recall that the prisoner's dilemma arises when it is in the best interest of all players to cooperate, but where each player has an incentive to adopt a dominant strategy that results in a less-than-optimal outcome. And yet, as the P&G/Unilever experience suggests, rival firms often enter into long-term arrangements that are mutually beneficial. What explains this seeming paradox? What was it about their relationship that allowed P&G and Unilever to escape the prisoner's dilemma?

One possible explanation may be found in the dynamics of mutually dependent advertising, pricing, new product development, research and development, and other business decisions. Unlike one-time, static games, the ongoing nature of business competition suggests that a more appropriate analytical approach is to consider static games that are played over and over again without end. Such static games are said to be **infinitely repeated**. It is somewhat paradoxical that business rivals locked in a fierce and ongoing competitive struggle for larger market share and greater profits often find that cooperation is mutually advantageous, although this is not guaranteed. In the next chapter we will consider **finitely repeated games**, which are played a limited number of times, and in which the end of the game may or may not be known with certainty. In the analysis that follows, we will continue to assume that all players have complete information.

> **Infinitely repeated game** A game that is played over and over without end.
>
> **Finitely repeated game** A game that is played a limited number of times. Finitely repeated games may have a certain or an uncertain end.

The P&G/Unilever relationship is an example of an infinitely repeated game, or at least a repeated game in which the end in unknown. While this informal relationship appears to be stable, other collusive arrangements, such as the Organization of Petroleum Exporting Countries (OPEC), have been subject to frequent breakdowns. What accounts for these differences? As we will see, the nature of the relationship between the players and inherent centrifugal economic forces are central to our understanding of the fragility or durability of cooperative business relationships.

CASE STUDY 4.1: EXTRA-CREDIT GAME PART I[1]

To illustrate the fragile nature of cooperation in a one-time, static game, consider the results of an experiment that was conducted by the author involving twenty-nine MBA students enrolled in a managerial economics class. The experiment was conducted during three class meetings over a three-week period. The experiment involved five rounds. The first two rounds were conducted in the first week, the third and fourth rounds were conducted in the second week, and the final round in the third week. Only the first two rounds of this experiment will be discussed in this case study. A discussion of the final three rounds will be deferred until Chapter 5.

The basic rules of the game were the same in each round of the experiment. Each student was instructed to write his or her name on a 3" × 5" index card. They were then asked to select from two colors (*orange* or *blue*), record their choice on the index card, and keep the card secret from the other students. After the students recorded their selections, the index cards were collected by the author, the colors tallied, and the results announced. The

TABLE C4.1
Payoffs for the Extra-Credit Game

Number of students selecting *blue*	Number of students selecting *orange*	Payoff to each student selecting *blue*	Payoff to each student selecting *orange*
0	29	—	2.0
1	28	0.1	2.1
2	27	0.2	2.2
3	26	0.3	2.3
4	25	0.4	2.4
5	24	0.5	2.5
6	23	0.6	2.6
⋮	⋮	⋮	⋮
23	6	2.3	4.3
24	5	2.4	4.4
25	4	2.5	4.5
26	3	2.6	4.6
27	2	2.7	4.7
28	1	2.8	4.8
29	0	2.9	—

students' identities and choice of strategies were known only to the author. Depending on their choices, the students were awarded extra-credit points that would be added to their final numerical grade. These payoffs are summarized in Table C4.1.

According to the payoff schedule, students adopting an *orange* strategy receive two more extra-credit points than students adopting a *blue* strategy. Clearly, *orange* is a strictly dominant strategy. On the other hand, adopting an *orange* strategy doesn't always result in the best payoff. If every student adopts the strictly dominated *blue* strategy, the payoff to each student is 2.9 extra-credit points. If all twenty-nine students adopt their strictly dominant *orange* strategy, the payoff to each student is a lesser 2.0 extra-credit points. As we will see in subsequent rounds of this experiment, the distribution of payoffs is important when choosing a strategy.

Since the results were not announced until after all moves were made, this experiment amounted to a noncooperative, static, one-time game. The fact that *orange* is a strictly dominant strategy was verified by the fact that every student selected *orange*. The important thing to note in the first round of this experiment is that even though the students recognized that it was in their collective best interest to adopt the strictly dominated *blue* strategy, they nonetheless chose to adopt the strictly dominant *orange* strategy.

In the second round, students were given the opportunity to formulate a common strategy. Given the payoff structure, it is clearly in the students' best interest to cooperate by adopting a *blue* strategy, which results in a better payoff. As before, the students' identities and choices of color were secretly recorded. What was the outcome? Although twelve students remained loyal to the agreement, seventeen students defected! The "honest" students were penalized by receiving 0.3 fewer extra-credit points (2.0 – 1.7) while the "cheaters" were awarded with an additional 1.7 extra-credit points. As one "honest" student later complained: "I guess that it is true that no good deed ever goes unpunished."

The above experiment underscores the fragility of collusive agreements. Recall, for example, the prisoner's dilemma discussed in Chapter 1 and depicted in Figure 1.1. Despite the better payoff of only six months by agreeing not to confess, each suspect had an incentive to double-cross the other to get a better deal. If it is that difficult to get two players to adhere to an agreement in which both players have a strictly dominant strategy to defect, imagine how much harder it will be to get twenty-nine players to remain loyal!

COLLUSION

Cooperation among players to achieve a mutually beneficial outcome may be modeled as a cooperative game. In our discussion of one-time, static games with complete information, we observed that when the incentives of individual players conflict with the best interest of the group (i.e., the prisoner's dilemma), cooperation is inherently unstable. Each player has an incentive to violate an agreement. But, when static games are played repeatedly, cooperation between and among rivals may constitute a Nash equilibrium strategy. For defection to be profitable, the gains from violating an agreement must outweigh possible future losses. Even when defection is profitable, the threat of retaliation may be sufficient to bind the players to an agreement.

When firms in an industry recognize that their actions are mutually interdependent, they might agree to coordinate their business decisions to maximize profits for the entire industry. **Collusion** is an agreement among firms for the purpose of increasing **market power**. Collusive behavior may include such activities as price fixing, production quotas, or other practices that stifle competition. Collusion need not be a formal, explicit agreement, but may result from a common understanding of the business environment. In fact, tacit collusion may constitute a focal-point equilibrium.

> **Collusion** When firms in an industry coordinate their activities to restrict competition to increase market power and profits.
>
> **Market power** The ability of a firm to influence the market-clearing price of its product by significantly changing industry output.

The precise details of a collusive arrangement depend on the particular circumstances and characteristics of the industry. Collusion frequently takes the form of a secret agreement, even though this activity is frequently illegal. Since 1890, for example, the U.S. government has endeavored to restrain anticompetitive business practices by enacting a host of antitrust legislation, the most significant of which include the Sherman Act (1890), the Clayton Act (1914), and the Federal Trade Commission Act (1914). In Europe, collusion is outlawed by Article 85 of the Treaty of Rome. A classic example of illegal collusion occurred in the U.S. electrical goods industry in the 1950s (see, for example, Fuller [1962] and Porter [1980]). This episode involved the sale of turbine generators by General Electric and Westinghouse to several government-owned utilities using a rigged, seal-bid auction. In this case, General Electric and Westinghouse secretly agreed which company would submit the winning and losing bids. After this conspiracy was uncovered and dismantled, prices of large turbines plummeted by 50 percent between 1958 and 1963.

Although business collusion typically aims to restrict supply to raise prices, it also involves many other business activities, such as reducing service quality, which until recently was practiced in the European airline industry, limiting advertising expenditures, and restricting firms' sales territory, such as occurred in the European chemical industry in the 1920s. The agreement identified ICI (Imperial Chemical Industries, PLC) as the exclusive supplier to Great Britain and the Commonwealth countries, Dupont to the United States, and German firms to the European continent. By the early 1930s this practice was declared illegal.

The most recognized manifestation of collusive behavior is the **cartel**, which is a formal arrangement among producers to allocate market shares and/or maximize industry profits. Cartel agreements coordinate the decisions of member firms, such as pricing and output policies. Perhaps the best-known example of a cartel is OPEC. (See Case Study 8.2). Members of OPEC have been able to influence world oil prices by jointly agreeing on production quotas for member states. Unlike P&G/Unilever, OPEC's joint production and pricing arrangements have historically proven to be fragile because some cartel members have an incentive to cheat. Venezuela, for example, has repeatedly violated almost every joint production agreement. The government in Caracas encourages fellow OPEC members, especially the cartel's "swing" producer, Saudi Arabia, to restrict output and raise prices, but then increases its own production to bolster profits. As word of Venezuela's cheating behavior spreads, the production-sharing agreement collapses.

> **Cartel** A formal agreement among firms in an industry to allocate market shares and/or increase industry profits.

Is "cheating" inevitable when cooperative, static games are played only once? Will the outcome be any different if the game is played more than once? Does it matter whether the game is played a certain, finite number of times? What if the game is played a finite number of times, but the end of the game is uncertain? What if the game is infinitely repeated? In business, for example, ongoing intense competition between and among rival firms is the rule rather than the exception. A common understanding of when, and if, a repeated game comes to an end, and the ability of rivals to detect and punish defectors, are important factors in determining whether long-term cooperation is possible.

THE PRISONER'S DILEMMA REVISITED

What was it about the ongoing relationship between P&G and Unilever that made it in each company's best interest not to break the peace? What was it about this infinitely repeated, static game that may have enabled the players to escape the prisoner's dilemma? An insight into the possible underlying dynamics of the P&G/Unilever relationship may be gleaned by considering the static game depicted in Figure 4.1, which involves two players with three pure strategies.

The reader should verify that if the game depicted in Figure 4.1 is played just once and larger payoffs are preferred, there are two pure-strategy Nash equilibrium strategy profiles: {A2, B2} and {A3, B3}. The payoffs for these Nash equilibria are highlighted in boldface. It should be equally apparent that it would be in both players' best interest to cooperate by playing the strategy profile {A1, B1}. Unfortunately, such an agreement is likely to collapse since each player has an incentive to double-cross the other by switching strategies. This, of course, is an example of the prisoner's dilemma since group incentives are superseded by individual incentives.

Would cooperation between the players be any more likely to endure if the game depicted in Figure 4.1 were played more than once? In this case, each player's strategy becomes the collection of moves made at each stage of the game. Does this repeated game have a Nash equilibrium? It turns out that there are several. One such Nash equilibrium is for player A to move A2 in the first and every subsequent stage, and for player B to move B2 in the first and every subsequent stage. The reason for this is that neither player can move differently at any stage without suffering a lower payoff.

FIGURE 4.1
Infinitely Repeated, Static Game in Pure Strategies

		Player *B*		
		B1	*B2*	*B3*
	A1	(20, 20)	(4, **25**)	(0, 0)
Player *A*	*A2*	(**25**, 4)	(**8, 8**)	(0, 0)
	A3	(0, 0)	(0, 0)	(**1, 1**)

Payoffs: (Player *A*, Player *B*)

TRIGGER STRATEGIES

An interesting question is whether the strategy profile {*A1, B1*} in Figure 4.1, which is not a Nash equilibrium when the game is played just once, is a Nash equilibrium if the game is infinitely repeated. To see if it is, suppose that player *A* publicly announces that he or she intends to adopt the following strategy:

1. Move *A1* in stage 1;
2. If player *B* moves *B1* in stage 1, move *A1* in stage 2;
3. If player *B* does not move *B1* in stage 1, move *A2* in stage 2 and thereafter.

Player *A*'s strategy of moving *A2* should player *B* violate the agreement is referred to as a **trigger strategy**. A trigger strategy is a move by one player in response to an unanticipated move by a rival. Once adopted, a trigger strategy will continue to be used until a rival initiates another unanticipated move. In the game depicted in Figure 4.1, suppose that players *A* and *B* agree to the strategy profile {*A1, B1*} in every stage of an infinitely repeated game. Violation of this agreement by player *B* "triggers" a change in player *A*'s strategy. In particular, if player *B* defects, player *A* will retaliate by moving *A2* in all subsequent stages. Is this trigger strategy sufficient to bind player *B* to the agreement?

> **Trigger strategy** A strategy adopted by one player in response to an unanticipated move by a rival. A trigger strategy will continue to be used until a rival makes another unanticipated move.

As it turns out, player *A*'s trigger strategy for binding player *B* to the agreement is unnecessary since player *B* has a private incentive to cooperate. To see why, suppose that player *A* moves *A1* in stage 1 and player *B* defects immediately. Player *B*'s payoff in stage 1 is 25, while player *A*'s payoff is 4. Player *B*'s defection, however, triggers a change in player *A*'s strategy. Player *A* "punishes" player *B* by switching to *A2* in stage 2. The strategy profile in stage 2 is thus {*A2, B2*}, which, as we have already established, is a Nash equilibrium. By violating the agreement in stage 1, the payoff to player *B* is 25 + 8 + 8 + . . . , while the payoff to player *A* will be 4 + 8 + 8 + ... By violating the agreement, the payoff to player *B* is less than the payoff of 20 + 20 + 20 + . . . by cooperating. Since player *B* has an incentive to cooperate, defection is a *strictly dominated* strategy. Neither player has an incentive to defect, so player *A*'s trigger strategy is unnecessary. Because of the symmetry of the game, maintaining the agreement also constitutes a Nash equilibrium strategy for player *A*.

What is interesting about the game depicted in Figure 4.1 is that while the strategy profile {*A1, B1*} does not constitute a Nash equilibrium for a one-time, static game, it does represent a

FIGURE 4.2

Revised Infinitely Repeated, Static Game in Pure Strategies

Player B

		B1	B2	B3
	A1	(20, 20)	(4, **50**)	(0, 0)
Player A	A2	(**50**, 4)	(**19, 19**)	(0, 0)
	A3	(0, 0)	(0, 0)	(**1, 1**)

Payoffs: (Player A, Player B)

Nash equilibrium strategy profile when the game is infinitely repeated. In this game, it is clearly in the best interest of rational players to cooperate. The threat of retaliation is not necessary to bind the players to this mutually beneficial arrangement. This happy state of affairs, however, is not guaranteed. To see why, consider the static game depicted in Figure 4.2.

The reader should verify that if the game depicted in Figure 4.2 is played just once there are two Nash equilibria strategy profiles: {A2, B2} and {A3, B3}. The payoffs for these Nash equilibria are highlighted in boldface. Although the strategy profile {A1, B1} results in a better payoff for both players, it is not a Nash equilibrium since both players have an incentive to violate the agreement. But, is the strategy profile {A1, B1} a Nash equilibrium if it is played in every stage of an infinitely repeated game? To see if it is, suppose that player A adopts the same trigger strategy outlined above. If player A moves A1 in stage 1 but player B defects and plays B2, player B's payoff is 50 while player A's payoff is 4. Player B's defection provokes player A's trigger strategy of moving A2 in the second and subsequent stages. The payoff to player B is now $50 + 19 + 19 + \ldots$ while the payoff for player A is $4 + 19 + 19 + \ldots < 20 + 20 + 20 + \ldots$. Player A is clearly worse off as a result of player B's defection, but what about player B? To answer this question, we need to evaluate the payoffs to player B.

EVALUATING PAYOFFS IN INFINITELY REPEATED GAMES

Does the knowledge that defection will result in retaliation eliminate the possibility of defection? Not necessarily. To begin with, if the threat of retaliation is not credible, it will be ignored. Even if threats are credible, a player may still defect if it is beneficial to do so. To see this, it is necessary to compare the present value of the stream of payoffs from violating an agreement to the present value of the stream of payoffs from cooperating.[2] The present value of a stream of future payoffs for player j is summarized in Equation (4.1).

$$PV^j = \frac{\pi_0^j}{(1+i)^0} + \frac{\pi_1^j}{(1+i)^1} + \cdots + \frac{\pi_n^j}{(1+i)^n} = \sum_{t=0}^{n} \frac{\pi_t^j}{(1+i)^t} \qquad (4.1)$$

In Equation (4.1), i is the **discount rate** and t the stage of the game. If the payoffs in each stage of the game are equal, Equation (4.1) becomes:

$$PV^j = \pi^j \sum_{t=0}^{n} \left(\frac{i}{1+i}\right)^t \qquad (4.2)$$

Equation (4.2) is referred to in the finance literature as the *present value of an annuity due.*[3] For an infinitely repeated game ($n = \infty$), Equation (4.2) becomes:[4]

$$PV^j = \frac{\pi^j(1+i)}{i} \tag{4.3}$$

> **Discount rate** The rate of interest used to discount to the present the flow of expected future payoffs.

In the game depicted in Figure 4.1, if we assume a discount rate of 5 percent, the present value of the payoffs to each player by cooperating is:

$$PV_C^j = \frac{20(1.05)}{0.05} = 420; \qquad j = A, B \tag{4.4}$$

In Equation (4.4), PV_C^j is the present value of the stream of payoffs from cooperating. By contrast, if player B violates the agreement in stage 1, cooperation will break down and each player will receive a Nash equilibrium payoff thereafter. The payoff to player B from violating the agreement is the one-time payoff from defection plus the present value of the subsequent stream of payoffs from the Nash equilibrium strategy profile $\{A2, B2\}$, which may be written as:

$$PV_V^B = 25 + \frac{8}{(1.05)^1} + \frac{8}{(1.05)^2} + \cdots = (25-8) + \frac{8(1.05)}{0.05} = 185 < PV_C^B \tag{4.5}$$

In Equation (4.5), PV_V^B is the present value of the payoffs to player B by violating the agreement. The value $25 - 8 = 17$ represents the additional payoff to player B from violating the agreement compared with the present value of the stream of payoffs by not cooperating. This value is the "reward" for being unfaithful. By contrast, the present value of the payoffs to player A is:

$$PV_V^A = (4-8) + \frac{8(1.05)}{0.05} = 164 < PV_C^A \tag{4.6}$$

In Equation (4.6), PV_V^A is the present value of the payoffs to player A when player B defects. The value $4 - 8 = -4$ represents the loss to player A by adhering to the agreement compared with the present value of the stream of payoffs by not cooperating. This value is the penalty to player A for being faithful. The above results indicate that both players will suffer lower payoffs when player B defects and player A remains faithful. Because of the symmetry of this game, the same must also be the case if player A defects and player B remains faithful.

In general, the benefit to each player by colluding is:

$$PV_C^j = \frac{\pi_C^j(1+i)}{i}; \qquad j = A, B \tag{4.7}$$

In Equation (4.7), π_C^j is the payoff to player j from colluding. The present value of the payoffs to a player violating the agreement is equal to the *net* gain from defection $(\pi_V^j - \pi_N^j)$ plus the present value of the Nash equilibrium payoffs (PV_N^j). In general, the benefit to the defecting player when a rival remains faithful is:

$$PV_V^j = \left(\pi_V^j - \pi_N^j\right) + PV_N^j = \left(\pi_V^j - \pi_N^j\right) + \frac{\pi_N^j(1+i)}{i}; \qquad j = A, B \tag{4.8}$$

When will it pay for a player to violate an agreement? Defection pays whenever the present value of the payoffs from violating the agreement is greater than the present value of the payoffs from colluding. For the two-player games in Figures 4.1 and 4.2, this condition is summarized in inequality (4.9):

$$PV_V^j = \left(\pi_V^j - \pi_N^j\right) + PV_N^j > PV_C^j; \qquad j = A, B \qquad (4.9)$$

Equation (4.9) may be rewritten as:

$$\left(\pi_V^j - \pi_N^j\right) + \frac{\pi_N^j\left(1+i\right)}{i} > \frac{\pi_C^j\left(1+i\right)}{i} \qquad (4.10)$$

In the game depicted in Figure 4.2, does player B have an incentive to violate the agreement? If we assume that the discount rate is 5 percent, the net gain to player B from violating the agreement is $\pi_V^B - \pi_N^B = 50 - 19 = 31$. The Nash equilibrium payoff in any subsequent state is $\pi_N^B = 19$ and the payoff in any stage from the agreement is $\pi_C^B = 20$. Substituting these data into Equations (4.7) and (4.8) we obtain:

$$PV_V^B = \left(50-19\right) + \frac{19\left(1.05\right)}{0.05} = 430 > 420 = \frac{20\left(1.05\right)}{0.05} = PV_C^B \qquad (4.11)$$

The payoff to faithful player A when player B defects is:

$$PV_V^A = \left(4-19\right) + \frac{19\left(1.05\right)}{0.05} = 384 < 420 = PV_C^A \qquad (4.12)$$

In Equation (4.12), PV_V^A is the present value of player A's trigger strategy.

The above calculations indicate that player B has an incentive to defect, while player A is harmed by remaining faithful to the agreement. Because of the symmetry of the game, both players have an incentive to violate the agreement in the first round. If this occurs then the payoff to both players is:

$$PV_N^j = \frac{19\left(1.05\right)}{0.05} = 399 < 420 = PV_C^j \qquad (4.13)$$

Inequality (4.13) verifies that the game depicted in Figure 4.2 is a prisoner's dilemma. It pays for both players to cooperate, but each has an incentive to defect.

The above examples demonstrate that despite the use of a trigger strategy, a strategy that does not constitute a Nash equilibrium in a one-time static game may be a Nash equilibrium in an infinitely repeated static game, but there is no guarantee of this. Depending on the payoffs and the discount rate, both players may have an incentive to defect. In the case of Figure 4.2, player A's trigger strategy was not sufficient to bind player B to the agreement.

─────────────────────── **Demonstration Problem 4.1** ───────────────────────

Consider the static game depicted in Figure 1.2, which is reproduced below. Suppose that the discount rate in this infinitely repeated game is 5 percent.

    a. What is the present value of the stream of payoffs to each firm from a no-collusion strategy?
    b. What is the present value of the stream of payoffs to each firm from cooperating?
    c. What is the present value of the stream of payoffs from defection?
    d. Based on your answers to parts a and b, is this cartel stable?

FIGURE 1.2
Pricing Game

Firm B

Firm A		High price	Low price
	High price	($1 million, $1 million)	($100,000, $5 million)
	Low price	($5 million, $100,000)	**($250,000, $250,000)**

Payoffs: (Firm A, Firm B)

Solution

a. A Nash equilibrium occurs when both firm A and firm B charge a low price. The present value of the stream of payoffs to each firm from a no-collusion strategy is:

$$PV_N^j = \frac{\pi_N^j(1+i)}{i} = \frac{\$250,000(1.05)}{0.05} = \$5,250,000 \qquad (D4.1.1)$$

b. The present value of the stream of payoffs from charging a high price is:

$$PV_C^j = \frac{\pi_C^j(1+i)}{i} = \frac{\$1,000,000(1.05)}{0.05} = \$21,000,000 \qquad (D4.1.2)$$

c. The present value of the stream of payoffs from defection is:

$$PV_V^j = \left(\pi_V^j - \pi_N^j\right) + \frac{\pi_N^j(1+i)}{i} = \$4,750,000 + \$5,250,000 = \$10,000,000 \quad (D4.1.3)$$

d. Since $PV_V^j < PV_C^j$ ($\$10,000,000 < \$21,000,000$), there is no incentive to violate the agreement.

──────────── **Demonstration Problem 4.2** ────────────

Refer, again, to the static game in Figure 1.2. Suppose that the discount rate is 20 percent.

a. What is the present value of the stream of payoffs from a no-collusion strategy?
b. What is the present value of the stream of payoffs from cooperating?
c. What is the present value of the stream of payoffs from defection?
d. Based on your answers to parts a and b, is this cartel stable?

Solution

a. The present value of the payoffs to each firm from a no-collusion, Nash equilibrium strategy for an infinitely repeated game is:

$$PV_N^j = \frac{\pi_N^j(1+i)}{i} = \frac{\$250,000(1.20)}{0.20} = \$1,500,000 \qquad (D4.2.1)$$

b. The present value of the stream of payoffs by cooperating is:

$$PV_C^j = \frac{\pi_C^j(1+i)}{i} = \frac{\$1,000,000(1.20)}{0.20} = \$6,000,000 \qquad \text{(D4.2.2)}$$

c. The present value of the payoffs from defection is:

$$PV_V^j = \left(\pi_V^j - \pi_N^j\right) + \frac{\pi_N^j(1+i)}{i} = \$4,750,000 + \$1,500,000 = \$6,250,000 \quad \text{(D4.2.3)}$$

d. In this case, since $PV_V^j < PV_C^j$ ($\$6,250,000 > \$6,000,000$), there is an incentive to defect.

CHEATING RULE FOR INFINITELY REPEATED GAMES

The results in the preceding section may be simplified to derive a cheating rule for two-player, infinitely repeated games. Rearranging Inequality (4.10) we obtain:

$$\frac{\pi_C^j - \pi_N^j}{\pi_V^j - \pi_C^j} < i \qquad (4.14)$$

Inequality (4.14) has a straightforward interpretation. The right-hand side is the discount rate that would make a player indifferent between cooperating and defecting. If Inequality (4.14) is satisfied, it will be in a player's best interest to violate the agreement. If the left-hand side of Inequality (4.14) is greater than the discount rate, it will pay for the player to collude. Finally, if the left-hand side of Inequality (4.14) is equal to the discount rate, a player will be indifferent between cooperating and violating the agreement.

Principle: For a two-player, infinitely repeated, static game, a cartel will be unstable if

$$\frac{\pi_C^j - \pi_N^j}{\pi_V^j - \pi_C^j} < i \qquad (4.14)$$

where $j = A$, B, π_C^j is the payoff from an agreement, π_N^j the Nash equilibrium payoff, π_V^j is the payoff from violating the agreement, and i is the interest rate at which payoffs can be reinvested, that is, the discount rate.

──────────────── **Demonstration Problem 4.3** ────────────────

Consider, again, the payoff matrix summarized in Figure 1.2. Suppose that each firm adopts the following trigger strategy: Charge a high price; if the other firm defects, charge a low price for all future moves. Below what discount rate can we expect the cartel to break down?

Solution

Substituting the values from demonstration problem 4.2 into the left-hand side of Inequality (4.14) we get:

$$\frac{\pi_C^j - \pi_N^j}{\pi_V^j - \pi_C^j} = \frac{\$1,000,000 - \$250,000}{\$5,000,000 - \$1,000,000} = 0.1875 \qquad \text{(D4.3.1)}$$

Ceteris paribus, if the discount rate is greater than 18.75 percent, each firm will have an incentive to violate the agreement, in which case the cartel will be unstable. If the discount rate is less than this value, it will be in the best interest of both firms to collude. Finally, if the discount rate is exactly 18.75 percent, each firm will be indifferent between colluding and defecting.

───────────────── **Demonstration Problem 4.4** ─────────────────

Consider the infinitely repeated, static game in Figure D4.4. Two firms must decide whether to charge $30 or $50 for their product. The payoffs (in millions of dollars) represent the firms' profits from each strategy profile.

    a. Suppose that the discount rate is 20 percent. Is the cartel stable?
    b. Suppose that the discount rate is 10 percent. Is the cartel stable?
    c. What is the discount rate at which the firms will be indifferent to violating the agreement to charge $50?

FIGURE D4.4
Static Game in Demonstration Problem 4.4

		Firm B	
		$30	*$50*
Firm A	*$30*	($60, $60)	($200, $30)
	$50	($30, $200)	($80, $80)

Payoffs: (Firm A, Firm B)

Solution

    a. The strictly dominant strategy of both firms is to charge $30. The Nash equilibrium payoff is $\pi_N^j = \$60$. The payoff to each firm by colluding is $\pi_C^j = \$80$. The payoff from defection is $\pi_V^j = \$200$. Substituting these values into the left-hand side of Inequality (4.14) we obtain:

$$\frac{\pi_C^j - \pi_N^j}{\pi_V^j - \pi_C^j} = \frac{\$80 - \$60}{\$200 - \$80} = 0.1667 \qquad \text{(D4.4.1)}$$

    At a discount rate of 20 percent, it will pay for a firm to defect.
    b. Since 0.10 < 0.1667, it pays for both firms to cooperate.
    c. If the discount rate is 16.67 percent, the players will be indifferent between defecting and cooperating.

MAKING THREATS CREDIBLE

Let us briefly review what we have learned thus far.[5] In the games depicted in Figures 4.1 and 4.2, player A's strategy was to play $A1$ until player B violates the agreement, and then moves to $A2$. The switch to $A2$ was player A's trigger strategy to bind player B to the agreement. Because of the symmetry of the payoffs, we assumed that player B used the same trigger strategy. In the game depicted in Figure 4.1, these trigger strategies were sufficient to bind the players to the agreement. In the game depicted in Figure 4.2, on the other hand, each player had an incentive to defect. The players' trigger strategies were not enough to cement the agreement.

Now, suppose that player A in the game depicted in Figure 4.2 publicly announces the following trigger strategy in the hope of preventing player B from violating the agreement:

1. Move $A1$ in stage 1;
2. If player B moves $B1$ in stage 1, move $A1$ in stage 2;
3. If player B does not move $B1$ in stage 1, move $A3$ in stage 2 and thereafter.

According to this strategy, player A will remain faithful to the agreement so long as player B remains faithful. On the other hand, if player B violates the agreement, player A will adopt the trigger strategy of moving $A3$ in all subsequent stages. This trigger strategy is sometimes referred to as a **grim strategy** because it involves the most punitive response to player B's defection. In the event of defection, the strategy profile in the second and subsequent stages is $\{A3, B3\}$, which we have already seen is a Nash equilibrium.

> **Grim strategy** A trigger strategy that involves the most punitive response to a rival's defection from a cooperative agreement.

If we assume a discount rate of 5 percent, the present value of the expected stream of payoffs to player B from violating the agreement is:

$$\left(\pi_V^B - \pi_N^B\right) + \frac{\pi_N^B(1+i)}{i} = (50-1) + \frac{1(1.05)}{0.05} = 70 < 420 \qquad (4.15)$$

Player A's payoff is:

$$\left(\pi_V^A - \pi_N^A\right) + \frac{\pi_N^A(1+i)}{i} = (4-1) + \frac{1(1.05)}{0.05} = 24 < 420 \qquad (4.16)$$

There is no question that both players are better off by keeping the agreement. Player B does not benefit from defecting. Thus, player A's trigger strategy of moving $A3$ in the event of a defection would appear to bind player B to the agreement. The symmetry of the payoffs suggests that a similar trigger strategy by player B will bind player A to the cartel. Does this mean that this trigger strategy has enabled the players to overcome the prisoner's dilemma? Not necessarily. The reason is that player A's trigger strategy is only effective if the threat of retaliation by moving $A3$ is credible. In this case, player A's trigger strategy is an empty threat because it results in a lower payoff (24) than moving $A2$ (384). Other things being equal, player B has no reason to believe that player A will follow through with his or her threat to move $A3$ in the event of defection.

A trigger strategy that punishes a defector can introduce an element of stability into the cartel if, and only if, the threat is credible. If the threat is not credible, a trigger strategy will not bind a player to an agreement if it is in that player's best interest to defect. This is often true of verbal commitments. Sam Goldwyn, the legendary Hollywood mogul, may have said it best: "A verbal contract is not worth the paper it's written on" (Berg 1989, p. 386). Credibility is an essential element of all trigger strategies.

A threat, commitment, or promise is an example of a **strategic move**, which is an attempt to alter the behavior of a rival in a way that favors the player making the threat. A strategic move will fail unless a rival believes that the player will follow through with the threat. What a player says and does, however, may be two different things. In situations involving strategic behavior, the player being threatened should be wary of bluffs. **Bluffing** is an attempt by a player to gain a strategic advantage over a rival through a display of bravado that has no basis in fact. Although a bluff is essentially an empty threat, it may be successful under the right conditions.

> **Strategic move** An attempt by a player to gain an advantage by altering the behavior of a rival.
>
> **Bluffing** An attempt by a player to gain a strategic advantage over a rival through a display of bravado that has no basis in fact.

For a threat to produce the desired strategic results, it must be irrevocable. Since the player making the threat stands to gain a strategic advantage, rivals will test for weaknesses. Because credibility will be tested, a player who makes a threat must be prepared to back up words with action. Retreat must be impossible. Dixit and Nalebuff (1991, Chapter 6) discuss eight methods for preventing retreat. Depending on the circumstances, any one of these methods may prove successful. The first two methods (reputation and contracts) work by changing a rival's payoffs. The remaining methods work by making it difficult to backtrack.

Reputation

A player's reputation may not mean too much in a game that is played only once, since betraying a promise will not affect future game play. On the other hand, if the game is played repeatedly, a player's credibility will be seriously tested. Establishing and maintaining a reputation for following through with threats, commitments, and promises strengthens a player's credibility.

Countries such as the United States and Israel have an official policy of not negotiating with terrorists. The rationale behind this policy, and the importance of standing by this commitment, is straightforward. Terrorists must believe that their actions will not yield any benefits. Because acquiescing to their demands will only encourage more terrorism, they should be steadfastly rebuffed.

Paradoxically, reneging on a commitment can sometimes make a commitment more credible. The reason for this is that failure to honor a commitment may rule out future actions that may not be in a rival's best interest. Suppose, for example, that terrorists highjack an airplane and threaten to kill the passengers unless their demands are met. The terrorists might reason that because the lives of innocent people are at stake, the government will have no choice but to negotiate. Is there a way out of this dilemma? One possibility is for the government to agree to the terrorists' demands, so as to secure the release of the hostages. But, once the hostages are released, the authorities renege and attack the terrorists. After being double-crossed, how much confidence will terrorists have in future government assurances? By breaking its commitment not to negotiate, the government may make its no-negotiation pledge more credible.

In the above example, we assumed that the government and the terrorists are rational. Establishing and maintaining a player's reputation to enhance credibility has important strategic implications. On the other hand, establishing a reputation for being irrational can have the same effect. Threats that might seem incredible for a rational person may have strategic value when coming from a crazy person. When a lunatic makes a threat, no matter how off the wall, it must be taken seriously. This is the dilemma that often confronts U.S. government officials when dealing with rogue states run by religious fanatics seeking nuclear weapons, or parents dealing with unruly young children. Being perceived as a nutcase may have strategic value.

Contracts

One of the most common methods of making commitments credible is to write a contract. A **contract** is a formal agreement that binds two or more players to a specific course of action in exchange for something of value. Failure by either party to adhere to the terms and conditions of the contract results in sanctions.

> **Contract** A formal agreement that obligates the parties to perform, or refrain from performing, some specified act in exchange for something of value.

Although the threat of punishment in the event of noncompliance is important, it is not enough. As we have seen, punishment must be sufficiently severe to ensure that compliance yields greater benefits than noncompliance. A sufficiently large penalty, however, may still not be enough to guarantee compliance. For one thing, contract renegotiation must be impossible. A player's commitment to behave in a particular manner will be compromised if it is possible to alter the terms of the contract. The ability to renegotiate weakens a credible commitment because it allows a player to backtrack.

Another requirement for a successful contract is that it must be enforceable. Moreover, the enforcement agent must be properly motivated. Thomas Schelling (1989) gave an example of this in a commencement address before the Rand Graduate School. Referring to a drug rehabilitation clinic in Denver that caters to wealthy clients, he noted that as a precondition for admission, each patient must sign a letter of self-incrimination that attests to his or her substance abuse and addiction. In the event that a patient fails a random drug test while at the clinic, it is understood that the letter will be made public. Many patients subsequently try to buy their way out of this agreement. The clinic has an incentive to follow through with this threat because its reputation is at stake. The moral of this story is that the mere existence of a contract will not make a commitment credible. More is needed, such as a mechanism to avoid renegotiation, an independent and motivated oversight and enforcement agent, or the loss of reputation. In fact, if the player's reputation is important enough, a formal contract may not be necessary.

Closing Doors

A useful credibility device is to cut off all communications once a commitment is made. This action may make a threat or commitment irreversible. An extreme example of closing doors is the final wishes of a dying person, especially in a last will and testament. Once an individual is dead, promises made in life are irreversible. Sending a registered letter is another example. Once sent, it cannot be retrieved. Moreover, the sender cannot claim ignorance of the letter's contents. Once the recipient has signed for the letter, he or she cannot claim ignorance of its existence. One problem with cutting off communications is that it becomes difficult or impossible to monitor compliance. For this reason, it may be necessary to employ the services of a disinterested monitoring agent to ensure compliance, such as an estate executor in the case of a last will and testament.

Burning Bridges

Cutting off escape routes can make a threat or commitment credible by making retreat impossible. Players need not literally burn bridges, or ships (see Application 4.1), to make a commitment credible. Edward Land, founder of Polaroid, refused to diversify out of the instant photography business. As a result, rivals such as Kodak were well aware of Land's messianic commitment to defend his company's territory against patent infringement at all costs. Walter Mondale knew that he would irreversibly alienate some voters in his 1984 presidential bid by pledging to raise taxes to stanch the flow of red ink in the federal budget. He hoped to cement his position among voters who favored his economic program. In 1989, Egon Krenz, the prime minister of the German Democratic Republic, dismantled portions of the Berlin Wall to lend credibility to his proposed economic reform program. This action worked because it denied the East German government the opportunity to retreat, lest they run the risk of a mass exodus to the West.

Brinksmanship

Thomas Schelling (1960, p. 200) describes brinksmanship as "the deliberate creation of a recognizable risk . . . that cannot be completely controlled." Dixit and Nalebuff (1991, p. 155) refer to brinksmanship as "leaving the outcome beyond your control." Brinksmanship lends credibility to a commitment by exposing a rival to a shared risk. An antagonist who violates the *status quo* may result in both players slipping over the brink "whether we want to or not, carrying him with us" (Schelling 1960, p. 200).

Brinksmanship was an element of U.S.–Soviet foreign policy during the Cold War (see Application 4.2). A cornerstone of this foreign policy was known by the acronym MAD, or "mutually assured destruction." Both sides knew that since both sides possessed massive stockpiles of nuclear weapons, a nuclear exchange would result in near total global annihilation. Perhaps because both sides recognized the futility of a third world war, the long-feared and anticipated conflagration never occurred.

Baby Steps

When the stakes are very high, the problem of credibility may be insurmountable. On the other hand, it might be possible to establish trust and reach an accommodation if the game is broken down into a series of smaller games. Finding solutions to smaller problems may lead to an outcome in which the whole may be greater than the sum of its parts. Taking small, incremental steps may overcome the credibility problem because players may be more willing to take a chance of being double-crossed when the associated loss is small. Once credibility is established, this tactic is repeated. Over time, as trust is established and reinforced, both sides may be willing to make larger commitments.

Moving in baby steps reduces the size of the threat by reducing the magnitude of the payoff. As an example, suppose a homeowner hires a contractor to rebuild his or her front porch. Unless both players have had prior dealings, neither will completely trust the other. The homeowner is unlikely to pay the full contract price up front for fear that the contractor might not use quality materials, do quality work, or complete the job in a timely manner. On the other side, the contractor may not completely trust the homeowner to pay once the job is finished. To overcome this problem, the homeowner might agree to pay the contractor on a periodic basis. In this way, the homeowner can monitor the contractor's progress, while the contractor can be assured of minimizing the risk associated with nonpayment. If both parties are satisfied upon completion, not only might this pave the way for future business dealings, it may also redound to the benefit of the contractor through enhanced reputation and satisfied customer referrals.

There is a potential problem with this approach to establishing credibility. It is called the **end-of-game problem**, which will be discussed at greater length in the next chapter. If either player believes they will be cheated in the last round, previously established trust may completely unravel. Suppose that the homeowner believes that the contractor will threaten not to complete the job unless he or she receives a substantial premium over the originally contracted price. This tactic may be successful if the cost of locating another contractor to complete the job is sufficiently high. To avoid being "sandbagged" on the last round, the homeowner will consider discontinuing the relationship one round earlier. This transforms the penultimate round into the last round. Since both parties will project forward and reason back, the end-of-game problem still exists. The contractor may decide to end the relationship in the next-to-penultimate round, and so on. One way to avoid this problem is to make the end of the game uncertain, in which case it is not easy for either player to identity the last round. This problem does not arise in an infinitely repeated game because a final round does not exist.

> **End-of-game problem** For finitely repeated games with a certain end, each stage effectively becomes the final stage, in which case the game reduces to a series of noncooperative one-time games.

Teamwork

An individual player who finds it personally difficult to avoid the temptation to violate an agreement may be able to draw strength from others. Teamwork and peer pressure can be powerful inducements, especially when a player places a high value on the opinion of others. In the midst of the horror and chaos of battle, a lone soldier may be tempted to desert his or her post. But, if the soldier is a member of a squad, platoon, company, or battalion, then "duty, honor, and country" can result in a strong commitment to a well-defined objective. Desertion by one soldier can lead to panic by all, which could result in a rout and extensive loss of life. For this reason, a soldier's military training is supplemented with credible assurances of death by firing squad in the event of desertion. The possibility of death resulting from an attack becomes more attractive than the certainty of death from desertion.

Agents

Another method for making threats and commitments credible is to use negotiating agents who lack the authority to compromise or make concessions. This last method helps explain why professional athletes and entertainment celebrities are willing to pay 10 percent agency commission to negotiate service contracts on their behalf. It also explains the role of union leadership in collective bargaining sessions with management. In both cases, the principals' agents can be dismissed if they exceed their negotiating mandate or do not act in the best interest of their clients.

Application 4.1   **The Hunt for Red October**

In the blockbuster movie *The Hunt for Red October,* which was based on Tom Clancy's best-selling novel of the same name, Captain Marco Ramius (portrayed by Sean Connery) and a group of Soviet naval officers conspire to defect and deliver the nuclear submarine *Red October* into the hands of the United States Navy. Captain Ramius is criticized by a co-conspirator for his decision to inform Soviet Admiral Yuri Padorin (Peter Zinner) of his intentions, which prompted a frantic search and destroy operation. In explaining his decision, Ramius described a famous tactic used by Hernando Cortés in early 1519 when he landed in Mexico with a force of only six hundred men, twenty horses, and ten small cannons to conquer an Aztec nation of more than five million people. "When Cortés landed in the New World," Ramius explained, "he burned his ships. As a result, his men were well motivated."

By denying his men the opportunity to retreat, Cortés gave them a stark choice: fight or die. This tactic also had a demoralizing effect on the Aztecs because it underscored the determination of Cortés and his soldiers. The Aztec leader, Montezuma, attempted to stop Cortés, but his efforts failed because his soldiers lacked unity and tenacity. In part, his people were frightened by Spanish horses and firearms, which they had never seen, but they also believed in the Quetzalcoatl legend that prophesied that the Aztecs would return to the "white god." For these reasons, they saw opposing Cortés as futile. By the end of the year, Cortés had entered the Aztec capital and imprisoned Montezuma.

Application 4.2   Fail-Safe

For brinksmanship to be effective, it must be measured. The reason for this is that an accident or miscalculation could cause a strategic situation to spiral out of control. An example of a measured response in the game of brinksmanship can be found in Eugene Burdick and Harvey Wheeler's 1962 best-selling novel and underappreciated 1964 movie, *Fail-Safe*.

In the novel and movie, a combination of human and computer errors resulted in an American strategic bomber being ordered to drop its nuclear payload on Moscow. The U.S. president, military, and congressional leaders must find a way to deal with the crisis. In spite of their best efforts, the bomber breaches Soviet defenses and is sure to complete its mission. Fearing an all-out nuclear confrontation, the American president (portrayed by Henry Fonda) offers a deal to the Soviet premier: He will order that a nuclear bomb be dropped on New York City to avoid a retaliatory strike that will almost certainly escalate into a full-scale, mutually destructive, nuclear conflagration. The Soviet premier accepts the president's offer and nuclear Armageddon is avoided. The movie concludes with the destruction of both cities.

Fail-Safe highlights both the benefits and dangers of a policy of brinksmanship. Although this policy lends credibility to a threat, small errors can lead to highly undesirable outcomes.

DETERMINANTS OF BUSINESS COLLUSION

The cartels discussed above involved only two firms. If the economic benefit of violating an agreement is greater than the economic benefit of cooperation, the cartel will probably collapse. In some cases, even if there is an incentive to defect, a cartel may endure if there exists a trigger strategy that binds players together by punishing defectors. The necessary conditions for the long-term viability of a cartel were the main focus of this chapter. But, what factors account for the emergence of a cartel in the first place? In this section we will discuss three important determinants of business collusion: Number of firms with similar interests, the size of the firms relative to the industry as a whole, and the visibility of the agreement.

Number of Firms with Similar Interests

Collusion is more likely the smaller the number of firms with similar interests. A coincidence of common interests is no guarantee that a cartel will be successful. Collusive agreements are difficult when there are a large number of firms. The reason for this is that it becomes increasingly difficult to monitor member compliance. To see this, suppose that there are n parties to the agreement. Each member of the cartel must monitor the behavior of the other $n - 1$ members. The total number of monitoring arrangements necessary to police the cartel is $n(n - 1)$. In the two-firm case, $2(2 - 1) = 2$ monitoring arrangements are needed to police the cartel. In the case of the Vienna-based Organization of Petroleum Exporting Countries (OPEC), which has 13 members (Algeria, Angola, Ecuador, Indonesia, Iran, Iraq, Kuwait, Libya, Nigeria, Qatar, Saudi Arabia, United Arab Emirates, and Venezuela), $13(13 - 1) = 156$ monitoring arrangements are necessary to police compliance.

Policing cartel agreements is made even more difficult in the case of OPEC because of the divergent cultural, historical, economic, and political characteristics of the members. Is there any wonder why OPEC meets as frequently as it does to hammer out new production-sharing agreements? The incentive to cheat, especially by members with low production quotas, is very strong. To make matters worse, as membership in the cartel increases, so do monitoring costs. Sanctions against offending members become increasingly more difficult to impose, which increases the possibility that the cartel will break down.

Firm Size Relative to the Industry

Economies of scale exist in the monitoring and policing of collusive agreements. It is less expensive for large firms to monitor the behavior of a few large rivals or a large number of relatively small rivals than it is for small firms to monitor the behavior of a relatively large number of small rivals or a small number of large rivals.

Visibility

An important factor contributing to the durability of a cartel is the manner in which such arrangements are entered into. Collusions are either explicit or tacit. A cartel is explicit if its members meet overtly to hammer out agreements. An explicit collusive agreement specifies the responsibilities of each member, such as member production quotas, collective pricing policies, common advertising strategies, market shares, sanctions, and so forth.

When explicit agreements are impossible or illegal, firms may engage in tacit or secret collusion. Firms may achieve a mutually beneficial understanding based on shared experiences or a common understanding of the business environment. This common understanding is sometimes referred to as "conventional wisdom." The resulting agreement may constitute a focal-point equilibrium. Firms develop this common understanding by observing the behavior of their rivals over time, and it may result in an outcome that would otherwise be possible only through explicit collusion. When collusion is illegal, firms may enter into explicit, but secret, agreements.

To understand how tacit collusion might emerge, consider a possible price war between two do-it-yourself superstores, such as Home Depot and Lowe's Home Improvement. Both firms are considering how to price a popular item in their hardware department. While both companies have a core of loyal customers, another group of customers is sensitive to price differences. Market research has given senior management of both firms a good idea how sales will be affected by a rival's pricing decisions. The problem confronting each firm is how to price its product and determine advertising expenditures without prior knowledge of its rival's pricing policy. The payoffs from alternative combinations of pricing and advertising strategies are summarized in the static game depicted in Figure 4.3.

The reader should verify that charging a lower price is a strictly dominant strategy for both firms. The payoffs for the Nash equilibrium strategy profile {*Lower price, Lower price*} are highlighted in boldface. It should also be clear from the payoff matrix that it is in both firms' best interest to adopt a *higher price* strategy. Since neither firm wants to risk losing its market share by charging a higher price, each firm has an incentive to adopt its strictly dominant strategy.

FIGURE 4.3
Home Improvement Pricing Game

Loew's

		Lower price	Higher price
Home Depot	Lower price	**(90, 80)**	(110, 50)
	Higher price	(60, **110**)	(100, 90)

Payoffs: (Home Depot, Loews)

Suppose that Home Depot and Lowe's Home Improvement tacitly agree to charge a higher price. The strategy profile {*Higher price, Higher price*} is not a Nash equilibrium since both firms have an incentive to violate this agreement. On the other hand, when the firms observe over time that their efforts to capture a larger market share at the expense of its rival are thwarted by corresponding price reductions, it is less likely that either firm will initiate a price war. If the threat of reprisal is credible, tacit price fixing agreements are more likely to endure.

There are other credible threats that a firm might employ to discourage other firms from defecting. Suppose, for example, that Home Depot promises its customers that "we will not be undersold" or provide "low price guarantees." Knowing that any price reduction will be matched, Lowe's has less of an incentive to cut prices. Of course, this deterrent is viable only if Home Depot's threat to match its competitor's price cuts is carried out.

Another method for fixing prices is the most-favored-customer policy. This policy guarantees that a customer will get the best price the firm is willing to offer any other customer. If another customer receives a lower price, all most-favored customers will receive the same low price as well. A firm that adopts a most-favored-customer policy is better able to withstand pressures to lower prices since the firm can always argue that "I'd really like to offer you a lower price, but if I do it for you then I'll have to do it for others as well."

Price leadership often emerges from this sort of price competition. Pricing decisions made by one firm are quickly imitated by other firms in the industry. Suppose that Home Depot is the industry's price leader. Once Home Depot sets the higher price, other firms in the industry follow in lockstep. In this way, a price war is avoided and all firms benefit from the higher price.

> **Price leadership** A form of price collusion in which a dominant firm initiates a price change that is matched by the rest of the industry.

The threat of retaliation if a firm violates an explicit or tacit collusive agreement is only credible if it is actually carried out. If member firms are unwilling or unable to punish violators, explicit and tacit collusions will be unstable and may break down. On the other hand, if the threat of punishment is sure and swift, collusive agreements are more likely to last. The cost of punishing a defector, however, can be quite high. Not only must firms lower the price charged to the defector's customers, they must also lower the price charged to their own customers. While each profit-maximizing firm in the industry may experience a small increase in unit sales resulting from the generally lower price, the rogue firm's attempt to enhance its market share will fail. In the end, the price war results in lower profits for the entire industry. This is precisely what happens in the retail and airlines industries whenever a single firm attempts to lure customers from rivals by offering discounts. This situation is significantly altered, however, if market conditions allow individual firms to engage in discriminatory pricing.

Price discrimination is the practice of charging different individuals, or groups of individuals, different prices for the same good or service. There are various degrees of price discrimination.

Third-degree price discrimination occurs when a firm is able to segment the market into easily identifiable groups, with each group being charged a different price. If discriminatory pricing is possible, it is possible to punish violators by charging a lower price to the defector's customers while continuing to charge a higher price to its own customers. In this case, the cost of policing a collusive agreement is considerably reduced.

> **Price discrimination** The practice of charging different consumers, or groups of consumers, different prices for the same good or service.

CHAPTER REVIEW

Static games that are played over and over again without end are said to be *infinitely repeated.* *Finitely repeated games* are played a limited number of times and may have certain or uncertain ends. Cooperation among players to achieve a mutually beneficial outcome may be modeled as a *cooperative game.*

The *prisoner's dilemma* illustrates one-time, static games when the incentives of individual players are in conflict with the best interest of the group. When this happens, cooperation is inherently unstable, since each player has an incentive to defect. When static games are played an infinite number of times, cooperation may constitute a Nash equilibrium. Even when defection is profitable, the threat of retaliation may be sufficient to bind the players to an agreement. For this to be so, the present value of the stream of payoffs from defection must be greater than the present value of the stream of payoffs from cooperation.

Collusion is cooperation among firms to increase *market power.* Collusive behavior in business includes price fixing, production quotas, or other practices that stifle competition and increase profits. Collusion may be explicit or tacit agreement. If it results from a common understanding of the business environment, collusion may be a *focal-point equilibrium.*

The most recognized manifestation of collusive behavior is the *cartel,* which is a formal arrangement among producers to allocate market shares and/or maximize industry profits. Cartel agreements coordinate the decisions of member firms, such as establishing joint pricing and output policies. The necessary conditions for the emergence of an industry cartel include the *number of firms* with similar interests, the *size of the individual firms* relative to the industry as a whole, and the *visibility* of the agreement.

In general, if the economic benefit of violating an agreement is greater than the economic benefit of cooperation, the cartel will probably collapse. Even if there is an incentive to defect, a cartel may endure if there exists a trigger strategy that binds players together by punishing defectors. In general, a *trigger strategy* is a move by one player in response to unanticipated moves by a rival. A trigger strategy will continue to be used until a rival makes another unanticipated move. Trigger strategies can sometimes be used to enforce a cartel by eliminating the incentive to defect. A trigger strategy that punishes a defector may introduce an element of stability into the cartel if the threat of retaliation is credible. If the threat is not credible, a trigger strategy will not by itself prevent defection.

A *threat, commitment,* or *promise* is an example of a *strategic move,* which is an attempt to alter a rival's behavior in a way that favors the player making the threat. A strategic move will fail unless a rival believes the threat is credible. *Bluffing* is an attempt by a player to gain a strategic advantage over a rival through a display of bravado that has no basis in fact. Although a bluff is essentially an empty threat, under the right conditions, it may be successful.

For a threat to produce the desired strategic results, it must be irrevocable. A threat that can be changed loses its strategic value. The chapter discussed eight techniques for making threats credible, including *reputation, contracts, closing doors, burning bridges, brinkmanship, baby steps, teamwork,* and *agents.*

CHAPTER QUESTIONS

4.1   Explain the prisoner's dilemma in a one-time static game.

4.2   Explain the circumstances whereby the prisoner's dilemma may be overcome in an infinitely repeated, pure-strategy, static game.

4.3   Explain the difference between a credible threat and a trigger strategy. How are they similar?

4.4   Under what conditions will trigger strategies be successful in maintaining the integrity of a collusive agreement?

4.5   The existence of a trigger strategy that punishes defectors bind players to an agreement in an infinitely repeated, static game. Do you agree? Explain.

4.6   Under what circumstances will collusion in repeated games be stable?

4.7   Tacit collusions are more stable than explicit collusions. Do you agree? Why?

CHAPTER EXERCISES

4.1   Consider again the static game depicted in Figure 4.2. Suppose that player A adopts the following strategy:

1. Move *A1* in stage 1;
2. If player B moves *B1* in stage 1, move *A1* in stage 2;
3. If player B does not move *B1* in stage 1, move *A3* in stage 3;
4. Continue to move *A3* unless player B agrees to move *B1* in stage 4; and so on.

   a. What are the players' payoffs if player B moves *B1* in stage 1, defects in stage 2, and continues to defect in the future?
   b. What are the players' payoffs if player B moves *B1* in stage 1, defects in stage 2, renegotiates the agreement in stage 3, defects in stage 4, and so on.
   c. Is player A's strategy sufficient to bind player B to the agreement?

4.2   Consider the payoff matrix summarized in Figure E4.2. The numbers in each cell represent the expected profits in thousands of dollars from alternative *aggressive* and *reactive* price strategy profiles. Suppose that the discount rate in this infinitely repeated game is 7 percent.

   a. What is the present value of the stream of payoffs to Orange Company and Blue Company by not cooperating in an infinitely repeated game?
   b. What is the economic benefit to Orange and Blue from cooperating?
   c. What is the economic benefit to Orange or Blue from defection?
   d. Based on your answers above, is collusion viable?

4.3   Suppose in chapter exercise 4.2 that the discount rate rises to 12 percent.

   a. What is the present value of the stream of payoffs to both firms in an infinitely repeated game by not cooperating?
   b. What is the present value of the stream of payoffs to both firms from cooperation?
   c. What is the present value of the stream of payoffs to either from defection?
   d. Based on your answers above, is the collusive agreement stable?

FIGURE E4.2
Static Game in Chapter Exercise 4.2

Blue

		Aggressive	Reactive
Orange	Aggressive	($500, $500)	($100, $600)
	Reactive	($600, $100)	($150, $150)

Payoffs: (Orange, Blue)

FIGURE E4.4
Static Game in Chapter Exercise 4.4

Firm B

		Don't cheat	Cheat
Firm A	Don't cheat	(10, 10)	(–5, 20)
	Cheat	(20, –5)	(5, 5)

Payoffs: (Firm A, Firm B)

4.4 Consider the cooperative, one-time, static game shown in Figure E4.4 involving two firms that have entered into a collusive agreement. The payoffs are in millions of dollars. Having entered into the agreement, both firms must decide whether to adhere to the agreement (*Don't cheat*), or to defect (*Cheat*).

   a. Does either firm have a dominant strategy in a noncooperative, one-time game?
   b. If both firms follow a maximin strategy, what is the strategy profile for this game?
   c. Suppose that this game is infinitely repeated and firm *B* defects. What is firm *A*'s best response?
   d. How might the use of a trigger strategy affect your answer to part c?

4.5 Consider the two-player, noncooperative, one-time, static pricing game depicted in Figure E4.5. The payoffs are in millions of dollars.

   a. Does either player have a dominant strategy? If so, what is the Nash equilibrium strategy profile for this game?
   b. If this game were repeated an infinite number of times, would either player adopt a nondominant strategy?

4.6 Consider the infinitely repeated, static game depicted in Figure E4.2. Suppose that each firm adopts the trigger strategy that if a rival defects, it will respond by reverting to its dominant strategy in all future stages. At what discount rate is a player indifferent between cooperating and defecting?

4.7 Consider the static game depicted Figure E4.7. Peridot Corporation and Amethyst Company must decide whether to charge a low price or a high price for their product. Payoffs are in thousands of dollars.

   a. For a noncooperative, one-time, static game, does either firm have a dominant strategy? If not, what is each firm's secure strategy? Does this game have a Nash equilibrium?

FIGURE E4.5
Static Game in Chapter Exercise 4.5

Firm B

Firm A		High price	Low price
	High price	(10, 10)	(−5, 20)
	Low price	(20, −5)	(5, 5)

Payoffs: (Firm A, Firm B)

FIGURE E4.7
Static Game in Chapter Exercise 4.7

Amethyst

Peridot		Low	High
	Low	($25, $25)	($200, $40)
	High	($40, $200)	($50, $50)

Payoffs: (Peridot, Amethyst)

   b. If this were a cooperative, one-time, static game, what pricing strategy should each firm adopt? Why?

   c. Suppose that the discount rate is 5 percent. What is the present value of the stream of payoffs to each company by not cooperating?

   d. What is the present value of the stream of payoffs to both companies by cooperating?

   e. What is the present value of the stream of payoffs to either company by defecting?

   f. Based on the answers to the above questions, is collusion stable?

   g. Suppose that the discount rate is 25 percent. Is collusion stable?

4.8   Suppose that Tsunami Corporation and Cyclone Company are considering a change in their pricing policies. This game is depicted in Figure E4.8.

   a. What is the Nash equilibrium strategy profile for this noncooperative, one-time, static game?

   b. Suppose that the discount rate is 7 percent. If this game is infinitely repeated, what is the present value of the stream of payoffs to both companies by not cooperating?

   c. What is the present value of the stream of payoffs to both companies by cooperating?

   d. What is the present value of the stream of payoffs to either firm from defection?

   e. Based on the answers to the above questions, will cooperation be stable?

4.9   Consider the static game depicted in Figure E4.9.

   a. Does this one-time game have a unique Nash equilibrium?

   b. Suppose this game was played just once. What strategy profile would result in the best payoff for both players?

   c. Suppose that this game is infinitely repeated. Below what discount rate can we expect cooperation to break down?

4.10 Suppose that profits earned in chapter exercise 4.9 are reinvested at an annual interest rate of 25 percent.

FIGURE E4.8
Static Game in Chapter Exercise 4.8

Cyclone

		High price	Low price
Tsunami	High price	($6, $6)	($4, $8)
	Low price	($8, $4)	($5.5, $5.5)

Payoffs: (Tsunami, Cyclone)

FIGURE E4.9
Static Game in Chapter Exercise 4.9

Player B

		X	Y
Player A	X	(5, 5)	(16, 1)
	Y	(1, 16)	(6, 6)

Payoffs: (Player A, Player B)

a. What is the present value of the stream of payoffs to both companies by not cooperating?
b. What is the present value of the stream of payoffs to both companies by cooperating?
c. What is the present value of the stream of payoffs to either company by defecting?
d. Based on the answer to the above questions, is cooperation viable?

ENDNOTES

1. This case study was inspired by an experiment involving a group of Texas A&M University students that was reported in the *Wall Street Journal,* December 4, 1986.
2. Discounting is necessary because of the time value of money. That is, $1 received or expended today is worth more than $1 received or expended tomorrow because the $1 may be invested and earn a rate of return of i.
3. An annuity is a series of equal payments, which are made at fixed intervals for a specified number of periods. Fixed payments made at the beginning of each period are referred to as an annuity due. Fixed payments made at the end of each period are referred to as an ordinary annuity.
4. For an infinitely repeated static game, Equation (4.1) may be rewritten as:

$$PV = \pi \left[1 + \frac{1}{(1+i)^1} + \frac{1}{(1+i)^2} + \frac{1}{(1+i)^3} \cdots \right] = \pi S \qquad \text{(F4.4.1)}$$

where

$$S = 1 + \frac{1}{(1+i)^1} + \frac{1}{(1+i)^2} + \frac{1}{(1+i)^3} + \cdots \qquad \text{(F4.4.2)}$$

After multiplying both sides of Equation (F4.4.2) by $1/(1+i)$ we obtain:

$$S\left(\frac{1}{1+i}\right) = \frac{1}{(1+i)^1} + \frac{1}{(1+i)^2} + \frac{1}{(1+i)^3} + \cdots \qquad \text{(F4.4.3)}$$

Subtracting Equation (F4.4.3) from Equation (F4.4.2) yields:

$$S - S\left(\frac{1}{1+i}\right) = 1 \qquad \text{(F4.4.4)}$$

Solving for S we obtain:

$$S = \frac{1+i}{i} \qquad \text{(F4.4.5)}$$

Substituting Equation (F4.4.4) into Equation (F4.4.1) yields:

$$PV = \frac{\pi(1+i)}{i} \qquad \text{(F4.4.6)}$$

5. This section is adapted from Avinash Dixit and Barry Nalebuff's highly informative and entertaining *Thinking Strategically: The Competitive Edge in Business, Politics, and Everyday Life,* New York: W. W. Norton, 1991, Chapter 6.

CHAPTER

5

FINITELY REPEATED, STATIC GAMES WITH COMPLETE INFORMATION

In this chapter we will:

- *Analyze pure strategy, cooperative, finitely repeated static games with certain and uncertain ends in which the players have complete information;*
- *Discuss the end-of-game problem;*
- *Develop a cheating rule for cooperative, finitely repeated, static games with an uncertain end;*
- *Discuss common enforcement mechanisms for cooperative, finitely repeated games.*

INTRODUCTION

We have thus far considered the two extreme versions of static games with complete information: those that are played just once, and those that are played an infinite number of times. We saw in Chapter 4 that in infinitely repeated, static games it may be possible to escape the prisoner's dilemma if the present value of the stream of future payoffs by cooperating is greater than by not cooperating. Even when it is not in the best interest of the players to cooperate, it may still be possible to introduce stability into a collusive arrangement using trigger strategies and credible threats. Even when threats are credible, however, cheating may still occur if defection is profitable. One of the questions that we will seek to answer in this chapter is whether it is still possible to solve the prisoner's dilemma when all players know that the game will be played more than once, but a finite number of times.

FINITELY REPEATED GAMES WITH A CERTAIN END

Finitely repeated static games in pure strategies differ fundamentally from the infinitely repeated games discussed in the previous chapter. There are two classes of finitely repeated games. In the first class, the players know that the game will come to an end, but are uncertain as to when that will occur. In the second class, the final stage is known with certainty. We will begin our discussion by examining finitely repeated, static games with a certain end.

> **Finitely repeated static game** A finitely repeated game is played a limited number of times. Finitely repeated static games may have certain or uncertain ends.

The one-time static game depicted in Figure 5.1 involves two players with three pure strategies. If larger payoffs are preferred, the reader should verify that this game has two Nash equilibria strategy profiles: {*A2, B2*} and {*A3, B3*}. The payoffs for these strategy profiles are highlighted in boldface. It should be obvious, however, that both players would be better off by agreeing to adopt the strategy profile {*A1, B1*}. This game is an example of a prisoner's dilemma since both

FIGURE 5.1

Two-Stage Game in Pure Strategies

Player B

		B1	B2	B3
	A1	(15, 15)	(4, **20**)	(0, 0)
Player A	A2	(**20**, 4)	(**8, 8**)	(0, 0)
	A3	(0, 0)	(0, 0)	(**1, 1**)

Payoffs: (Player A, Player B)

players have an incentive to violate such an agreement. Player A has an incentive to switch to A2 and player B has an incentive to switch to B2. As in all such games, individual incentives trump the incentive of the cooperative. As a result, both players earn a payoff of 8, instead of a payoff of 15 by cooperating.

We saw in the previous chapter that it may be possible to escape the prisoner's dilemma if a static game is infinitely repeated. For this to be the case, the present value of the stream of expected future payoffs from violating the agreement must be less than the present value of the stream of expected future payoffs from the agreement. The present value of the stream of future payoffs depends on the discount rate. From Inequality (4.14), the reader should verify that for the game depicted in Figure 5.1, it will be in the players' best interest to violate the agreement for any discount rate that is greater than 140 percent. But, what if this game is played a finite number of times? Will it still be in both players' best interest to form a cartel? We will explore this question by considering the simplest example of a finitely repeated game. We will assume that the game in Figure 5.1 is played just two times.

The first thing to note about Figure 5.1 is that the strategy profiles {A2, B2} and {A3, B3} are Nash equilibria if the game is played twice. Consider, for example, the strategy profile {A2, B2}. This combination of moves constitutes a Nash equilibrium strategy because neither player can improve his or her payoff by switching strategies in either stage. Player A can do no better by switching from A2 in either period, regardless of the strategy adopted by player B. The same can also be said for player B. If both players adopt the strategy profile {A2, B2} in both periods, the undiscounted payoff to each player is 8 + 8 = 16. On the other hand, both players will be better off by agreeing to the combination of moves {A1, B1} in both stages, since the undiscounted payoff to each player is 15 + 15 = 30 > 16. But, does this combination of moves constitute a Nash equilibrium strategy in a two-stage game?

Although both players are better off by moving {A1, B1} in both stages, this strategy cannot be part of a Nash equilibrium. The reason is that once the first stage has been played, the second and final stage becomes a one-time game, and the strategy profile {A1, B1} is not a Nash equilibrium since both players will have an incentive to violate the agreement. In the second stage, the strategy profile {A2, B2} constitutes a Nash equilibrium. We are left to determine whether the combination of moves {A1, B1} in the first stage and {A2, B2} in the second stage constitutes part of a Nash equilibrium strategy. If it does, the undiscounted payoffs to both players will be 15 + 8 = 23, which is better than the payoffs from playing {A2, B2} in both stages.

Unfortunately, the combination of moves {A1, B1} in the first stage and {A2, B2} in the second stage does not constitute a Nash equilibrium strategy. To see why, consider the game from player A's perspective. If player B violates the agreement in the first stage while player A moves A1, the undiscounted payoff to player B for the entire game is 20 + 8 = 28 > 15 + 8 = 23. By contrast, the undiscounted payoff to player A is 4 + 8 = 12, which is less than 8 + 8 = 16 by not cooperating in either stage. Because of the symmetry of the game, this is also true for player *B*. Thus, it is in both players' best interest to violate the agreement in the first and second stages of the game. The only strategy that constitutes a Nash equilibrium is for player *A* to move *A2* and for player *B* to move *B2* in both stages of this game.

Since both players will obviously benefit from the combination of moves {*A1, B1*} in the first stage, is it possible for either player to devise a tactic that will keep his or her rival from defecting in the first stage? Let us consider this problem from player *A*'s perspective. Suppose that player *A* announces to player *B* that he or she intends to adopt the following strategy:

1. Move *A1* in stage 1;
2. If player *B* moves *B1* in stage 1, move *A2* in stage 2;
3. If player *B* does not move *B1* in stage 1, move *A3* in stage 2.

The rationale underlying this strategy is as follows. Player *A* moves *A1* in stage 1 because this will result in the best payoff. If player *B* defects in stage 1, player *A* will punish player *B* by moving *A3* in stage 2. Regardless of how player *B* moves in stage 1, player *A* will defect and move *A2* in stage 2 in the hope that player *B* will continue to move *B1*. The third part of player *A*'s strategy is critical to the survivability of the agreement in the first stage.

By announcing that he or she will retaliate by playing *A3* in stage 2 if player *B* defects in stage 1, player *A* hopes to bind player *B* to the agreement. Suppose that player *B* adopts *B2* in the first stage while player *A* sticks to the agreement by moving *A1*. If player *B* moves *B2* in stage 2 after defecting in stage 1, the undiscounted payoff to player *B* will be $20 + 0 = 15 < 15 + 8 = 23$. It should be clear that it will be in player *B*'s best interest to stick to the agreement, at least in the first stage. Since the game is symmetrical, it seems reasonable that player *B* will adopt the same strategy as player *A*, which is:

1. Move *B1* in stage 1;
2. If player *A* moves *A1* in stage 1, move *B2* in stage 2;
3. If player *A* does not move *A1* in stage 1, move *B3* in stage 2.

Although the combination of moves {*A1, B1*} in the first stage and {*A2, B2*} in the second stage appears to constitute a Nash equilibrium strategy, this is an illusion. This combination of moves is a Nash equilibrium only if each player's threat to retaliate by moving *A3* or *B3* in the event of defection is credible, which it is not. To see why, recall that if player *B* defects and player *A* does not defect, the resulting strategy profile is {*A1, B2*}. If player *B* continues to move *B2* in stage 2, the undiscounted payoff to player *A* by moving *A3* is $4 + 0 = 4$. On the other hand, if player *A* moves *A2*, his or her undiscounted payoff is $4 + 8 = 12$. Since player *A*'s payoff is greater by moving *A2*, moving *A3* in the second stage is an empty threat. Only if player *B* believes that player *A* is serious about following through with the threatened retaliation will player *A*'s strategy constitute a Nash equilibrium. Because of the symmetrical nature of this game, the same line of reasoning applies to player *B*. Thus, the only combination of moves that constitutes a Nash equilibrium is {*A2, B2*} in both stages.

END-OF-GAME PROBLEM

We have seen how the use of trigger strategies in infinitely repeated games may introduce an element of stability into collusive agreements. By contrast, the use of trigger strategies in static games with a certain end will fail. This observation is known as the **end-of-game problem**. The end-of-game problem transforms finitely repeated games with a certain end into a series of one-time, static games, which makes it difficult for players to escape the prisoner's dilemma.

In the game depicted in Figure 5.1 the strategy profile {*A1, B1*} cannot be part of a Nash equilibrium strategy in a two-stage game. The reason was that the threat of retaliation in the second stage lacked credibility. But, what if the threat was credible? Would such a threat also work in the second stage to prevent defection? No, for the simple reason that there is no further stage in which to carry out a threat. But, what if the game depicted in Figure 5.1 had three stages? Would the threat of retaliation work in the second stage? Does the combination of moves {*A1, B1*} constitute part of a Nash equilibrium in the second stage, but not the third stage? What if the game is played four, five, or *n* times? Will an agreement to move {*A1, B1*} constitute part of a Nash equilibrium strategy in every stage except the last? To answer this question, consider again the game in Figure 5.1. Suppose that this game comes to an end after, say, the fifth stage. Let us further assume that no matter how implausible, the threat by player *A* to move *A3* and the threat by player *B* to move *B3* in the event of defection is credible. Is this threat sufficient to bind both players to an agreement to play {*A1, B1*} in each stage of this game?

To answer this question, we will employ the technique known as **backward induction**, also known as the **fold-back method**. Backward induction simply involves the process of projecting forward and reasoning backward. We begin by projecting the game to the fifth, and final, stage of this game. The threat of retaliation in the fifth stage is not credible because there are no further stages in which to carry out the threat. Regardless of whatever else transpired in the previous four stages, the combination of moves {*A2, B2*} will be played in stage 5. Is the threat of retaliation in the fifth stage enough to bind the players to {*A1, B1*} in the fourth stage?

> **Backward induction (fold-back method)** A method for determining the outcome of a game by projecting forward and reasoning backward.

Unfortunately, the answer to this question is also "no." By moving {*A2, B2*} in stage 5, stage 4 effectively becomes the last stage in which cooperation is possible. Thus, a trigger strategy that threatens retaliation in the fifth stage becomes meaningless. Defection in stage 5 is a foregone conclusion. Thus, retaliation in stage 5 lacks credibility. By "folding back" the process to stage 4, stage 3 now effectively becomes the last stage, and so on all the way back to the first stage. Just as in the case when the game depicted in Figure 5.1 was played twice, the search for a Nash equilibrium in a multistage static game reduces to a series of noncooperative, one-time, static games. Thus, the Nash equilibrium strategy for player *A* is to move *A2,* and for player *B* to move *B2* in each and every stage of this game.

When a finitely repeated game in pure strategies has a certain end, the use of trigger strategies to enforce cooperation is doomed to failure. The reason for this is relatively straightforward. Since the players understand that they cannot be punished for their actions in subsequent stages, each has an incentive to adopt a strategy that is consistent with a one-time noncooperative game. This is the end-of-game problem. For the game in Figure 5.1, the end-of-game problem means that each player will adopt his or her dominant strategy in the first and all subsequent stages. Even if player *A* believes that player *B* will move *B1*, it is still in player *A*'s best interest to move *A2* since player *B* cannot retaliate. The same line of reasoning, of course, holds true for player *B*. Since no one wants to be left "holding the bag" when someone else cheats, no one cooperates. Thus, the Nash equilibrium strategy for this game is for both players to adopt their dominant one-time game strategies.

> **End-of-game problem** For finitely repeated, pure-strategy games with complete information with a certain end, each stage effectively becomes the final stage, in which case the game reduces to a series of noncooperative, one-time, static games.

What is remarkable about finitely repeated static games with a certain end is that regardless of the number of stages, the entire game is nothing more than a collection of noncooperative, one-time static games. Once the last stage is identified, each player has an incentive to view the next-to-the-last stage as a one-time game, which transforms that stage into the last stage, and so on. This unraveling renders cooperative agreements unworkable. How the players move in the last stage determines how the players move in the first stage, and all subsequent stages. This leads to the following general principle for playing finitely repeated, static games in pure strategies with complete information and a certain end.

> ***Principle:*** A player's Nash equilibrium strategy in a finitely repeated, static game with pure strategies, complete information, and a certain end is to play each stage as if it was a one-time static game.

Two interesting examples of the end-of-game problem are found in Baye (2003, Chapter 10). The first example involves a worker who announces that he or she plans to quit. In general, it is reasonable to assume that people work hard partly because they fear being fired if they are caught "goofing off." In fact, if the net benefit of being diligent is greater than the net benefit of shirking, it is in the worker's best interest to do a good job.

Now, suppose that a worker announces on Monday morning that Tuesday will be his or her last day on the job. Will he or she work diligently on Tuesday? Probably not. Since the worker has no intention of showing up for work on Wednesday, any threat by management to dismiss the employee for goofing off on Tuesday will not carry much weight. The worker's choice between working hard and shirking on Tuesday is a noncooperative, one-time game in pure strategies. Since management cannot retaliate on Wednesday, this trigger strategy is rendered impotent.

Will the employee's attitude toward work on Monday be affected by his or her decision to quit at the end of Tuesday? The worker has identified Tuesday as the final stage of this noncooperative game. Thus, there is an incentive to view Monday as a one-time game as well, which suggests that the worker will shirk his or her responsibilities on that day as well. If the employee gives two weeks' notice, each day on the job becomes a noncooperative, one-time game. Does this scenario sound unrealistic? If it does, consider how your attitude toward work might change if you submitted your two weeks' notice. Would you be disposed to work just as hard as you did before, or would you slack off while counting down your final days on the job?

What, if anything, can management do to make you work hard? Management could, of course, fire the worker on the spot, but this move could be counterproductive since the worker would change his or her strategy from giving two weeks' notice to quitting at the close of business on the planned day of resignation. This would make the task of finding replacement workers at short notice, without disrupting the production process, extremely difficult. Moreover, firing an employee without cause (there being no evidence that the worker will shirk his or her responsibilities) could have negative legal implications.

The above discussion of a finitely repeated game with a certain end, however, suggests a possible solution to management's dilemma. The answer lies in extending the game beyond the resignation date. For example, management could offer the employee assistance in identifying new employment opportunities, or perhaps provide letters of recommendation to potential future employers. By extending the employee/management relationship into the future, it may be in the worker's best interest to avoid "burning bridges" by goofing off during the final days of his or her employment.

Baye's second example of the end-of-game problem deals with the so-called "snake oil" salesman. During the American westward expansion of the late nineteenth century, snake-oil salesmen traveled from one frontier town to another selling bottles of elixir promising to cure everything from toothaches to baldness. Of course, these claims were bogus, but by the time their "patsies" realized that they had been "had," the snake-oil salesman would be long gone. By contrast, what if a local merchant attempted to pull a similar scam? It is precisely because the local merchant is playing a finitely repeated game with an uncertain end that the threat of retaliation for unethical behavior, such as loss of business or jail time, helps to ensure that his or her products are of reliable quality. For the snake-oil salesman there is no tomorrow, and so the transaction is played as a noncooperative, one-time game.

CASE STUDY 5.1: EXTRA-CREDIT GAME PART II

In the previous chapter, the fragile nature of a cooperative strategy profile in a simultaneous-move, one-time game, was illustrated in Application 4.1: Extra-Credit Game Part I. In that game, a class of twenty-nine MBA students was instructed secretly to choose between an *orange* or *blue* strategy. The students were awarded extra-credit points based on the number of students choosing a particular color. The payoffs from alternative strategy profiles were summarized in Table C4.1. In that game, students adopting an *orange* strategy received two more extra-credit points than students adopting a *blue* strategy. Thus, *orange* was a strictly dominant strategy. On the other hand, adopting an *orange* strategy did not always result in the best payoff. If every student adopted a strictly dominated *blue* strategy, the payoff to each student was 2.9 extra-credit points. If every student adopted their strictly dominant *orange* strategy, each would only receive two extra-credit points.

In the second stage of the experiment, the students were given the opportunity to form a cartel by collectively adopting a *blue* strategy. This transformed the extra-credit game into a cooperative, simultaneous-move, one-time game in which each student continued to have a strictly dominant *orange* strategy. Despite the agreement to adopt a mutually beneficial *blue* strategy, there were seventeen defections. Students who remained faithful to the agreement received 1.7 extra-credit points, while defectors received 3.7 extra-credit points. Students who did not defect were penalized by receiving 0.3 fewer extra-credit points, while those students who violated the agreement were awarded with an additional 1.7 extra-credit points. This example underscores the fragility of collusive agreements. The incentive to cheat undermined the viability of the cartel.

In the third stage of this experiment, the students were asked to agree on a common strategy and, as before, secretly record their choice. As before, only the instructor knew the identities and strategy choices. Unlike the second stage, however, the third stage included a system of rewards and punishments for violating or not violating the agreement. Students were informed that defectors would be penalized 0.5 extra-credit points, while students who did not defect would be awarded 0.5 extra-credit points. What were the results of this third stage of this game? A substantial number of students continued to violate the agreement, although the number of defections declined somewhat. Seventeen students did not defect while twelve defected. Nondefectors received 2.2 extra-credit points and the violators received 3.2 extra-credit points. Why did this system of rewards and punishments not bind the players to the cartel? The reason is that the penalty for defection was too small. The difference in payoffs from adopting an *orange* strategy was still uniformly better by 1 extra-credit point. As a result, *orange* remained a strictly dominant strategy. The incentive to violate the agreement was not eliminated.

The fourth, and final, stage of the experiment duplicated the third stage except that the transfer of extra credit from defectors to nondefectors was increased to 2 points. This transformed *blue* into a strictly dominant strategy. Not surprisingly, when the results were tabulated, *blue* was unanimously adopted. In this case, the penalty was sufficient to deter defection and bind the students to the agreement. But this unanimous selection was not the best! The optimal cooperative agreement for the *group* would have been for twenty-eight students to adopt a *blue* strategy and for one student to defect by adopting an *orange* strategy. The reason for this is that twenty-eight students would each receive 4.8 extra-credit points, while the defector would receive only 2.8 points, which is only marginally less than 2.9 extra-credit points he or she would receive when the group unanimously adopted a *blue* strategy.

The problem with this cartel is finding a volunteer to be the "sacrificial lamb." This problem may not be as insurmountable as it appears. A student with a high grade point average will receive less marginal benefit from two additional extra-credit points than a student with a low grade point average. In fact, a student with the high grade point average might even derive greater marginal utility from accolades received from other students for "taking one for the team."

FINITELY REPEATED GAMES WITH AN UNCERTAIN END

In the previous section we saw that a finitely repeated game with a certain end collapses into a series of noncooperative, one-time, static games. If the end of the game is uncertain, however, the outcome is quite different. To see how, let us once again consider the finitely repeated game depicted in Figure 5.1. The only difference is that while the players know the game will eventually come to an end, they just don't know when. Because the players do not know when the last stage will occur, it may be possible to use a trigger strategy that threatens retaliation to bind rivals to an agreement. In fact, this scenario is not much different from the infinitely repeated, static games examined in the previous chapter. The main distinction lies in our evaluation of the present value of the expected stream of future payoffs in the presence of this uncertainty.

Let us begin by denoting θ as the probability that the game will come to an end in the second stage where $0 \leq \theta \leq 1$. Conversely, $(1 - \theta)$ is the probability that the game will be played in stage 2. If the game is played in the second stage, the probability that the game will be played in the third stage is $(1 - \theta)^2$, and so on. The probability that the game will be played in the nth stage is $(1 - \theta)^{n-1}$. As before, we will assume that both players will revert to their strictly dominant strategy in the event of a defection by a rival. To demonstrate that $\{A2, B2\}$ is a Nash equilibrium for the game in Figure 5.1, it is necessary to show that neither player gains by violating the agreement to move $\{A1, B1\}$ in each stage. Since the end of the game is uncertain, the undiscounted expected payoff to each player for the entire game by cooperating is given by the expression:

$$15\Big(1 + (1-\theta) + (1-\theta)^2 + (1-\theta)^3 + \cdots\Big) \tag{5.1}$$

On the other hand, if a player defects in the first stage while his or her rival remains faithful, but reverts to a dominant strategy in the second and subsequent stages, the undiscounted expected present value of the stream of future payoffs is given by the expression:

$$20(1) + 8\Big((1-\theta) + (1-\theta)^2 + (1-\theta)^3 + \cdots\Big) \tag{5.2}$$

For defection to be profitable, Expression (5.2) must be greater than Expression (5.1), that is:

$$20(1) + 8\Big((1-\theta) + (1-\theta)^2 + (1-\theta)^3 + \cdots\Big)$$
$$> 15(1) + 15\Big((1-\theta) + (1-\theta)^2 + (1-\theta)^3 + \cdots\Big) \tag{5.3}$$

Since the left-hand side of Inequality (5.3) is the sum of a geometric progression, this expression may be abbreviated as:[1]

$$\frac{(1-\theta)}{\theta} > \frac{5}{7} \tag{5.4}$$

Solving Inequality (5.4) we find that $\theta < 0.583$. The interpretation of this result is that it is in the players' best interest to defect if the probability that the game will end in the next stage is less than 58.3 percent. In general, the greater the probability that a game will move to the next stage, the less the likelihood that a trigger strategy that punishes defectors will bind players to an agreement. In the extreme case where $\theta = 0$, a finitely repeated game becomes an infinitely repeated game. At the other extreme, when $\theta = 1$, a finitely repeated game becomes a one-time game, or what is almost the same thing, a finitely repeated game with a certain end, in which case the threat of retaliation in the event of defection is ineffective.

Application 5.1   Groundhog Day

In the darkly humorous movie *Groundhog Day*, Phil Connors (portrayed by Bill Murray) plays an unpleasant television weatherman from Pittsburgh who finds little pleasure in his life, except perhaps for making his colleagues miserable. In the movie, Phil is sent to the sleepy Pennsylvania town of Punxsutawney to cover an annual event known as *Groundhog Day*. The centerpiece of the festivities occurs when locals and tourists gather together in the center of town at a place called Gobbler's Knob. At the appointed time, a groundhog called Punxsutawney Phil is extracted from an ersatz tree stump to predict whether or not there will be an early end to the winter.

Unlike in previous years, Phil is accompanied by his television producer Rita (Andie MacDowell) and cameraman Larry (Chris Elliott). After cynically covering the event, Phil and colleagues are forced to spend the night in a Punxsutawney hotel because a snowstorm has blocked the roads. Next morning, Phil awakens to a rendition of "Pennsylvania Polka" on the clock radio and the announcer proclaiming that it is Groundhog Day. Phil remembers having already covered this event, but nobody else in the town seems to, including Rita and Larry. At first, Phil reacts as anyone might by reliving the events of the day in much the same way as previously. Once again, the same snowstorm strands him in Punxsutawney. The next morning he awakens again to the "Pennsylvania Polka" and another Groundhog Day.

Caught in this seemingly endless repetition, Phil alters his behavior in self-indulgent and self-destructive ways, such as eating, smoking, and drinking adult beverages to excess, driving on railroad tracks, insulting the police, dropping a plugged-in radio in his own bath, jumping into the path of a truck, and diving off a building. At one point in the movie, Phil even kidnaps Punxsutawney Phil and, after a high-speed chase with the police in pursuit, drives off a cliff. Unfortunately, none of these actions make the slightest bit of difference; after each demise, Phil wakes up in the same hotel to the sounds of "Pennsylvania Polka" and another Groundhog Day. Phil is stuck living the same day over and over again, seemingly with no end in sight. In the vernacular of game theory, Phil's trip to Punxsutawney was transformed from a one-time game to a finitely repeated game with an uncertain end.

Was Phil's strategic behavior altered by the change in the game? It would appear so. Had he been able to return to Pittsburgh as scheduled, he most likely would have continued to be the same obnoxious, overbearing character with few friends and no close relationships. By the end of the movie, however, he manages to discover a better side of his personality, after which his nightmare in Punxsutawney comes to an end. The viewer is left with the impression that Phil's future is better off from the experience. Or, as a game theorist might put it, the present value of the stream of future payoffs by cooperating and being pleasant with others was greater than the present value of the stream of future payoffs by repeatedly playing the Nash equilibrium strategy profile of a one-time game.

CHEATING RULE FOR FINITELY REPEATED GAMES WITH AN UNCERTAIN END

Analytically, the only difference between infinitely repeated games and finitely repeated games with an uncertain end is the probability that the game will end after each stage of play. The undiscounted expected stream of payoffs for the jth player may be written as:

$$E(\pi) = \pi_1^j + (1-\theta)\pi_2^j + (1-\theta)^2\pi_3^j + (1-\theta)^3\pi_4^j + \cdots \tag{5.5}$$

The present value of the expected stream of future payoffs for player j may be written as:

$$PV^j = \pi_1^j\left(\frac{1-\theta}{1+i}\right)^0 + \pi_2^j\left(\frac{1-\theta}{1+i}\right)^1 + \cdots + \pi_n^j\left(\frac{1-\theta}{1+i}\right)^{n-1} = \sum_{t=0}^{n}\pi_t^j\left(\frac{1-\theta}{1+i}\right)^{t-1} \tag{5.6}$$

In Equation (5.6), i is the discount rate and $(1-\theta)$ is the probability that the game will be played in the next period. If we assume that the payoff is the same in each period, it may be demonstrated that the present value of the stream of payoffs by not cooperating is:[2]

$$PV_N^j = \frac{\pi_N^j(1+i)}{i+\theta} \tag{5.7}$$

Similarly, the present value of the stream of expected payoffs by cooperating is:

$$PV_C^j = \frac{\pi_C^j(1+i)}{i+\theta} \tag{5.8}$$

Finally, the present value of the stream of expected payoffs from violating the agreement is the one-time, net payoff from defecting $(\pi_V^j - \pi_N^j)$ plus the present value of the stream of expected Nash equilibrium payoffs (PV_N^j), that is:

$$PV_V^j = \left(\pi_V^j - \pi_N^j\right) + PV_N^j \tag{5.9}$$

When will it pay to defect in a finitely repeated game with an uncertain end? A player has an incentive to defect when the present value of the stream of expected payoffs from violating the agreement is greater than the present value of the stream of expected payoffs from cooperating. This condition is summarized in Inequality (5.10).

$$\left(\pi_V^j - \pi_N^j\right) + PN_N^j > PV_C^j \tag{5.10}$$

or

$$\left(\pi_V^j - \pi_N^j\right) + \frac{\pi_N^j(1+i)}{i+\theta} > \frac{\pi_C^j(1+i)}{i+\theta} \tag{5.11}$$

──────────────── **Demonstration Problem 5.1** ────────────────

Consider the finitely repeated, static game depicted in Figure 5.1. Suppose that the discount rate is 5 percent and the probability that the game will end is 50 percent.

    a. What is the present value of the stream of payoffs to each player from a *no collusion* strategy?

    b. What is the present value of the stream of payoffs to each player by cooperating?

    c. What is the present value of the stream of payoffs from defection if the other player remains faithful?

    d. Based on your answers to parts a and b, is this agreement stable?

Solution

    a. A Nash equilibrium occurs when both players adopt an *A2* strategy. The present value of the payoffs to each firm from a no collusion strategy is:

$$PV_N^j = \frac{\pi_N^j(1+i)}{i+\theta} = \frac{8(1.05)}{0.55} = 15.27 \qquad (D5.1.1)$$

    b. The present value of the stream of payoffs to both players by adopting a cooperative strategy is:

$$PV_C^j = \frac{\pi_C^j(1+i)}{i+\theta} = \frac{15(1.05)}{0.55} = 28.64 \qquad (D5.1.2)$$

    c. The present value of the stream of payoffs to either player by defecting when his or her rival remains faithful is:

$$PV_V^j = \left(\pi_V^j - \pi_N^j\right) + PV_N^j = (15-8) + 15.27 = 27.27 \qquad (D5.1.3)$$

    d. Since $PV_V{}^j < PV_C{}^j$, there is no incentive for either player to violate the agreement.

──────────────── **Demonstration Problem 5.2** ────────────────

Consider the finitely repeated, static game depicted in Figure 1.2, which is reproduced below. Suppose that the discount rate is 5 percent and the probability that the game will end is 1 percent.

    a. What is the present value of the stream of payoffs to each player from a *no collusion* strategy?

    b. What is the present value of the stream of payoffs to each player by cooperating?

    c. What is the present value of the stream of payoffs from defection if the other player remains faithful?

    d. Based on your answers above, is this agreement stable?

Solution

    a. The present value of the payoffs to each firm from a no collusion strategy is:

FIGURE 1.2
Pricing Game

Firm B

		High price	Low price
Firm A	High price	($1 million, $1 million)	($100,000, $5 million)
	Low price	($5 million, $100,000)	**($250,000, $250,000)**

Payoffs: (Firm A, Firm B)

$$PV_N^j = \frac{\pi_N^j(1+i)}{i+\theta} = \frac{\$250,000(1.05)}{0.06} = \$4,375,000 \qquad (D5.2.1)$$

b. The present value of the stream of payoffs to both players by cooperating is:

$$PV_C^j = \frac{\pi_C^j(1+i)}{i+\theta} = \frac{\$1,000,000(1.05)}{0.06} = \$17,500,000 \qquad (D5.2.2)$$

c. The present value of the stream of payoffs to either player by defecting when his or her rival remains faithful is:

$$PV_V^j = \left(\pi_V^j - \pi_N^j\right) + PV_N^j = \$4,750,000 + \$4,375,000 = \$9,125,000 \qquad (D5.2.3)$$

d. Since $PV_V^j < PV_C^j$, there is no incentive for either player to violate the agreement.

Inequality (5.11) may be rearranged to yield the cheating rule for finitely repeated games with an uncertain end. The interpretation of Inequality (5.11) is similar to the interpretation of Inequality (4.10). If the left-hand side of Inequality (5.11) is greater than the discount rate, it will pay for a player to defect, in which case the agreement will be unstable. If the left-hand side of Inequality (5.11) is equal to the discount rate, a player will be indifferent between cooperating and defecting. Inequality (5.11) may be rearranged to yield a cheating rule for finitely repeated static games with an uncertain end. This rule is summarized in Inequality (5.12).

$$\frac{\pi_C^j - \pi_N^j - \theta\left(\pi_V^j - \pi_N^j\right)}{\pi_V^j - \pi_C^j} < i \qquad (5.12)$$

It should also be noted that Inequality (5.12) differs from the cheating rule in Inequality (4.14) by the addition of $-\theta(\pi_V^j - \pi_N^j)$ in the numerator. This term captures the uncertainty about the gains from violating an agreement when the end of the game is unknown.

Principle: For a two-player, cooperative, non-zero-sum, static, finitely repeated game with an uncertain end, a collusive agreement will be unstable if:

$$\frac{\pi_C^j - \pi_N^j - \theta\left(\pi_V^j - \pi_N^j\right)}{\pi_V^j - \pi_C^j} < i \qquad (5.12)$$

where $0 < \theta \leq 1$ is the probability that the game will end after each play.

─────────────── **Demonstration Problem 5.3** ───────────────

Consider the game depicted in Figure 1.2.

    a. Suppose there is a 1 percent chance that the game will come to an end in each stage. For what values of the discount rate can we expect this agreement to break down?

    b. Suppose that the discount rate is 5 percent. If the agreement breaks down, what are the probabilities that the game will come to an end in each stage?

Solution

    a. Substituting the data from Figure 1.2 into the left-hand side of Inequality (5.13) we obtain:

$$\frac{\pi_C^j - \pi_N^j - \theta\left(\pi_V^j - \pi_N^j\right)}{\pi_V^j - \pi_C^j} = \frac{\$750,000 - 0.01(\$4,750,000)}{\$4,000,000} = 0.1756 \qquad (D5.3.1)$$

We can expect this agreement to break down for any discount rate greater than 17.56 percent.

    b. Substituting the data from Figure 1.2 into the left side of Inequality (5.12) we get:

$$\frac{\$750,000 - \theta(\$4,750,000)}{\$4,000,000} > 0.05 \qquad (D5.3.2)$$

Solving for θ, this agreement will break down if the probability that the game will come to an end is less than 11.58 percent.

──

COMMON ENFORCEMENT MECHANISMS

Even in games where players have an incentive to defect, many explicit and tacit cooperative arrangements manage to survive. In those cases where a cartel does break down, they are subsequently reformed, even though the conditions that led to its disintegration still exist. One possible explanation for this is the use of enforcement mechanisms that bind players to an agreement and do not rule out future cooperation. Two common enforcement mechanisms are the *tit-for-tat* and *preemption* trigger strategies.

Tit-for-Tat

A player who adopts a **tit-for-tat strategy** will not knowingly violate an agreement, but neither will that player allow defection to go unpunished, nor will that player quietly accept punishment. Tit-for-tat is a variation of the eye-for-an-eye rule, which is found in the Bible (Exodus 21:22). This enforcement mechanism begins with players cooperating in the first stage, but they subsequently mimic their rivals' behavior. According to Robert Axelrod (1984), tit-for-tat exhibits four general characteristics of an effective trigger strategy: transparency, affability, provocability, and forgiveness. Tit-for-tat is transparent because it is simple and easy to understand. It is affable because it never initiates defection. But, tit-for-tat is also provocative because it will never let a defection go unpunished. Finally, it is forgiving because it allows players to return to the *status quo ante,* that is, it does not rule out the possibility of cooperation in the future.

The problem with a tit-for-tat strategy is that once initiated, tit-for-tat could lead to a chain reaction of strike and counterstrike, which results in a complete breakdown of an agreement. In its most naive version, a player will react to a real or perceived insult by striking back. At no point does tit-for-tat accept defection without retaliation. There is no mechanism for short-circuiting this downward spiral.

> **Tit-for-tat strategy** An enforcement mechanism in which a player does not knowingly violate an agreement, but neither will he or she allow defection to go unpunished nor quietly accept punishment.

To illustrate the tit-for-tat trigger strategy, consider two neighborhood gas stations, PETROX and GLOMAR. Both gas stations have a core of loyal customers who are not sensitive to price changes. There is, however, a third group of customers that is very sensitive to price changes. Price competition between the gas stations targets members of this third group.

Suppose that both gas stations initially charge a price of $3 per gallon for 87-octane gasoline. The wholesale price of gasoline paid by each gas station is $2.70 per gallon. Suppose that each gas station adopts the following tit-for-tat strategy: If my rival cuts price, I will respond by cutting price by twice that amount. Since neither gas station wants to lose money, the lowest that either gas station will charge is $2.80 per gallon ($2.70 per gallon wholesale plus an additional $0.10 per gallon to cover operating expenses). In the first two stages, PETROX and GLOMAR charge the same price. Then, in the third stage, PETROX believes that GLOMAR will cut its price by $0.01 to $2.99 per gallon. In response, PETROX doubles GLOMAR's expected price cut by reducing its price to $2.98. GLOMAR retaliates in the fourth stage by cutting its price to $2.96. PETROX strikes back by lowering its price in the fifth stage to $2.94, and so on. As a result of PETROX's miscalculation of GLOMAR's pricing plans, the price of regular gasoline charged by both gas stations eventually declines to $2.80 per gallon. This process is illustrated in Figure 5.2.

One reason why tit-for-tat is so compelling is that it tends to perform well against a variety of alternative strategies. Axelrod (1984) invited several game theorists to participate in a computer tournament. They were asked to submit strategies for playing finitely repeated games of the prisoner's dilemma type. These strategies were then pitted one-on-one against every other strategy. Although tit-for-tat never beat another strategy, the match-ups did occasionally result in a tie. Nevertheless, tit-for-tat won the tournament by accumulating the highest overall score. The reason, according to Axelrod, is that although tit-for-tat never initiates conflict, neither does it allow defection to go unpunished. Moreover, it does not close the doors to future cooperation.

The most severe shortcoming of a tit-for-tat strategy is the possibility that a player will misinterpret a rival's actions or intentions. This miscalculation may take one of two forms. Either a player believes that a rival has made a certain move when, in fact, a completely different move was made, or a player correctly observes a rival's moves, but misinterprets a rival's motives. With a tit-for-tat strategy, this miscalculation will immediately result in a punishing countermove, which will set in motion a cycle of retaliation by both players. The price war depicted in Figure 5.2, for example, may result less from a deliberate attempt by GLOMAR to capture a larger market share, and more from PETROX's miscalculation of GLOMAR's intentions.

Preemption

In contrast to tit-for-tat, which is a reactive trigger strategy, **preemption** is proactive. Preemption is an example of a strategic move, which is an attempt by one player to alter a rival's behavior in a direction that is favorable to the player initiating the action. Preemption attempts to limit a rival's strategy options in favor of the player initiating the action. Consider, for example, the two-player, static game in Figure 5.3 involving two breweries, Pie Eye Beer and Red Nose Lager.

FIGURE 5.2
Tit-for-Tat Pricing Strategy

Stage	PETROX	GLOMAR
1	$3.00	$3.00
2	$3.00	$3.00 (2.99?)
3	$2.98	$3.00
4	$2.98	$2.96
5	$2.94	$2.96
6	$2.94	$2.92
7	$2.90	$2.92
⋮	⋮	⋮
12	$2.82	$2.80
13	$2.80	$2.80
⋮	⋮	⋮
n	$2.80	$2.80

FIGURE 5.3
Preemption Game

		Red Nose Magazines	Red Nose Television
Pie Eye	Magazines	(5, 1)	(1, 0)
	Television	(1, 1)	(2, 4)

Payoffs: (Pie Eye, Red Nose)

Each brewery is considering whether to advertise on television or in magazines. The payoffs represent an expected increase in profits in millions of dollars. The reader should verify that neither company has a dominant strategy and that this game has two Nash equilibria strategy profiles: {*Magazines, Magazines*} and {*Television, Television*}. Although Pie Eye considers the Nash equilibrium strategy profile {*Magazines, Magazines*} the best outcome, Red Nose considers it the worst. Conversely, Red Nose considers the Nash equilibrium strategy profile {*Television, Television*} the best outcome while Pie Eye considers it the worst. What, if anything, can Pie Eye do to turn this game to its advantage?

> **Preemption** An example of a strategic move in which a player moves first in an attempt to limit a rival's strategy options.

One thing that Pie Eye can do is preempt Red Nose by moving first and committing itself to a *Magazine* strategy. By moving first, Pie Eye has transformed a static game into a dynamic (sequential-move) game. By announcing its intention to advertise in magazines, Red Nose's best response is to advertise on television. The resulting Nash equilibrium strategy profile {*Magazine, Magazine*} results in a payoff of $5 million for Pie Eye and only $1 million for Red Nose.

In the game in Figure 5.3, preemption was successful because Pie Eye enjoyed a **first-mover advantage**, which will be discussed in greater detail when we discuss sequential-move games in Chapter 11. In fact, Red Nose could also have exploited this situation to its advantage by pre-empting Pie Eye. Of course, the key to Pie Eye's success is convincing Red Nose that its strategic move is credible. If Pie Eye's commitment to magazine advertising was little more than a public announcement, Red Nose might be able to trump Pie Eye by immediately signing advertising contracts with several television outlets. As we saw in the previous chapter, contracts are an ef-fective way to make a commitment credible. In this case, Pie Eye would have little choice but to respond by advertising on television as well.

> **First-mover advantage** When a player is able to obtain a higher payoff by credibly committing to a strategy before his or her rivals commit to their strategies.

A **scorched-earth policy** is a preemptive strategy that occurs when a player commits to de-stroying his or her assets in an effort to make his or her commitment credible, thereby altering the behavior of his or her rivals. An example of a scorched-earth policy involved Houghton Mifflin's effort to thwart a takeover bid from Western Pacific by threatening to release its best-selling authors from their contractual obligations. Many of the publisher's authors also pledged not to re-sign with the company. Houghton Mifflin's efforts were so successful that Western Pacific withdrew its takeover bid.

> **Scorched-earth policy** A preemptive strategy in which a player commits to destroying his or her assets in an effort to alter a rival's behavior.

CASE STUDY 5.2: PHILIP MORRIS VERSUS BAT

We have seen that a tit-for-tat strategy may be used as an enforcement mechanism to keep a firm from cutting its prices to capture a larger market share at the expense of its rivals. Firms will reason that any gains from price-cutting will be transitory. Since lower prices translate into lower per-unit margins above cost with no change in market shares, there is little incentive for firms to engage in a price war.

An example of the consequences of firms engaging in a price war is provided in Be-sanko, Dranove, and Shanley (2000, pp. 296–297). In the 1990s, Philip Morris and British American Tobacco (BAT) dominated the Costa Rican tobacco market, with market shares of around 30 percent and 70 percent, respectively. The market consisted of three segments. Philip Morris dominated the premium and mid-priced segments with its Marlboro and Derby brands, respectively. BAT dominated the low-priced segment with its Delta brand. A prosperous Costa Rican economy in the 1980s allowed both companies to increase prices faster than the rate of inflation. By the end of the decade, industry per-unit margins exceeded 50 percent.

Storm clouds in the cigarette industry began to form in the early 1990s as the health dangers of smoking became more widely known. The effect in Costa Rica was a greater reduction in the demand for premium and mid-priced cigarettes than for low-priced brands. For the first time since the early 1980s, BAT's market share began to increase at Philip Morris's expense. How would Philip Morris respond to this threat?

Early on Saturday, January 16, 1993, Philip Morris cut the price of Marlboro and Derby cigarettes by 40 percent. The timing of the price cuts was not coincidental: Philip Morris reasoned that should BAT retaliate with a price cut of its own, it would not be able to satisfy the predicted increase in demand because of low, year-end holiday inventories. Philip Morris also reasoned that local management would be unable to respond immediately to the price cut without first consulting senior management in London.

Philip Morris had miscalculated. By Saturday afternoon, BAT had slashed prices by 50 percent. According to industry analysts, the new price of Delta cigarettes was only slightly higher than their marginal cost of production. The ensuing price war lasted two years, by the end of which cigarette sales had increased by 17 percent because of lower prices—but market shares remained unchanged. By the time the price war ended, Philip Morris's and BAT's profits were, respectively, $8 million and $20 million lower than previously.

What prompted Philip Morris to cut prices in Costa Rica? One reason was that management did not expect BAT to retaliate. Philip Morris had successfully used this tactic to increase its market share at BAT's expense in other Central American countries. Had Philip Morris been able to anticipate BAT's rapid response in Costa Rica, it might not have initiated the price war. This case study underscores the importance of correctly anticipating a rival's countermove. A rapid, retaliatory strike can serve as an effective enforcement mechanism to keep rivals from using price cuts to increase its market share.

CHAPTER REVIEW

In this chapter we examined *finitely repeated static games* in pure strategies. There are two classes of finitely repeated games: those in which the final stage of the game is known with certainty, and those in which the end of the game is uncertain.

When a finitely repeated game in pure strategies has a certain end, the use of trigger strategies to enforce cooperation will fail because a player is unable to punish a rival's defection beyond the last stage. This transforms the next-to-the-last stage into the last stage. By "folding back" the process yet another stage, finitely repeated games unravel into a series of noncooperative one-time games. This unraveling is known as the *end-of-game problem.*

If the end of the game is uncertain, the outcome will be quite different. Since the players do not know when the last stage will occur, it may be possible to use the threat of retaliation to bind players to an agreement. This outcome is similar, but not identical, to infinitely repeated, static games in pure strategies. The main distinction lies in the evaluation of the present value of the stream of expected future payoffs in the presence of this uncertainty.

In general, the greater the probability that the game will come to an end, the less the likelihood that a trigger strategy that punishes defectors will bind players to an agreement. When the probability that the game will come to an end is zero, a finitely repeated game is transformed into an infinitely repeated game. At the other extreme, when the probability that the game will come to an end is unity, a finitely repeated game becomes a one-time game or, what is almost the same thing, a finitely repeated game with a certain end. In this game, the threat of retaliation in the event of defection is completely ineffective.

Two common methods for penalizing cheaters in situations involving the end-of-game problem are the *tit-for-tat* and *preemption*. A player who adopts a tit-for-tat strategy will never initiate cheating, but neither will that player allow cheating by rivals to go unpunished. At no point does tit-for-tat accept defection without retaliation. There is no mechanism for short-circuiting this downward spiral.

The most severe shortcoming of a tit-for-tat strategy is the possibility that a player will misinterpret a rival's actions or intentions. This miscalculation may take one of two forms. Either a player believes that a rival has made a certain move when, in fact, a completely different move was made, or a player correctly observes a rival's moves, but misinterprets a rival's motives. With a tit-for-tat strategy, this miscalculation will immediately result in a punishing countermove, which will set in motion a cycle of retaliation by both players.

In contrast to tit-for-tat, which is reactive, *preemption* is proactive. Preemption is an example of a *strategic move,* which is an attempt by one player to alter a rival's behavior in a direction that is favorable to the player initiating the action. Preemption attempts to limit a rival's strategy options in favor of the player initiating the action. It transforms a static (simultaneous-move) game into a dynamic (sequential-move) game.

Preemption is successful if it gives the initiating player a *first-mover advantage.* This is when a player is able to obtain a larger payoff by credibly committing to a strategy before a rival is able to commit to a strategy. An example of preemption is a *scorched-earth policy.* This occurs when a player commits to a strategy by destroying his or her assets. This alters a rival's behavior by making the player's commitment credible.

CHAPTER QUESTIONS

5.1   Analytically, what is the difference between a repeated game with an uncertain end and an infinitely repeated game?

5.2   Consider a static game that has two Nash equilibria. Explain carefully why a strategy profile that may not be a Nash equilibrium in a one-time game may be part of a Nash equilibrium strategy in a game that is played twice.

5.3   What is meant by the end-of-game problem?

5.4   Does the end-of-game problem apply to repeated games with an uncertain end? Explain.

5.5   Explain what is meant by the term "backward unraveling" in finitely repeated games with a certain end.

5.6   A computer software analyst with IBM has given notice that she plans to resign in two weeks to take a job with Microsoft. The analyst is currently a member of a team that is developing an innovative business application to be used with Microsoft's Excel spreadsheet. Explain how the analyst's job performance in her final two weeks might be affected by her decision to resign. Can you think of any other problems associated with the analyst's decision to join Microsoft? What, if anything, can management do to avoid a change in the analyst's job performance, and also to avoid any other potential problems that may arise in the move to Microsoft?

5.7   What is it about tit-for-tat that makes it an effective trigger strategy? Explain how Phillip Morris used a tit-for-tat strategy to counter the effort by British American Tobacco to dominate the market for cigarettes in Costa Rica.

5.8   What is the main weakness of a tit-for-tat trigger strategy?

5.9   Explain what is meant by a strategic move? What is it about preemption that qualifies it as a strategic move?

5.10   Explain why a scorched-earth policy qualifies as a preemptive strategy.

CHAPTER EXERCISES

5.1   Consider the finitely repeated, static game depicted in Figure E5.1.

    a.   If this game is played just once, does either player have a dominant strategy?

FIGURE E5.1
Static Game in Chapter Exercise 5.1

Player B

		B1	B2
Player A	A1	(5, 5)	(2, 6)
	A2	(6, 2)	(3, 3)

Payoffs: (Player A, Player B)

FIGURE E5.2
Static Game for Chapter Exercise 5.2

Firm Y

		A	B	C
	A	(2, 2)	(0, 0)	(1, 1)
Firm X	B	(0, 0)	(5, 5)	(8, 4)
	C	(1, 1)	(4, 8)	(7, 7)

Payoffs: (Firm X, Firm Y)

b. What is the Nash equilibrium strategy profile, if any, if the game is played one time?

c. Could a Nash equilibrium in a one-time game not be part of a Nash equilibrium in a game that is played twice?

d. Are there any credible threats that could be made by either player that would result in the largest payoff for either player if this game is played three times?

5.2 Consider the finitely repeated, static game depicted in Figure E5.2.

a. If this game is played just once, does either player have a dominant strategy?

b. If this game is played just once, what is the Nash equilibrium strategy profile for this game?

c. Assuming that all threats are credible, what strategy profile will result in the largest payoff for both players if this game is played three times?

5.3 For ten years, Argon Airlines and Boron Airways have been engaged in fierce competition for passengers along the lucrative Boston–New York–Washington, D.C. corridor. While consumers have benefited enormously from this price competition, the airlines have suffered losses, which threaten continued service by both airlines along this route. As a result, Argon and Boron have entered into an agreement to discontinue offering discount airfares. After years of bad feelings, however, neither company completely trusts the other. The payoffs for this game, which are summarized in Figure E5.3, are in millions of dollars.

a. Does either firm have a dominant strategy?

b. Because of distrust between the airlines, the probability that the game will end in the next stage is 15 percent. Below what discount rate can we expect cooperation to break down?

c. If the discount rate is 10 percent, at what uncertainty level will the players in this game be indifferent between cooperation and defection?

FIGURE E5.3

Static Game in Chapter Exercise 5.3

Boron

		Discount	No discount
Argon	Discount	(−1, −1)	(30, −5)
	No discount	(−5, 30)	(5, 5)

Payoffs: (Argon, Boron)

5.4   Consider the static game depicted in Figure E5.1. Suppose that the discount rate is 25 percent.

  a. What is the present value of the stream of payoffs from a *no collusion* strategy?
  b. What is the present value of the stream of payoffs by colluding?
  c. What is the present value of the stream of payoffs from defection?
  d. If the probability that the game will come to an end in the next stage is 50 percent, at what discount rate are the players indifferent between colluding and defecting?
  e. If the discount rate is 25 percent, at what probability that the game will end in the next stage will a player be indifferent between colluding and defecting?
  f. If the discount rate is 65 percent, at what probability that the game will end in the next stage will a player be indifferent between colluding and defecting?

5.5   Consider the static game depicted in Figure E5.2. Suppose that the discount rate is 10 percent.

  a. What is the present value of the stream of payoffs by not cooperating?
  b. What is the present value of the stream of payoffs by colluding?
  c. What is the present value of the stream of payoffs from defection?
  d. If the probability that the game will come to an end in the next stage is 20 percent, at what discount rate are the firms indifferent between colluding and defecting?
  e. If the discount rate is 30 percent, at what probability that the game will end in the next stage will a firm be indifferent between colluding and defecting?
  f. If the discount rate is 40 percent, at what probability that the game will end in the next stage will a firm be indifferent between colluding and defecting?

ENDNOTES

1. $20 + 8\left[(1-\theta)+(1-\theta)^2+(1-\theta)+\cdots\right]$

$$> 15 + 15\left[(1-\theta)+(1-\theta)^2+(1-\theta)^3+\cdots\right] \qquad \text{(F5.1.1)}$$

$$5 > 7\left[(1-\theta)+(1-\theta)^2+(1-\theta)^3+\cdots\right] \qquad \text{(F5.1.2)}$$

$$5 > 7S \qquad \text{(F5.1.3)}$$

Let:

$$S = (1-\theta)+(1-\theta)^2+(1-\theta)^3+\cdots \qquad \text{(F5.1.4)}$$

Multiply both sides of (F5.1.3) by $(1-\theta)$:

$$S(1-\theta) = (1-\theta)^2 + (1-\theta)^3 + (1-\theta)^4 + \cdots \qquad \text{(F5.1.5)}$$

Subtract (F5.1.4) from (F5.1.3):

$$S - S(1-\theta) = (1-\theta) \qquad \text{(F5.1.6)}$$

Solving for S:

$$S = \frac{1-\theta}{\theta} \qquad \text{(F5.1.7)}$$

Substituting (F5.1.6) into (F5.1.2):

$$5 > 7\left(\frac{1-\theta}{\theta}\right) \qquad \text{(F5.1.8)}$$

$$\frac{5}{7} > \frac{1-\theta}{\theta} \qquad \text{(F5.1.9)}$$

2.
$$PV^j = \pi^j\left[\left(\frac{1-\theta}{1+i}\right)^0 + \left(\frac{1-\theta}{1+i}\right)^1 + \left(\frac{1-\theta}{1+i}\right)^2 + \cdots\right] \qquad \text{(F5.2.1)}$$

$$PV^j = \pi^j S \qquad \text{(F5.2.2)}$$

$$S = 1 + \left(\frac{1-\theta}{1+i}\right)^1 + \left(\frac{1-\theta}{1+i}\right)^2 + \left(\frac{1-\theta}{1+i}\right)^3 + \cdots \qquad \text{(F5.2.3)}$$

Multiplying both sides of Equation (F5.2.3) by $(1-\theta)/(1+i)$ we get:

$$S\left(\frac{1-\theta}{1+i}\right) = \left(\frac{1-\theta}{1+i}\right) + \left(\frac{1-\theta}{1+i}\right)^2 + \left(\frac{1-\theta}{1+i}\right)^3 + \cdots \qquad \text{(F5.2.4)}$$

Subtracting Equation (F5.2.4) from Equation (F5.2.3):

$$S - S\left(\frac{1-\theta}{1+i}\right) = 1 \qquad \text{(F5.2.5)}$$

Solve for S:

$$S = \left(\frac{1+i}{i+\theta}\right) \qquad \text{(F5.2.6)}$$

Substituting Equation (F5.2.6) into Equation (F5.2.2):

$$PV^j = \frac{\pi^j(1+i)}{i+\theta} \qquad \text{(F5.2.7)}$$

CHAPTER

6

MIXING PURE
STRATEGIES

In this chapter we will:

- *Demonstrate that pure-strategy Nash equilibria do not exist for zero-sum games;*
- *Discuss the minimax theorem and its application to zero-sum games;*
- *Show how mixing pure strategies guarantees that zero-sum and non-zero-sum games have a unique, mixed-strategy Nash equilibrium;*
- *Examine optimal mixing rules and when to use them;*
- *Analyze the use of mixed strategies in games involving bluffing.*

INTRODUCTION

In our discussion of noncooperative, one-time, static games with complete information we were confronted with a bit of a dilemma. We found that in games involving pure strategies, a unique Nash equilibrium may not exist, or there may be multiple Nash equilibria. When this occurs, it is difficult to predict the outcome of the game. In a previous chapter we learned that in games involving multiple Nash equilibria, a unique focal-point equilibrium may exist if the players share a common "understanding" of the problem. Unfortunately, there is no guarantee that this will be the case. In such cases, the decision maker may adopt a strategy based on some arbitrary selection criteria, such as the maximin or minimax regret decision rules. The reader may have found this somewhat arbitrary approach less than satisfying. Are we to conclude from this that game theory is of limited usefulness when analyzing real-world strategic situations?

Game theory would long ago have been relegated to the intellectual backwaters if its application were limited to situations involving just pure strategies. Game theory is a powerful tool for analyzing strategic situations precisely because a unique Nash equilibrium does, in fact, exist for all games, although not always in pure strategies. Recall that a pure strategy is a complete and nonrandom game plan. By contrast, a mixed strategy involves randomly mixing pure strategies, which expands the usefulness of game theory to a much broader range of strategic applications. We will begin our discussion of mixed strategies by examining noncooperative, static games with zero-sum payoffs. The reader may recall that such games do not have pure-strategy Nash equilibria.

ZERO-SUM GAMES

Nearly a quarter of a century before John Nash demonstrated the existence of a "fixed-point" equilibrium in noncooperative games, John von Neumann was developing his own theories of strategic behavior. As a junior faculty member at Princeton University, von Neumann became intrigued with the strategic behavior of poker players. Poker is an example of a zero-sum game in which one player's gain is another player's loss.

FIGURE 6.1
Matching Pennies Game

Zoe

		Heads	Tails
Chloe	Heads	(1, −1)	(−1, 1)
	Tails	(−1, 1)	(1, −1)

Payoffs: (Chloe, Zoe)

An important aspect of poker games is the bluff, which occurs when a player bets big on a bad hand. The objective of bluffing is to get opponents to "fold" (quit) by persuading them that they are sitting on losing hands, when, in fact, the opposite may be true. Not surprisingly, von Neumann found that for bluffing to be successful, big bets must be made randomly on both good and bad hands. The underlying logic is obvious. Rivals will eventually see through a player who repeatedly bets big on inferior hands. When this happens, not only will bluffing become ineffective, but rivals will exploit the attempt. This observation led von Neumann (1928) to publish the proof of his famous *minimax theorem,* which will be discussed at greater length below. Unfortunately, von Neumann's discovery had few real-world applications because it was limited to zero-sum games.

Example: Matching Pennies Game

To illustrate the concept of a zero-sum game, consider the children's game of matching pennies. In this game, Chloe and Zoe each hold a penny in a closed hand with either the obverse ("heads") or the reverse ("tails") side face up. Chloe and Zoe then simultaneously open their hands. If the sides match (i.e., two heads or two tails), Zoe pays Chloe a penny. Since Chloe wins a penny and Zoe loses a penny, the payoffs may be written (1, −1). If the sides do not match, Chloe pays Zoe a penny, in which case the payoffs are (−1, 1). The normal form of this game is depicted in Figure 6.1. This is a zero-sum game because the sum of the payoffs from any strategy profile is zero. The reader should verify that this game does not have a pure-strategy Nash equilibrium.

Example: Sink-the-Battleship Game

Our second example of a zero-sum game is the sink-the-battleship game.[1] Like the matching pennies game, this is a finitely repeated, static game with an uncertain end. The game involves two countries, Acre and Ilia, which are engaged in a naval battle. Each country's battleship has both offensive and defensive capabilities. In Figure 6.2, Ilia's battleship, which is located at point *I* in the grid, launches a missile at Acre's battleship, which is located at point *A*. This information is immediately fed into Acre's computer-controlled defense system, which instantly counter attacks by launching an antimissile missile.

The flight path of each country's missile is programmed at launch. Although Acre does not know the flight path of Ilia's missile, it does know the missile's technical capabilities. Acre knows, for example, that Ilia's missile will not follow a direct flight path from point *I* to point *A* so that it cannot easily be intercepted. Ilia will fire its missile in a zigzag pattern with possible flight paths illustrated along the line segments in Figure 6.2.

We will attempt to identify the optimal flight paths (strategies) of both combatants. The reader may recall from Chapter 2 that it is sometimes possible to find a pure-strategy Nash equilibrium by eliminating dominated strategies. If an iterated, dominant strategy exists, the game will have a unique, pure-strategy Nash equilibrium. Even if a unique Nash equilibrium does not exist,

FIGURE 6.2
Sink-the-Battleship Game

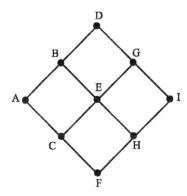

eliminating all dominated strategies will simplify the game and make the decision-making process more tractable.

In this game, each country's missile travels at the same speed, which is 20 seconds per line segment. The missiles can travel in straight lines or make 90-degree turns. The major difference between the capabilities of the combatants' missiles is total flight time. While Ilia's missile has enough fuel to reach its target along any flight path, such as *IGDBA,* Acre's antimissile missile can stay aloft for only 60 seconds due to the extra weight required to carry enough explosives to guarantee a killing hit. Thus, Acre's antimissile missile is restricted to a flight path such as *ACEG.* For this reason, it is only necessary to consider each combatant's strategies along three line segments of the grid in Figure 6.2. If Ilia's missile can avoid being intercepted during the first 60 seconds, it will destroy Acre's battleship.

Each combatant in this game has eight possible strategies involving three line segments. These strategies are summarized in Figure 6.3. Since each combatant has eight possible strategies, this game has 64 possible strategy profiles. Our objective is to determine the optimal strategy profile for each combatant. An optimal strategy for Acre is one that leads to a hit, which is designated in the payoff matrix by the symbol ☼. An optimal strategy for Ilia results in a miss, in which case the cell in the payoff matrix is left blank. The sink-the-battleship game is an example of a zero-sum game since a "win" by one combatant is a "loss" for the other. For example, the strategy profile {*ABDG, IGEH*} will result in Ilia destroying Acre's battleship. The reason for this is that as we move from point to point every 20 seconds, at no time will Acre's antimissile missile and Ilia's missile arrive at the same place at the same time. On the other hand, the strategy profile {*ABEC, IHEB*} results in a win for Acre since Ilia's missile will be destroyed after 40 seconds. To determine hits and misses in the matrix, we need only compare the last two letters in each player's strategy. If either of these letters appears in the same position for both strategies, Ilia's missile will be intercepted. If not, Ilia's missile will destroy Acre's battleship.

The game depicted in Figure 6.3 is rather complicated, and selecting an optimal strategy appears to be a rather daunting task. But, eliminating dominated strategies will dramatically simplify this game by narrowing down the combatants' choices. Consider the game from Acre's perspective. Strategy *ACEB,* for example, dominates strategies *ABDG, ABEG, ABEH, ACEH,* and *ACEG.* The reason is that this strategy results in the same hits and fewer misses. After eliminating these dominated strategies we are left with the simplified game depicted in Figure 6.4.

In the revised normal-form game depicted in Figure 6.4, Acre's strategy *ABEC* dominates *ACFH* because it results in the same hits and fewer misses. Eliminating *ACFH* results in the simplified game in Figure 6.5.

FIGURE 6.3
Normal Form of the Sink-the-Battleship Game

Ilia

		IGDB	IGEB	IGEC	IGEH	IHFC	IHEC	IHEB	IHEG
	ABDG	☆							☆
	ABEG		☆	☆	☆		☆	☆	☆
	ABEH		☆	☆	☆		☆	☆	☆
Acre	ABEC		☆	☆	☆	☆	☆	☆	☆
	ACFH				☆	☆			
	ACEH		☆	☆	☆		☆	☆	☆
	ACEG		☆	☆	☆		☆	☆	☆
	ACEB	☆	☆	☆	☆		☆	☆	☆

FIGURE 6.4
First Iterated Payoff Matrix for the Sink-the-Battleship Game

Ilia

		IGDB	IGEB	IGEC	IGEH	IHFC	IHEC	IHEB	IHEG
	ABEC		☆	☆	☆	☆	☆	☆	☆
Acre	ACFH				☆	☆			
	ACEB	☆	☆	☆	☆		☆	☆	☆

FIGURE 6.5
Second Iterated Payoff Matrix for the Sink-the-Battleship Game

Ilia

		IGDB	IGEB	IGEC	IGEH	IHFC	IHEC	IHEB	IHEG
Acre	ABEC		☆	☆	☆	☆	☆	☆	☆
	ACEB	☆	☆	☆	☆		☆	☆	☆

Now, let us consider the game from Ilia's perspective. Rather than looking for hits, we will focus on strategies that lead to misses. Strategies *IGEB, IGEC, IGEH, IHEC, IHEB,* and *IHEG* are dominated by strategies *IGDB* and *IHFC* because they will never result in a miss. After these dominated strategies are eliminated, we are left with the game depicted in Figure 6.6. It is not possible

FIGURE 6.6
Final Iterated Payoff Matrix for the Sink-the-Battleship Game

Ilia

		IGDB	IHFC
Acre	ABEC		☆
	ACEB	☆	

to reduce the sink-the-battleship game any further. This is a zero-sum game without a unique Nash equilibrium strategy profile in pure strategies because a win by one player is a loss by the other.

Although this game does not have a pure-strategy Nash equilibrium, we have accomplished quite a lot. In the original game, there was a one in eight chance of selecting the correct strategy. In the reduced game depicted in Figure 6.6, there is a one in two chance of choosing the correct strategy. Moreover, the remaining strategies tell us something about the combatants' preferred tactics. Acre's best strategy involves programming its antimissile missiles to patrol in small loops near its battleship, while the optimal strategy for Ilia is to program its missiles to travel along the frontier of the grid.

Although we dramatically simplified the strategy options confronting the players in the sink-the-battleship game, the game in Figure 6.6 is similar to the matching pennies game. Both combatants are still confronted with the challenge of adopting an optimal strategy in a game without a pure-strategy Nash equilibrium. The best result comes not from choosing one pure strategy over another, but from randomly mixing pure strategies. Before discussing the best way to do this, we will briefly review one of the most important contributions to the development of game theory and the search for a Nash equilibrium in games with zero-sum payoffs.

MINIMAX THEOREM

The first formal proof of the **minimax theorem** was presented by John von Neumann in a paper delivered to the Gottingen Mathematical Society on December 7, 1926 (Dimand and Dimand, 1992), and published two years later (von Neumann, 1928). Von Neuman was trying to come up with a way to predict a player's moves in zero-sum games. What he found was that a rational player will adopt a strategy that simultaneously attempts to maximize gains and minimize losses. When all players adopt this strategy, the maximum of the minimum (maximin) payoffs for all players will be equal. Later, Professor John Nash attempted to predict a player's strategy in *any* game, not just zero-sum games. Nash argued that a rational player will always adopt a strategy that is the best response to a rival's strategy. It was subsequently demonstrated for zero-sum games that the maximin and Nash equilibrium strategies are one and the same.

> **Minimax theorem** In a zero-sum, noncooperative game, a player will attempt to minimize a rival's maximum (minimax) payoff, while maximizing his or her own minimum (maximin) payoff. The minimax and the maximin payoffs of all players will be equal.

To illustrate the rationale underlying the minimax theorem, consider the following situation. Two young children, Andrew and Adam, accompany their mother, Rosette, on a trip to a shopping mall. At the mall, Andrew asks his mother to buy him a candy bar. Rosette knows that if she buys a candy bar for Andrew, Adam will want a candy bar as well. Since Rosette is concerned about her children eating too many sweets, she comes up with the following solution. She tells Andrew that he can have the candy bar, which we will assume is uniform throughout, but must share it with his brother. She also tells Andrew that he can divide the candy bar in any way he wants, but that Adam gets to choose the first piece. Andrew accepts his mother's offer and divides the candy bar exactly in half. Although Andrew knows nothing about game theory, he has behaved precisely the way the minimax theorem would have predicted. He minimized the maximum size of the piece that Adam gets, while at the same time maximizing the minimum size of his own piece. The minimax and the maximin payoffs are exactly the same; both Adam and Andrew get exactly one-half of the candy bar.

The groundbreaking work of John von Neumann and Oscar Morgenstern, while singularly significant, was of little use when analyzing most "real-world" situations. Von Neumann demonstrated that every two-person, zero-sum game had a maximin solution. Unfortunately, most situations with this property are limited to parlor games and sporting events. Economic competition in the marketplace, for example, is very often a "win-win" situation. Indeed, Adam Smith argued more than two centuries ago that individual initiative serves the common good. Admittedly, Smith's metaphor of the "invisible hand," and its laissez-faire corollary, has its limitations. Environmental pollution and the unbridled exploitation of natural resources, for example, have prompted intense debates about the possible dangers and welfare losses of global warming. In many cases, cooperation, not competition, results in socially optimal outcomes.

Example: Roach Coach Game

Why do fast-food restaurants, gasoline stations, and street vendors tend to cluster around a central location? At first blush, this behavior might appear to be counterintuitive since each vendor is competing for the same group of customers. Might not the business prospects of each vendor be improved if they were more widely dispersed? By segmenting the market, would not each vendor become a local monopolist—exploiting market power by charging higher prices and earning greater profits? One possible explanation for the concentration of commercial activity around a central hub is the minimax theorem. To see this, consider the roach coach game.

The term "roach coach" is a generic and derogatory term to describe the mobile refreshment stands that cluster at sporting events, beaches, parks, and other recreational venues. In the roach coach game, profit-maximizing vendors A and B sell refreshments at an adolescent soccer tournament. The only place to set up business is along a north–south road that abuts the western boundary of the soccer fields. The question facing each vendor is where to park? In this game, the spectators, who represent the vendors' potential customer base, are assumed to be randomly distributed throughout the tournament site. Suppose that the vendors initially consider parking at the location depicted in Figure 6.7. Does this represent a Nash equilibrium for this game?

To answer this question, let us consider the problem from vendor A's perspective. By moving a few feet further north, vendor A can reasonably expect to sell to all of the spectators south of his position, and half the spectators to vendor B's location. Reasoning the same way, by moving a few feet further south, vendor B can expect to sell to all of the spectators north of her position and half the spectators to vendor A's location. A countermove by vendor A to vendor B's countermove is to move a little further north. Vendor B's countermove is to move a little further south, and so on. What is the outcome of this circular reasoning of move and countermove? Both vendors will end up parking right next to each other at the center of the soccer fields. This pure-strategy Nash equilibrium is depicted in Figure 6.8.

The roach coach game is an application of the minimax theorem. Each vendor attempts to minimize the other vendor's maximum payoff, while maximizing his or her own minimum payoff. The minimax theorem accurately predicts that each vendor sells to exactly half the spectators at the soccer tournament.

FIGURE 6.7
Roach Coach Game

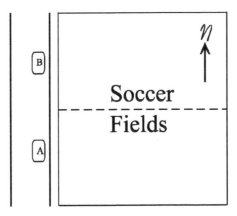

FIGURE 6.8
Solution to the Roach Coach Game

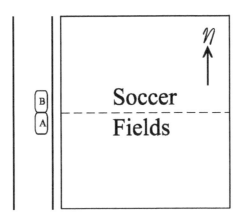

——————————————— **Demonstration Problem 6.1** ———————————————

Consider the two-player, zero-sum, one-time, static game depicted in Figure D6.1. Two companies, Avian and Alice Springs, sell mineral water. Each company is considering charging a high price of $2 per bottle or a low price of $1 per bottle. Each company incurs total fixed costs of $20,000 per period, regardless of whether or not they sell any water. At a price of $1 per bottle, both companies together sell 20,000 bottles, with total revenues of $20,000. At a price of $2 per bottle, both companies together sell 10,000 bottles, with total revenues amounting to $20,000. If both companies charge the same price, sales will be divided equally between them. If the two companies charge different prices, the company charging $1 will sell 20,000 bottles and the company selling $2 will sell nothing. If one company charges $1 per bottle while the other company charges $2, the profit earned by the first company will be $1(40,000) − $20,000 = $20,000, while the profit earned by the second company will be $2(0) − $20,000 = − $20,000.

FIGURE D6.1
Static Game in Demonstration Problem 6.1

Alice Springs

		Price = $1	Price = $2
Avian	Price = $1	($0, $0)	($20,000, −$20,000)
	Price = $2	(−$20,000, $20,000)	($0, $0)

Payoffs: (Avian, Alice Springs)

a. Suppose that both companies attempt to minimize the other company's maximum profits, while maximizing their own minimum profit. What is the solution profile for this game?
b. What, if anything, can you say about the payoffs to each company?
c. Is the solution profile for this game a Nash equilibrium?

Solution

a. If Avian adopts a *price = $1* strategy, Alice Springs' maximum payoff is $0. If Avian adopts a *price = $2* strategy, Alice Springs' maximum payoff is $20,000. Since the minimum of the maximum payoffs is $0, Avian will adopt a *price = $1* strategy. If Avian adopts *price = $1* strategy, Avian's minimum payoff is $0. If Avian adopts a *price = $2* strategy, Avian's minimum payoff is −$20,000. Since the maximum of the minimum payoffs is $0, Avian will again adopt a *price = $1* strategy. Since the payoff matrix is symmetrical, Alice Springs will also adopt a *price = $1* strategy. Thus, the strategy profile for this game is {*Price = $1, Price = $1*}.
b. Both companies' payoffs from the strategy profile {*Price = $1, Price = $1*} is $0 profits. This result represents a maximin and minimax payoff.
c. The strategy profile {*Price = $1, Price = $1*} for this game is a Nash equilibrium because it represents the best response by each company to the strategy adopted by its rival, and because neither company can improve its payoff by switching strategies.

MIXED STRATEGIES

One of the problems with the minimax theorem is that it does not apply equally to all zero-sum games. To see this, consider again the matching pennies game depicted in Figure 6.1. This game does not have a unique, pure-strategy Nash equilibrium because the minimum payoff for each strategy profile is always the same. Both *heads* and *tails* represent **pure strategies** because they are nonrandom. A pure-strategy Nash equilibrium does not exist for zero-sum games. Fortunately, this is not the end of the story. It turns out that by randomizing our choice of strategies, we can transform a zero-sum game with just two pure strategies into a game with an infinite number of strategies with different probabilities assigned to each. Randomizing pure strategies is referred to as mixing strategies. Our decision to mix strategies is itself a strategy.

A mixed strategy is the probability distribution of the set of all pure strategies such that the sum of the probabilities is equal to unity. Suppose, for example, that we have a game in which each player has a choice of three pure strategies: S_A, S_B, and S_C. Suppose that the probabilities attached to the randomized choice of strategies are 0.50, 0.30, and 0.20, respectively. This randomized collection of pure strategies is inclusive because the sum of the probabilities of their occurrence is unity. We could have assigned some other probability distribution to the choice of these pure strategies. In fact, there are an infinite number of such probability distributions.

Although the matching pennies game does not have a pure-strategy Nash equilibrium, it may be demonstrated that this game has a unique, mixed-strategy Nash equilibrium provided that Chloe and Zoe randomize their choice of *heads* and *tails,* with equal probabilities. One way for Chloe and Zoe to choose a strategy is to secretly flip a fair coin. Why does this result in a unique, Nash equilibrium? Because neither player can obtain a strictly better payoff by playing the game in any other way. To see this, suppose that Zoe randomly chooses *heads* 50 percent of the time. Recall that if the sides match, Chloe wins a penny. If the sides do not match, she loses a penny. If Chloe strictly prefers, say, heads then her expected payoff is $0.5(1) + 0.5(-1) = 0$. Is there any other mixing strategy that Zoe can adopt that will improve her expected payoff?

Suppose that Zoe randomly plays *heads* 80 percent of the time and *tails* 20 percent of the time. In this case, Chloe will strictly prefer to play *heads* as a pure strategy because she can expect to earn $0.80(1) + 0.20(-1) = 0.6$ cents. By contrast, by always playing *tails* Chloe can expect to earn $0.80(-1) + 0.20(1) = -0.6$ cents. Thus, Chloe will strictly prefer *heads* to playing a mixed strategy involving both *heads* and *tails*. By contrast, Zoe's random 80–20 strategy will result in an expected loss of –0.6 cents. It should be clear that Zoe can do no better with anything other than a 50–50 mixing rule. In fact, a 50–50 mixing rule strictly dominates all other mixing rules. By analogous reasoning, the same must also be true for Chloe. All other mixing rules will result in a payoff that is worse for both players, and no other rule will result in a payoff that is strictly better. In the matching pennies game, by eliminating all dominated strategies (that is, any strategy other than a random 50–50 mix), we are left with a unique, **mixed-strategy Nash equilibrium**.

Mixed-strategy Nash equilibrium A unique Nash equilibrium that occurs when players adopt their optimal mixing rules.

Demonstration Problem 6.2

Consider the interaction between labor and management depicted in Figure D6.2. The payoff matrix reflects management's desire for labor to work hard and labor's desire to take it easy. Management has two options: either secretly monitor worker performance (*observe*), or trust labor to work hard (*don't observe*). Labor also has two options: *work hard* or *goof off*. The payoff matrix reads as follows. Management "loses" if it observes because of the expense and loss of worker goodwill by monitoring diligent workers. On the other hand, the workers "win" because hard work is rewarded with extra pay, benefits, and so on. The payoffs from the strategy profile {*Observe, Work hard*} are (–1, 1). The same is true of the strategy profile {*Don't observe, Goof off*} because management loses money by employing "goldbricks" who get paid for goofing off. When the strategy profile is {*Don't observe, Work hard*}, management wins because it does not incur the expense of monitoring hard-working employees, while the workers lose because they now have to

FIGURE D6.2
Static Game in Demonstration Problem 6.2

		Labor	
		Work hard	Goof off
Management	Observe	(–1, 1)	(1, –1)
	Don't observe	(1, –1)	(–1, 1)

Payoffs: (Management, Labor)

earn their pay. Finally, the strategy profile {*Observe, Goof off*} results in payoffs of (1, −1) because management discovers, and presumably fires, shirkers.

    a. Does either player in this game have a pure-strategy, dominant strategy? Explain.
    b. Does this game have a pure-strategy Nash equilibrium? If not, why not?
    c. What does the absence of pure-strategy Nash equilibrium suggest for optimal management/employee relations?

Solution

    a. Neither player in this game has a pure-strategy, dominant strategy. If management *observes,* labor will *work hard.* If management does not observe, labor will *goof off.* Thus, labor's strategy depends on the strategy adopted by management. If labor works hard, management's best response is *don't observe.* If labor *goofs off,* management's best response is to *observe.* Thus, the strategy adopted by management depends on the strategy adopted by labor, and *vice versa.*
    b. This game does not have a Nash equilibrium in pure strategies. Since both players can always improve their payoffs by switching strategies, a Nash equilibrium in pure strategies does not exist for this game.
    c. The absence of a pure-strategy Nash equilibrium suggests that it is in the best interest of both management and labor to behave unpredictably by adopting a randomized 50–50 mixing rule.

OPTIMAL MIXING RULES

We saw in the matching pennies game that it may be possible for players to improve their payoffs in games by randomizing strategies. Unlike the matching pennies game, however, being unpredictable does not always mean that a player should randomize strategies by, say, flipping a coin. To see this, consider the following variation of the touchdown game, which was discussed in Chapter 2. Unlike the touchdown game, which involved a single play, the first-down game is a finitely repeated game with a certain end. The reader may recall that this class of games reduces to a series of one-time games. The challenge confronting both teams is to adopt a game plan with the greatest probability of victory.

Example: First-Down Game

The first-down game considers just one aspect of each team's game plan. At any time during the game, one team will be on offense and the other team will be on defense. The New York Giants and Baltimore Ravens are assumed to have different offensive and defensive skills. We will continue to assume a game with complete information in that both head coaches are aware of their rival's relative strengths and weaknesses. Figure 6.9 summarizes the probabilities that the Giants will gain first-down yardage for any combination of offensive and defensive strategies. The figure also summarizes the probabilities that the Ravens will be able to thwart the Giants' efforts.

    The Giants' passing game is somewhat stronger than its running game. If the Giants correctly anticipate that the Ravens will defend against the run, passing the ball will result in a first down 80 percent of the time. If the Giants correctly predict that the Ravens will defend against the pass, running the ball will result in a first-down yardage of a lesser, but still high, 70 percent of the time. Not surprisingly, the Giants' ability to gain first-down yardage is considerably worsened if they incorrectly anticipate the Ravens' defense. If the Giants' incorrectly predict a run defense

FIGURE 6.9

First-Down Game

Ravens

Giants		Pass defense	Run defense
	Pass offense	(10, 90)	(80, 20)
	Run offense	(70, 30)	(40, 60)

Payoffs: (Giants, Ravens)

FIGURE 6.10

Percentage of Times the New York Giants Successfully Gain First-Down Yardage

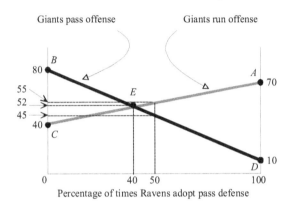

and pass the ball, they will be successful only 10 percent of the time. If they incorrectly predict a pass defense and run the ball, they will be successful only 40 percent of the time.

Since the Giants do not know what the Ravens are planning, how can they maximize their chances of gaining first-down yardage? Similarly, since the Ravens do not know what the Giants are planning, how can they minimize the Giants' chances of gaining first-down yardage? We will attempt to answer these questions by first considering the game from the Giants' perspective. Suppose that the Ravens adopt a pass defense 100 percent of the time (a run defense 0 percent of the time). If the Giants pass the ball, they will gain first-down yardage 10 percent of the time. If the Giants run the ball, they will gain first-down yardage 70 percent of the time. By contrast, if the Ravens adopt a run defense 100 percent of the time (a pass defense 0 percent of the time), the Giants will gain first-down yardage 80 percent of the time by passing the ball. If the Giants run the ball, they will gain first-down yardage 40 percent of the time. These conditions are depicted in Figure 6.10. Only by mixing defense plays can the Ravens reduce the Giants' chances of gaining first-down yardage. The Ravens must keep the Giants guessing.

One way for the Ravens' head coach to accomplish this is to flip a fair coin on every defensive play call. The reader will recall that this was the method used by Chloe and Zoe for randomizing pure strategies in the matching pennies game. If the Giants limit their play calls to passing the ball, the Ravens will adopt the correct defensive strategy 50 percent of the time. The Giants' expected success at gaining first-down yardage is 0.5(80) + 0.5(10) = 45, or 45 percent of the time. Likewise, by always running the ball, the Giants' expected success is 0.5(70) + 0.5(40) = 55, or 55 percent. It should be obvious that if the Ravens use a 50–50 mixing rule, it will be in the Giants' best interest always to run the ball, since this will gain first-down yardage 55 out of every 100 offensive plays. The Ravens' 50–50 defensive-strategy mixing rule is clearly superior to a game plan involving a single (pure) strategy. Always defending against the pass will result in a Giants' success rate of 70 percent. Always defending against the run will result in a Giants' success rate of 80 percent.

While the Ravens are better off by using a 50–50 mixing rule than always defending against the Giants' offensive strength, can they do better by using a different mixing rule? As we move along the horizontal axis in Figure 6.10 from left to right, the percentage of times that the Ravens adopt a *pass defense* increases from 0 to 100 percent of the time. As we move along the horizontal axis from right to left, the percentage of times that the Ravens adopt a *run defense* increases from 0 to 100 percent of the time (not shown). In other words, if the Ravens adopt a *pass defense* 100 percent of the time, this means that they will adopt a *run defense* 0 percent of the time. Conversely, if the Ravens adopt a *pass defense* 0 percent of the time, they will adopt a *run defense* 100 percent of the time.

As the percentage of times the Ravens adopt a *pass defense* increases from 0 to 100 percent, the Giants' success rate of running the ball increases along the straight line from point *C* to point *A*, although the success rate of passing the ball decreases along the straight line from point *B* to *D*. Alternatively, as the percentage of times the Ravens adopt a *run defense* increases from 0 to 100 percent, the Giants' success rate in passing the ball increases along the straight line from point *D* to point *B*, although the success rate of running the ball decreases along the straight line from point *A* to point *C*. Now, consider what happens when the Ravens adopt a 50–50 mixing rule. The reader should verify from Figure 6.10 that the Giants will successfully gain first-down yardage by running the ball 55 percent of the time, and 45 percent of the time by passing the ball.[2] From the Ravens' perspective, a 50–50 mixing rule is superior to a game plan that focuses on either the pass or the run.

Clearly, the Ravens will be better off by randomizing its defensive strategies by flipping a coin, but can it do better? An **optimal mixing rule** is a randomized mix of pure strategies that maximizes a player's minimum payoff, while minimizing a rival's maximum payoff. This outcome occurs at point *E* in Figure 6.10. The reader should verify that the optimal mixing rule for the Ravens is to adopt a 40–60 randomized mixing strategy. In other words, the Ravens should randomize its plays in such a way that four times in ten they defend against the pass, and six times in ten they defend against the run. With a 40–60 optimal mixing rule, the Ravens can reduce the Giants' first-down yardage success rate to a minimum of 52 percent.[3] The Ravens' optimal mixing rule of 40–60 minimizes the Giants' maximum payoff, while maximizing their own minimum payoff. Of course, the manner in which the plays are mixed is important. If the Ravens' defensive plays become predictable, the Giants will exploit nonrandom play calls to its advantage.

> **Optimal mixing rule** A randomized mix of pure strategies that maximizes a player's minimum payoff, while minimizing a rival's maximum payoff.

Now, let us consider the first-down game from the Giants' perspective, which is depicted in Figure 6.11. We want to determine the Giants' optimal mixing rule, which minimizes the Ravens' chances of stopping a first-down attempt, while maximizing its own chances of gaining first-down yardage. As we move along the horizontal axis of Figure 6.11 from left to right, the percentage of times that the Giants adopt a *pass offense* increases from 0 to 100 percent. As we move along the horizontal axis from right to left, the percentage of times that the Giants adopt a *run offense* (not shown) increases from 0 to 100 percent. If the Giants adopt a *pass offense* 100 percent of the time, the Ravens will successfully stop the Giants 90 percent of the time by defending against the pass, but only 20 percent of the time by defending against the run. If the Giants adopt a run offense 100 percent of the time, the Ravens will successfully stop the Giants 60 percent of the time by defending against the run, but only 30 percent of the time by defending against the pass.

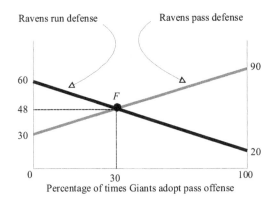

FIGURE 6.11
Percentage of Times the Baltimore Ravens Keep the New York Giants from Gaining First-Down Yardage

The lowest possible success rate for the Ravens occurs at point *F* in Figure 6.11. The reader should verify that the optimal mixing rule for the Giants is a 30–70 randomized mixing strategy. That is, the Giants should randomize its plays in such a way that three in ten times they pass the ball, and seven in ten times they run the ball. With a 30–70 optimal mixing rule, the Giants will be able to reduce the Ravens' success rate to a minimum of 48 percent. An interesting aspect of zero-sum games is that the sum of the success rates must equal 100 percent. Thus, the Ravens' success rate is completely consistent with the Giants' success rate of 52 percent using its optimal mixing strategy.

COMPUTING OPTIMAL MIXING RULES

A computationally more economical way of calculating a player's optimal mixing rule was suggested by J. D. Williams (1966).[4] Consider, again, the first-down game in Figure 6.9. Let *g* be the probability that the Giants adopt a *pass offense* and $(1 - g)$ the probability that they adopt a *run offense*. Similarly, let *r* be the probability that the Ravens adopt a *pass defense* and $(1 - r)$ the probability that they adopt a *run defense*. Determining the values of *g* and *r* will give us the optimal mixing rule for each team.

The Giants' expected payoff from a *pass offense* strategy is *r* multiplied by 10 percent (the Giants' payoff from the strategy profile {*Pass offense, Pass defense*}), plus $(1 - r)$ (the probability that the Ravens adopt a *run defense* strategy) multiplied by 80 percent (the Giants' payoff from the strategy profile {*Pass offense, Run defense*}), or $r(10) + (1 - r)(80) = 80 - 70r$. Similarly, the Giants' expected payoff from adopting a *run offense* strategy is *r* multiplied by 70 percent (the Giants' payoff from the strategy profile {*Run offense, Pass defense*}), plus $(1 - r)$ times 40 percent (the Giants' payoff from the strategy profile {*Run offense, Run defense*}), or $r(70) + (1 - r)(40) = 40 + 30r$. These calculations are summarized in the third column of Figure 6.12. The Giants will strictly prefer a *pass offense* strategy ($g = 1$) if $80 - 70r > 40 + 30r$, or $r > 4/10$. The Giants will strictly prefer a *run offense* ($g = 0$) if $80 - 70r < 40 + 30r$, or $r < 4/10$. Finally, the Giants will be indifferent between the two strategies when the probability that the Ravens will adopt a *run defense* strategy is $80 - 70r = 40 + 3r$, or $r = 4/10$. The Ravens' optimal mixing rule is to randomly defend against the pass 40 percent of the time, and to defend against the run 60 percent of the time. Assuming that the Ravens adopt its optimal mixing rule, the Giants' expected success rate is $80 - 70(0.4) = 40$

FIGURE 6.12
Calculating the Ravens' Optimal Mixing Rule

Ravens

		Pass defense	Run defense	
Giants	Pass offense	(<u>10</u>, 90)	(<u>80</u>, 20)	$r(10) + (1 - r)(80)$ $= 80 - 70r$
	Run offense	(<u>70</u>, 30)	(<u>40</u>, 60)	$r(70) + (1 - r)(40)$ $= 40 + 30r$

FIGURE 6.13
Calculating the Giants' Optimal Mixing Rule

Ravens

		Pass defense	Run defense
Giants	Pass offense	(10, <u>90</u>)	(80, <u>20</u>)
	Run offense	(70, <u>30</u>)	(40, <u>60</u>)
		$g(90) + (1 - g)(30)$ $= 30 + 60g$	$g(20) + (1 - g)(60)$ $= 60 - 40g$

+ 30(0.4) = 52 percent. The Ravens' optimal 40–60 mixing rule and success rate were precisely the results derived in Figure 6.10.

We can also verify the optimal mixing rule of the Giants using the same method. These calculations are summarized in the third row of Figure 6.13. The Ravens' expected payoff from adopting a *pass defense* strategy is equal to g multiplied by 90 percent (the Ravens' payoff from the strategy profile {*Pass offense, Pass defense*}), plus $(1 - g)$ multiplied by 30 percent (the Ravens' payoff from the strategy profile {*Run offense, Pass defense*}), or $g(90) + (1 - g)(30) = 30 + 60g$. Similarly, the expected payoff from adopting a *run defense* strategy is equal to g multiplied by 20 percent (the Ravens' payoff from the strategy profile {*Pass offense, Run defense*}), plus $(1 - g)$ times 60 percent (the Ravens' payoff from the strategy profile {*Run offense, Run defense*}), or $g(20) + (1 - g)(60) = 60 - 40g$. The Ravens will strictly prefer a *pass defense* strategy ($r = 1$) whenever $30 + 60g > 60 - 40g$, or $g > 3/10$, and strictly prefer a *run defense* ($r = 0$) if $30 + 60g < 60 - 40g$, or $g < 3/10$. Assuming that the Giants adopt its optimal mixing rule, the Ravens' expected success rate at gaining first-down yardage is $30 + 60(0.3) = 60 - 40(0.3) = 48$ percent. The Giants' 30–70 optimal mixing rule and success rate are precisely the results in Figure 6.11.

The Giants' 52 percent success rate and the Ravens' 48 percent success rate is a direct consequence of the minimax theorem, which asserts that whenever the interests of two rational players are diametrically opposed, both players will attempt to maximize their own payoff and minimize their rival's payoff. In zero-sum games, this leads to the rather remarkable conclusion that the sum of the success rates for the players must equal 100 percent. We first saw this when Andrew was asked to share a candy bar with Adam. Regardless of the division, the sum of the shares equals 100 percent of the candy bar. Thus, for zero-sum games, to determine the gains and losses, it is only necessary to compute the optimal mixing rule for one of the players. Finally, the minimax theory guarantees that since neither player can improve its payoff by adopting a different mixing strategy, the outcome must be a Nash equilibrium.

FIGURE 6.14
Revised Probabilities in the First-Down Game

Ravens

		Pass defense	Run defense
Giants	Pass offense	(10, 90)	(80, 20)
	Run offense	(90, 10)	(40, 60)

Payoffs: (Giants, Ravens)

FIGURE 6.15
Ravens' Optimal Mixing Rule After the Giants Improve Their Running Game

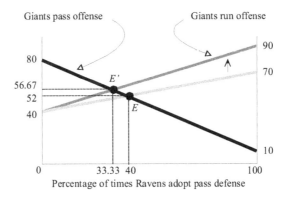

Now, suppose the Giants improve their offensive capabilities by trading for one of the league's premier running backs. Figure 6.14 summarizes the revised probabilities that the Giants will successfully gain a first-down yardage. In this revised game, the probability that the Giants will run the ball for first-down yardage should the Ravens adopt a *pass defense* strategy has increased from 70 percent to 90 percent. Given this change in the Giants' offensive capabilities, how should the teams modify their optimal mixing rules?

Using William's method, it is left as an exercise for the reader to verify that $r = 4/12$ and $(1 - r) = 5/12$. That is, the Ravens should randomly defend against the pass four out of twelve times, and randomly defend against the run the other eight times. In other words, the Ravens should randomly adopt a pass defense 33.33 percent of the time and a run defense 66.67 percent of the time. If the Giants adopt their optimal mixing rule, their success rate increases from 52 percent to 56.67 percent. The change in the Ravens' optimal mixing rule and Giants' success rate reflects the improvement in the Giants' running game. This new situation is depicted in Figure 6.15.

In a similar manner, the reader should also verify that $g = 5/12$ and $(1 - g) = 7/12$. The Giants should now randomly pass five out of twelve times, and randomly run the ball the other seven times. In other words, the Giants should randomly defend against the pass 41.67 percent of the time, and randomly defend against the run 58.33 percent of the time. If the Ravens adopt their optimal mixing rule, their success rate falls from 48 percent to 43.33 percent. The change in the Giants' optimal mixing rule and the Ravens' success rate reflects the improvement in the Giants' running game. This new situation is depicted in Figure 6.16.

FIGURE 6.16
Giants' Optimal Mixing Rule After Improving Their Running Game

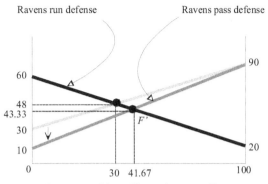

Percentage of times Giants adopt pass offense

Demonstration Problem 6.3

Consider the touchdown game depicted in Figure 2.11.

a. What are the optimal mixing rules for the Giants and Ravens?
b. Suppose that both teams adopt their optimal mixing rule and randomize their offensive and defensive plays. What is the Ravens' success rate at scoring the winning touchdown? What is the Giants' success rate at minimizing the Ravens' chances of scoring the winning touchdown?

Solution

a. Figures D6.3.1 and D6.3.2 summarize the calculations for computing each team's optimal mixing rules. Using William's method, the optimal mixing rule for the Ravens is 40–70. The Ravens should randomly pass the ball four out of seven times and randomly run the ball the three out of seven times. The optimal mixing rule for the Giants is 60–70, that is, the team should randomly defend against the pass six out of seven times and randomly defend against the run one out of seven times.
b. Assuming that both teams adopt their optimal mixing rule, the Giants' expected success rate is $40 + 10(4/7) = 80 - 60(4/7) = 45.7$ percent and the Ravens' expected success

FIGURE D6.3.1
Calculating the Ravens' Optimal Mixing Rule

		Ravens		
		Pass offense	*Run offense*	
Giants	*Pass defense*	(50, 50)	(40, 60)	$r(50) + (1 - r)(40)$ $= 40 + 10r$
	Run defense	(20, 80)	(80, 20)	$r(20) + (1 - r)(80)$ $= 80 - 60r$

FIGURE D6.3.2
Calculating the New York Giants' Optimal Mixing Rule

Ravens

		Pass offense	Run offense
Giants	Pass defense	(50, <u>50</u>)	(40, <u>60</u>)
	Run defense	(20, <u>80</u>)	(80, <u>20</u>)
		$g(50) + (1 - g)(80)$ $= 80 - 30g$	$g(60) + (1 - g)(20)$ $= 20 + 40g$

rate is $80 - 30(6/7) = 20 + 40(6/7) = 54.3$ percent of the time. As expected, since this is a zero-sum game, the sum of the two teams' success rates is 100 percent.

Unfortunately, the methods discussed in this section for determining a player's optimal mix of strategies is only appropriate for two-player, two-strategy games.[5] This was done so as to convey the essential ideas of strategic behavior without loss of generality. Although the method described above breaks down as the number of players and/or strategies is greater than two, in most cases the increased complexity associated with introducing more players and strategies are purely computational.

We have shown that by mixing strategies, even a zero-sum game can have a unique Nash equilibrium. In fact, this phenomenon is not limited to zero-sum games. It is possible to find at least one Nash equilibrium for every game with a finite number of players and pure strategies, although mixing strategies may be required to accomplish this. This phenomenon is referred to as the **Nash equilibrium existence theorem**.

> **Nash equilibrium existence theorem** Randomly mixing pure strategies using an optimal mixing rule is sufficient to guarantee that every game with a finite number of players and pure strategies has a unique Nash equilibrium.

WHEN TO USE OPTIMAL MIXING RULES

In the preceding section we learned that whenever players fail to use their optimal mixing rule, rivals can exploit this deviation. Suppose, for example, that the coaching staff of the Giants in Figure 6.9 does not adopt their optimal 30–70 mixing rule, but instead plays to their offensive strength by passing the ball 100 percent of the time. As we saw, the Ravens can exploit this by abandoning their optimal 40–60 mixing rule by defending against the pass 100 percent of the time. This will reduce the Giants' success rate from 52 percent to 10 percent. But, what if the Giants are not as naive as they appear? What if the Giants' head coach is "setting up" the Ravens for a really big play, such as when gaining first-down yardage means winning the game? There is always the danger that the head coach of the Giants is a superior strategist who adopts a less-than-optimal mixing rule for relatively unimportant plays, only to switch strategies when it really matters. Once the Ravens deviate from their optimal mixing rule to exploit a perceived deviation by the Giants, they become vulnerable to the "bait and switch." To extend the metaphor, the hunters become the hunted.

To avoid falling into this trap, the Ravens should never deviate from its optimal mixing rule. The same advice is also true for the Giants. Any deviation from a team's optimal mixing rule will increase a rival's success rate. In general, a player can never be exploited by sticking to his or her optimal mixing rule. A mixed-strategy Nash equilibrium is only possible when both players adopt their optimal mixing rule.

But, it is not enough for a player to use its optimal mixing rule to minimize a rival's maximum payoffs, while maximizing its own minimum payoffs. To be effective, the mix of strategies must also be random. On third down with a yard to go, for example, the best percentage play may be for the Giants' quarterback to hand the ball off to the halfback for a run up the middle. But, it is vitally important that the Giants avoid calling this play in all similar situations. Otherwise, the Ravens will key their defense to thwart the Giants' predictable choice of plays. Passing the ball downfield now and again will keep the Ravens' defense "honest," thereby increasing the likelihood that the run will be successful. Any systematic pattern of play calls, even if the plays are called in the right proportion, can be exploited. How can each team properly randomize its plays?

RANDOMIZING PURE STRATEGIES

How should the Giants properly randomize their offensive plays? One possibility is to flip a fair coin. Pass the ball if the coin comes up "heads" and run the ball if the coin comes up "tails." Unfortunately, this only works if the optimal mixing rule is 50–50. Even if the optimal mixing rule were 50–50, flipping a fair coin might not be the best way to randomize strategies because there is a tendency for people to look for patterns in random events. If a person flips a fair coin and "heads" comes up 20 times in a row, it is natural to guess that the next coin toss will come up heads as well, even though the next toss is an independent event. In order to keep rivals guessing, it is necessary to come up with a selection mechanism that keeps rivals from attempting to impose order on an otherwise random process.

To avoid the problem of linking outcomes, it is not only necessary to come up with a randomizing mechanism that is objective and independent, but one that is sufficiently complicated, and secret, to discourage rivals from anticipating a player's next move. A modest improvement over flipping a fair coin might be to roll a fair die. This approach not only has the advantage of making it more difficult to link outcomes, since one out of six repeated outcomes are far less likely than one out of two, but it also allows for more diverse mixing rules. For example, if the optimal mixing rule is 30–70, the Giants can call a pass play if the numbers 1 or 2 come up, and a run play otherwise. One problem with using a single die to make play calls is that it does not lend itself well to more complicated rules. Suppose, for example, that the optimal mixing rule is 25–75. Since 25 percent of 6 (the largest number on the die) is 1.5, what play should be called on a roll of 1 or 2?

An improvement to this randomizing mechanism is to use a pair of dice. There are 21 possible outcomes. If we ignore the pair 6–6, there are 20 possible outcomes. If the optimal mixing rule is 30–70, the Giants can decide to call a pass play for any combination 1–x, where x goes from 1 to 6, and a run play otherwise. If the optimal mixing rule is 40–60, the Giants can call a pass play on any toss of 1–x, 2–2, 2–3, or 2–4, and a run play otherwise. If the optimal mixing rule is 25–75, the Giants can call a pass play for any combination 1–1, 1–2, 1–3, 1–4, and 1–5, and a run play otherwise. Even if the optimal mixing rule does not result in a clear-cut combination of rolls, such as, say, 21–69, not only do the problems associated with rounding become more tractable, but certain combinations of dice rolls can be excluded as circumstances dictate.

Of course, we are not likely to see the head coach of the Giants rolling a pair of dice on the sidelines to determine the next play call. Moreover, certain circumstances, such as field position, defensive alignments, weather, field conditions, and so on, may influence play calls, even when such adjustments violate the team's optimal mixing rule. It will be in the head coach's best interest to have a randomized game plan ready at the kickoff.

Suppose, for example, that the head coach of the Giants expects to run 20 offensive plays per quarter. Using a 30–70 optimal mixing rule, the author generated the following sequence of offensive play calls using the method outlined above: R, R, R, P, P, R, R, P, R, R, P, P, P, R, P, P, R, R, P, R. Another mechanism for randomizing play calls is suggested in Dixit and Nalebuff (1991, Chapter 7). The head coach could take a quick glance at the second hand of his or her watch before each play. If the optimal mixing rule is 30–70, he or she might call a pass play if the second hand is between 1 and 20, and a run play if the second hand is between 21 and 60. If the optimal mixing rule is 40–60, he or she can call a pass play if the second hand is between 1 and 24, and a run play if the second hand is between 25 and 60, and so on. Even in the unlikely event that the head coach of the Ravens is aware of this method of randomizing plays, this knowledge is not likely to be of much help unless the watches of both coaches are perfectly synchronized.

Even if head coaches in the National Football League are remotely aware of the concept of optimal mixing rules, how likely is it that these or similar methods of randomizing play calls will be used? More often than not, experience, intuition, or hubris leads head football coaches to dismiss such ideas out of hand. After all, head coaches are hired to manage football games, not to play games of chance. Nevertheless, the logic of randomizing play calls to keep the opposition off balance is inescapable.

Principle: A player who randomizes his or her optimal mixing rule can never be exploited by a rival, provided that the rival does not know the player's game plan in advance.

The lesson to be learned from all of this is that it really does not matter if a rival knows your optimal mixing rule. Provided that your rival does not have access to your randomized game plan, there is nothing that he or she can do to exploit this information. On the other hand, if a player deviates from his or her optimal mixing rule, secrecy is essential. Otherwise, rivals will exploit this knowledge to the player's detriment. Lies, deceit, and misdirection are essential elements of effective game play. You may know that it is in your rival's best interest to mislead you, but unless you have some way to separate truth from fiction, a player's best course of action is to adopt his or her optimal mixing rule, and be unpredictable.

BLUFFING

Suppose that the head coach of the Giants positions his or her players in a pass-offense alignment, and then runs the ball? If the head coach of the Ravens calls for a pass defense, he has fallen for the bluff. As we saw in Chapter 4, bluffing is an example of a strategic move, which is an attempt by one player to gain an advantage by changing the behavior of a rival through a display of bravado that has no basis in fact. But, while a player cannot simply take a rival's display of strength at face value, it may reveal something important about his/her intentions.

An analysis of bluffing involves two important considerations. First, bluffing cannot be divorced from expected payoffs. Second, bluffing must be unpredictable to be successful. To see this, consider betting strategies in poker. "The poker hand must at all times be concealed behind the mask of inconsistency. The good poker player must avoid set practices and act at random, going so far, on occasion, as to violate the elementary principles of correct play" (McDonald 1950, p. 30). A "tight" player who never bluffs seldom wins a large pot because rivals will not "see" (match) or raise his or her bet. A "loose" player who always bluffs will always be "called" on a higher bet. If we have learned anything from the above discussion, the best strategy is to randomize bets on good and bad hands. To see why, consider a game of five-card stud-poker between "Wild Bill" Hickok and "Doc" Holiday. Wild Bill plays poker with Doc on a regular basis. Wild Bill has observed that Doc raises 75 percent of the time on a good hand, and calls 25 percent of the time. On the other hand, when Doc raises 25 percent of the time on a bad hand, and folds 75 percent of the time. The probabilities of Doc's betting strategies are summarized in Figure 6.17.

FIGURE 6.17
Probabilities that "Doc" Holiday Bets on Good and Bad Hand's

		Hand	
		Good	*Bad*
	Raise	75	25
Bet	*Call*	25	0
	Fold	0	75

Before the cards are dealt, Wild Bill and Doc believe there is a 50 percent chance of being dealt a good or bad hand. Doc's mixing rule depends on the hand that he is dealt. Because of this, Wild Bill can learn something about the probability that Doc has a good hand based on his betting strategy. If Doc folds, Wild Bill can be certain that Doc is holding a bad hand. If Doc calls, Wild Bill can be sure that Doc is holding a good hand. In either case, the betting is over and the next hand is dealt. But, if he raises, Wild Bill cannot be certain whether Doc has a good hand or a bad hand. The odds are 3 to 1 that Doc has a good hand. While Doc's decision to raise does not reveal whether he has a good hand or a bad hand, Wild Bill knows more than he did before the cards were dealt.

The process by which Wild Bill revises his probability estimates that Doc has a good hand is referred to as **Bayesian updating**.[6] Bayesian updating is the process by which players revise their expectations based on a rival's prior moves. Bayesian updating is based on **Bayes' theorem**, which is a statistical concept dealing with conditional probabilities. According to Bayes's theorem, the probability that Doc has a good hand after Wild Bill observes his decision to raise, call, or fold, is equal to the probability that he has a good hand and behaves in a certain way, divided by the total probability that he bets in a certain way. For example, if Doc folds, he must have a bad hand. If he calls, his hand must be good. If Doc raises, the probability that he has a good hand is $0.5(0.75) = 0.375$, or 37.5 percent. If Doc raises, the probability that he has a bad hand is $0.5(0.25) = 0.125$, or 12.5 percent. Thus, the total probability of Doc raising is 0.5, or 50 percent. Using Bayes' theorem, the probability that Doc has a good hand based on his decision to raise is equal to $0.375/0.5 = 0.75$, or 75 percent. Thus, based on his betting strategy, Wild Bill has revised his probability estimate that Doc has a good hand from 50 percent to 75 percent.

> **Bayesian updating** A process by which players revise their expectations based on a rival's prior moves.

> **Bayes' theorem** A statistical relationship that enables a player to update the probability of an event after learning that some other event has occurred.

CHAPTER REVIEW

The focus of this chapter was *zero-sum games*. In a zero-sum game, one player's gain is exactly offset by the other players' losses. John von Neumann discovered that in two-player, zero-sum, noncooperative, static games, it is rational for each player to adopt a maximin strategy. According to the *minimax theorem,* when both players adopt a maximin strategy, the game will have a Nash equilibrium.

A *pure strategy* for a player is a complete and nonrandom game plan. A pure-strategy Nash equilibrium does not exist for a zero-sum game. By randomizing strategies, a zero-sum game can have an infinite number of strategy profiles, one for each of an infinite number of probability distributions. Such randomized pure strategies are called *mixed strategies*. An *optimal mixing rule* is a randomized mix of pure strategies that maximizes a player's minimum payoff, while minimizing a rival's maximum payoff.

By randomly mixing pure strategies, the Nash equilibrium existence theorem asserts that it is possible to find at least one Nash equilibrium for every game with a finite number of players and pure strategies.

Whenever a player fails to use his or her optimal mixing rule, it is possible for a rival to exploit this deviation by deviating from his or her own optimal mixing rule. The danger is that a player may attempt to deceive a rival by adopting a less-than-optimal mixing rule for relatively unimportant moves, only to switch strategies when it really matters. In general, it does not matter if a player knows a rival's optimal mixing rule. Provided that a player does not know a rival's moves in advance, there is nothing that he or she can do to exploit this knowledge.

Bayesian updating, which is based on *Bayes' theorem*, is the process by which players revise their expectations based on a rival's prior moves. Bayes' theorem is a statistical relationship that enables a player to update the probability of an event after learning that some other event has occurred.

CHAPTER QUESTIONS

6.1   Explain the difference between zero-sum and non-zero-sum games.

6.2   Zero-sum games sometimes have a unique Nash equilibrium in pure strategies. Do you agree with this statement? Explain.

6.3   The matching pennies game is an example of a non-zero-sum game because someone always ends up with a penny. Do you agree? Why not?

6.4   Explain why the sink-the-battleship game does not have a pure-strategy Nash equilibrium.

6.5   Under what conditions will it be possible to identify a unique Nash equilibrium for the sink-the-battleship game?

6.6   Explain, in your own words, the rationale underlying the minimax theorem.

6.7   What is the difference between a pure strategy and a mixed strategy?

6.8   Is it possible to have a unique, pure-strategy Nash equilibrium in zero-sum games? In mixed strategies?

6.9   Is it possible to have a unique Nash equilibrium in non-zero-sum games when neither player has a dominant strategy? Explain.

6.10   What do we mean when we say that a player has an optimal mixing rule?

6.11   A good way to randomize strategies is to flip a fair coin. Is this statement true, false, or uncertain? Why?

6.12   A player who randomizes his or her optimal mixing rule can never be exploited by a rival. Do you agree? Why or why not?

6.13   It is never possible for a player to exploit a rival by knowing his or her optimal mixing rule. Do you agree with this statement? Explain.

6.14   In a zero-sum game, it doesn't matter if a player knows a rival's optimal mixing rule as long as he or she does not have access to the player's randomized game plan. Do you agree with this statement? Why?

6.15 What is a strategic move?

6.16 Give an example of bluffing that is not mentioned in the text.

6.17 In poker, explain how a rival's bluff can be used to improve a player's chances of winning.

CHAPTER EXERCISES

6.1 Consider the matching pennies game in Figure 6.1. If both players flip a fair coin to determine their moves, verify that Chloe's expected payoff is zero if she adopts any of the following strategies:

  a. A pure strategy of *heads.*
  b. A pure strategy of *tails.*
  c. A mixed strategy playing both *heads* and *tails* one-half of the time.
  d. A mixed strategy of *heads* one-quarter of the time and *tails* three-quarters of the time.

6.2 Suppose two candidates are running for the same office. The winner is the candidate receiving the majority of votes. The electorate's preferences on campaign issues are assumed to be uniformly and randomly distributed. If we assume that the voters will select that candidate nearest to their position, where will the two candidates locate themselves along the circle in Figure E6.2?

6.3 Consider the touchdown game depicted in Figure E6.3, which summarizes the probabilities that the New England Patriots or the Carolina Panthers will win on the last play of the game from alternative offensive and defensive strategy profiles.

  a. What is the optimal mixing strategy for the Patriots and the Panthers?
  b. Suppose that both teams adopt their optimal mixing rule and both teams randomize their offensive and defensive plays. What is the Patriots' expected success rate scoring the winning touchdown? What is the Panthers' expected success rate keeping the Patriots from scoring the winning touchdown?

6.4 Suppose that two individuals are playing a game of tennis. The situation depicted in Figure E6.4 summarizes the probabilities that the receiver will successfully return a serve to his or her backhand or forehand.

  a. What is the optimal mixing rule of the receiver? Using this mixing rule, what is the server's expected success rate?
  b. What is the optimal mixing rule of the server? Using this mixing rule, what is the receiver's expected success rate?
  c. Your answers to parts a and b illustrate what important theorem in game theory?

6.5 Suppose with practice that the receiver in chapter exercise 6.4 improves his or her ability to return a serve to the backhand side. The revised probabilities of a successful return are summarized in Figure E6.5. How will your answers to chapter exercise 6.4 differ as a result of these changed probabilities?

6.6 Verify that 50–50 is the optimal mixing rule for the sink-the-battleship game depicted in Figure 6.3.

6.7 Recall the battle-of-the-sexes game in Figure 3.1. Calculate the mixed-strategy Nash equilibrium for this game.

FIGURE E6.2
Issue Positions of the Electorate

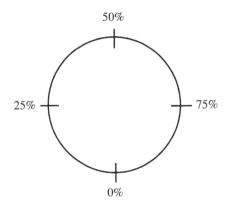

FIGURE E6.3
Static Game in Chapter Exercise 6.3

		New England Patriots	
		Pass offense	*Run offense*
Carolina Panthers	*Pass defense*	(65, 35)	(27, 73)
	Run defense	(26, 74)	(70, 30)

Payoffs: (Carolina Panthers, New England Patriots)

FIGURE E6.4
Probabilities that Receiver Successfully Returns a Serve

		Server	
		Serve to the forehand side	*Serve to the backhand side*
Receiver	*Forehand return*	90	20
	Backhand return	30	60

FIGURE E6.5
Probabilities that Receiver Successfully Returns a Serve

		Server	
		Serve to the forehand side	*Serve to the backhand side*
Receiver	*Forehand return*	90	20
	Backhand return	30	70

FIGURE E6.8
Chicken Game in Chapter Exercise 6.8

Paulie

		Chicken	Cool dude
Rocky	Chicken	(3, 3)	(1, 6)
	Cool dude	(6, 1)	(0, 0)

6.8   Figure E6.8 illustrates the chicken game. In this game, Rocky and Paulie agree to drive their cars at high speed directly at each other. The first to bail out is a "chicken," while the other is a "cool dude." If neither bail out, both die, but both will be remembered as cool dudes.

  a.   Does this game have a pure-strategy Nash equilibrium strategy profile?
  b.   Calculate Rocky and Paulie's optimal mixing rule.

6.9   Consider the oil-drilling game depicted in Figure 2.8.

  a.   What is the optimal mixing strategy for PETROX and GLOMAR?
  b.   What is the success rate of both companies if they adopt their optimal mixing rules?

ENDNOTES

  1.   This example was adapted from Dixit and Nalebuff (1991, Chapter 3).
  2.   These results may be verified by noting in Figure 6.12 that the success of a Giants *pass offense* (PO) is a function of the Ravens *pass defense* (PD), which may be summarized by the equation:

$$PO = 80 - 70PD \qquad (F6.2.1)$$

Likewise, the success of the Giants' *run offense* (RO) is a function of the Ravens *run defense* (RD), which may be expressed by the equation:

$$RO = 70 - 30RO = 70 - 30(1 - PD) = 40 + 30PD \qquad (F6.2.2)$$

By flipping a fair coin, $PD = RD = 0.5$. Substituting this value into Equations (F6.2.1) and (F6.2.2) we obtain:

$$PO = 80 - 70(0.5) = 45 \text{ percent} \qquad (F6.2.3)$$

$$RO = 40 + 30(0.5) = 55 \text{ percent} \qquad (F6.2.4)$$

  3.   Determining the mixing rule that minimizes the Ravens' chances of stopping the Giants while maximizing the Giants' chances of gaining first-down yardage can be determined by simultaneously solving Equations (F6.2.1) and (F6.2.2). Setting $PO = RO$ we get:

$$80 - 70PD = 40 + 30PD \qquad (F6.3.1)$$

Solving Equation (F6.3.1) we get $PD^* = 40$ percent, and $RD^* = 1 - PD^* = 60$ percent. Substituting this result into the right-hand or left-hand side of Equation (F6.2.1) or (F6.2.2) we obtain $PD^* = 40$ percent. Substituting this result into Equation (F6.3.1) we get:

$$80 - 70(0.4) = 40 + 30(0.4) = 52 \text{ percent} \qquad (F6.3.2)$$

FIGURE F6.4

Probabilities of the Payoffs for Player *B*

Player *B*

		B1	B2
Player *A*	A1	C	D
	A2	E	F

4. The simple arithmetic method described here is due to J. D. Williams in *The Compleat Strategyst* (1966). Consider Figure F6.4, which summarizes the probabilities of the payoffs for player *B*. The equilibrium ratio of strategy *B1* to strategy *B2* is $(F-D):(C-E)$. Player *B* will select the probability p of adopting strategy *B1* such that player *A* is indifferent between strategy *A1* and strategy *A2*, which is $pC + (1-p)D = pE + (1-p)F$. This may be rearranged to yield the equation $p/(1-p) = (F-D)/(C-E)$, or $(F-D):(C-D)$. By analogous reasoning, the optimal mix for player *A* is $(F-E):(C-D)$.

5. Linear programming is a more general method for finding mixed-strategy Nash equilibrium strategy profiles.

6. Bayesian updating will be discussed at greater length when we discuss multistage games with incomplete information in Chapter 16.

CHAPTER

7 STATIC GAMES WITH CONTINUOUS STRATEGIES

In this chapter we will:

- *Explain the difference between static games involving discrete and continuous pure strategies;*
- *Introduce the concept of a reaction (best-response) function;*
- *Analyze noncooperative static games with complete information involving pure continuous strategies;*
- *Discuss the possible benefits from coordinating strategies in games involving pure continuous strategies;*
- *Introduce the "tragedy of the commons";*
- *Explore the Coase theorem as a possible solution to the tragedy of the commons.*

INTRODUCTION

Throughout our discussion of strategic behavior and static games, players were confronted with a choice of discrete pure strategies. In the prisoner's dilemma, for example, the suspects had to decide whether to *confess* or remain *silent*. While this was perfectly reasonable, there are many other games in which the players must choose from among an infinite number of possible strategies. In the pricing game depicted in Figure 1.2, for example, firms *A* and *B* had to choose from just two strategies: *high price* or *low price*. But what does it mean to charge a "high price" or a "low price"? Should we define a *high price* strategy as any price that is higher than the price charged by a rival? If so, there are still an infinite number of possible prices to choose from, which means that there are an infinite number of strategy profiles and payoffs. In most business and economic applications, such as pricing and output decisions, expenditures on alternative advertising strategies, product differentiation, capital investment, auction bids, bargaining positions, and so on, it makes more sense to allow players to choose from a continuum of strategies rather than from a discrete few. In this chapter we will expand our search for a Nash equilibrium to include games involving continuous pure strategies.

CONTINUOUS STRATEGIES

To appreciate the challenges associated with finding a Nash equilibrium for games involving continuous pure strategies, consider the following price war between two neighborhood gas stations, PETROX and GLOMAR. While each gas station has a core of loyal customers, there exists a third group of customers that is sensitive to price changes. Suppose that PETROX adopts the following pricing strategy: If GLOMAR cuts its price for regular gasoline by $0.10 per gallon, PETROX will lower its price by 80 percent of that, or $0.08 per gallon. This will allow PETROX to generate some profit from its loyal customers, although price-sensitive customers will buy from

FIGURE 7.1
PETROX's Reaction Function

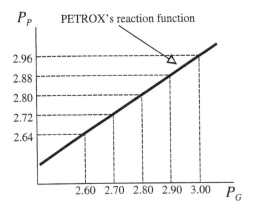

GLOMAR. On the other hand, if GLOMAR raises its price by $0.10 per gallon, PETROX will respond by raising its price by only $0.08 per gallon, in which case price-sensitive customers will buy from PETROX. The response by PETROX to price changes by GLOMAR is referred to as its reaction function. PETROX's reaction function is illustrated in Figure 7.1, where P_P and P_G are the prices charged by PETROX and GLOMAR, respectively. Suppose that the two gas stations are alike in every respect, including operating costs, the size of their loyal customer bases, and so on. For this reason, it is reasonable to assume that GLOMAR will adopt an identical pricing strategy. GLOMAR's reaction function is illustrated in Figure 7.2.

We will begin our analysis of the gasoline-price-setting game by assuming that both GLOMAR and PETROX initially charge $2.80 for a gallon of regular gasoline. Now, suppose that GLOMAR is contemplating raising its price by $0.10 to $2.90 per gallon. GLOMAR reasons that PETROX will react to such a price increase by raising its price by only $0.08, or 80 percent, to $2.88 per gallon. GLOMAR reasons that if this happens, it should consider raising its price by 80 percent of that, or $0.064 to $2.864 per gallon. If GLOMAR raises its price by $0.064, PETROX will counter with a price increase of $0.0512 to $2.8512 per gallon. GLOMAR will react by raising its price by $0.041 to $2.841 per gallon, and so on. This circular reasoning will continue until the price charged by both gas stations converges to $2.80 per gallon.

The reader is cautioned that this is not a sequential-move game. Both players will move just once. We are describing a process of circular reasoning whereby both players are attempting to decide exactly what that move will be. In this game, $2.80 per gallon is a unique Nash equilibrium strategy. Thus, the best strategy for both players is to continue to charge a price of $2.80 per gallon.

Suppose that instead of a price increase, GLOMAR contemplates lowering its price by $0.10 to $2.70 per gallon. GLOMAR reasons that PETROX will respond by lowering its price by only $0.08 to $2.72 per gallon. If PETROX lowers its price by $0.08, GLOMAR will counter by lowering its price by $0.064 to $2.736 per gallon, in which case PETROX is expected to react with a price cut of $0.0512, to $2.7488 per gallon, and so on. As in the case of a price increase, this back and fourth reasoning will eventually lead right back to the initial price of $2.80 per gallon. The Nash equilibrium strategy profile for the gasoline-price-setting game is {*$2.80, $2.80*}, which is depicted in Figure 7.3, where R_P and R_G are the reaction functions of PETROX and GLOMAR, respectively.

The Nash equilibrium for the gasoline-price-setting game is for both gas stations to charge a price of $2.80 per gallon. Unless there is a change in the underlying conditions of the game that

FIGURE 7.2
GLOMAR's Reaction Function

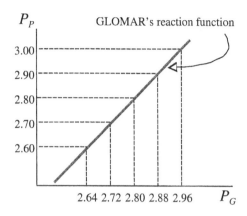

FIGURE 7.3
Nash Equilibrium in the Gasoline-Price-Setting Game

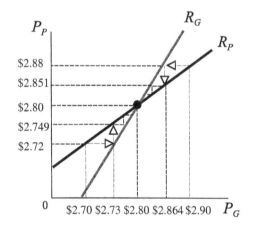

cause either or both reaction functions to shift, such as a change in the size and distribution of the gas stations' loyal customer base, a change in the price of crude oil or the wholesale price of gasoline, the number of competitors, the gas station's operating costs, and so on, the Nash equilibrium strategy profile {*$2.80, $2.80*} will continue undisturbed since neither gas station can improve its payoff by charging a different price. If PETROX believes that GLOMAR will charge a price of $2.80, the best response is to charge a price of $2.80 as well, and vice versa. This combination of pricing strategies represents the best response by each gas station to the pricing strategy of its rival.

It cannot be emphasized strongly enough that the gasoline-price-setting game is a one-time, static game. The circular reasoning just described is a mental process for determining each player's optimal strategy. An equally important point is that both players have an infinite number of strategies (prices) to choose from. These strategy choices lie along a continuum of prices. The players are not constrained by a finite number of choices.

FIGURE 7.4
First Elimination of Strictly Dominated Strategies

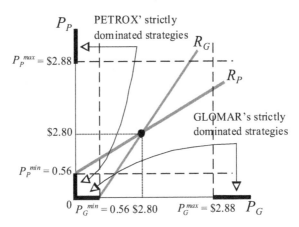

REACTION (BEST RESPONSE) FUNCTIONS

In the gasoline-price-setting game we introduced the idea of a player's **reaction function**. What exactly is a reaction function, and how important is it for finding a Nash equilibrium for games involving continuous strategies? In general, a reaction function (also known as a **best-response function**) is a formal statement of the optimal response by one player to the moves made by rivals. In the gasoline-price-setting game, the reaction functions of PETROX and GLOMAR are summarized by Equations (7.1) and (7.2), respectively. While the Nash equilibrium strategy profile for the gasoline-price-setting game can be found by solving these equations simultaneously, the underlying process involves eliminating dominated strategies.

$$P_P = 0.56 + 0.8P_G \qquad (7.1)$$

$$P_G = 0.56 + 0.8P_P \qquad (7.2)$$

> **Reaction (best-response) function** A relationship that expresses the best response by a player to the moves of a rival.

The Nash equilibrium for the gasoline pricing game is the strategy profile {*$2.80, $2.80*}, which occurs at the intersection of the two reaction functions in Figure 7.3. Is it possible to identify some other Nash equilibrium for this game? To answer this, we will begin by assuming that rational gas station owners would never wittingly adopt a dominated strategy. Consider Figure 7.4.

In searching for a Nash equilibrium we will iteratively eliminate dominated strategies. Given the reaction function in Equation (7.1), it is clear that PETROX would never charge a price less than $0.56, which is the price that PETROX would charge if GLOMAR gave its gasoline away. This is the minimum price (P_P^{min}) that PETROX would charge its loyal customers. The minimum price $P_P^{min} = \$0.56$ strictly dominates any lower price. These strictly dominated strategies should be eliminated from further consideration. Let P_P^{max} represent the price of $2.90 that PETROX would like to charge if it did not have to worry about retaliation. Prices greater than P_P^{max} are strictly dominated strategies and should be eliminated. By analogous reasoning, any price below P_G^{min} and greater than P_G^{max} are strictly dominated strategies for GLOMAR, and should be eliminated for precisely the same reasons. These strictly dominated pricing strategies are illustrated as

FIGURE 7.5

Simplified Gasoline-Price-Setting Game After Eliminating Strictly Dominated Strategies

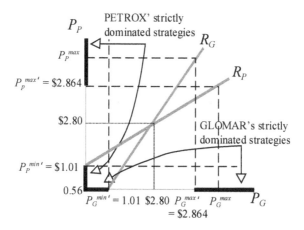

the heavily shaded lines in Figure 7.4. After eliminating these strictly dominated strategies, the resulting simplified gasoline-price-setting game is illustrated in Figure 7.5.

In the simplified game in Figure 7.5, the new origin is the minimum price of $0.56 that would be charged by both gas stations. It is left as an exercise for the reader to show that the new intercept for each reaction function now occurs at a price of $1.01. In the simplified gasoline-price-setting game, prices below $P_P^{min'} = \$1.01$ and above $P_P^{max'} = \$2.88$ are now the strictly dominated strategies, and must be eliminated as well. By analogous reasoning, GLOMAR prices below $P_G^{min'} = \$1.01$ and above $P_G^{max'} = \$2.88$ must also be eliminated. The resulting simplified game (not shown) has as its new origin a price of $1.01. We repeat this process until we have eliminated all strictly dominated strategies and are left with the Nash equilibrium strategy profile {*$2.80, $2.80*}.

———————————— **Demonstration Problem 7.1** ————————————

Suppose that an industry consists of two firms producing an identical product. The market demand for the combined output of both firms is:

$$(Q_1 + Q_2) = 201 - 0.2P \qquad (D7.1.1)$$

The total cost of production for each firm is:

$$TC_i = 100 + 5Q_i; \qquad i = 1, 2 \qquad (D7.1.2)$$

a. Calculate each firm's reaction function.
b. Determine the Nash equilibrium output strategy profile for this game. Calculate the corresponding Nash equilibrium price, and the profit for each firm.

Solution

a. Solving Equation (D7.1.1) for price we obtain:

$$P = 1,005 - 5(Q_1 + Q_2) \qquad (D7.1.3)$$

The total revenue equation for firm 1 is:

$$TR_1 = 1,005Q_1 - 5Q_1^2 - 5Q_1Q_2 \qquad (D7.1.4)$$

Combining Equations (D7.1.2) and (D7.1.4) we obtain firm 1's total profit function:

$$\pi_1 = TR_1 - TC_1 \qquad\qquad (D7.1.5)$$

$$\pi_1 = -100 + 1{,}000Q_1 - 5Q_1^2 - 5Q_1Q_2 \qquad\qquad (D7.1.6)$$

To determine the output level that maximizes firm 1's total profits, take the partial derivative of Equation (D7.1.6) with respect to Q_1 and set the resulting equation equal to zero.

$$\frac{\partial \pi_1}{\partial Q_1} = 1{,}000 - 10Q_1 - 5Q_2 = 0 \qquad\qquad (D7.1.7)$$

Analogously for firm 2:

$$\frac{\partial \pi_2}{\partial Q_2} = 1{,}000 - 10Q_2 - 5Q_1 = 0 \qquad\qquad (D7.1.8)$$

The reaction function for each firm is determined by solving Equations (D7.1.7) and (D7.1.8) for Q_1 and Q_2, respectively.

$$Q_1 = 100 - 0.5Q_2 \qquad\qquad (D7.1.9)$$

$$Q_2 = 100 - 0.5Q_1 \qquad\qquad (D7.1.10)$$

b. The Nash equilibrium output for each firm is determined by solving simultaneously Equations (D7.1.9) and (D7.1.10).

$$Q_1{}^*(Q_2^*) = Q_2{}^*(Q_1^*) = 66.67 \qquad\qquad (D7.1.11)$$

Thus, the Nash equilibrium output strategy profile for this game is {66.67, 66.67}. The market price for the combined output of the two firms is:

$$P^* = 1{,}005 - 5(66.67 + 66.67) = \$338.39 \qquad\qquad (D7.1.12)$$

The total profit for each firm is found by substituting the Nash equilibrium output levels for each firm into their respective profit functions.

$$\pi_1^* = -100 + 1{,}000(66.67) - 5(66.67)^2 - 5(66.67)(66.67) = \$22{,}121.12 \quad (D7.1.13)$$

$$\pi_2^* = -100 + 1{,}000(66.67) - 5(66.67)^2 - 5(66.67)(66.67) = \$22{,}121.12 \quad (D7.1.14)$$

The total combined profits of the two firms are \$44,242.24.

TRAGEDY OF THE COMMONS

A well-known example of a two-person, static game involving continuous strategies is the "tragedy of the commons." This phrase is used to describe competition over scarce resources and the conflict between individual self-interests and social welfare. The tragedy of the commons derives its name from a parable by William Forster Lloyd (1833) that was popularized by Garrett Hardin (1968). This version of the tragedy of the commons involves two shepherds using the village "commons" (a communal area open to public use) for sheep grazing. Because the size of the commons is fixed, the more sheep that are allowed to graze, the less grass will be available for each. This is not necessarily a problem if the number of sheep is very small, but if the shepherds' flocks grow, they will reach a point beyond which nutrition suffers and the amount of wool and of wool and mutton produced declines. More specifically, suppose that s_A and s_B represent the number of sheep grazed by shepherds A and B, respectively. Suppose that the benefit from grazing each sheep is:

$$B = 150 - s_A - s_B \tag{7.3}$$

Shepherd A's total benefits are given by the equation:

$$TB_A = s_A \times B = 150s_A - s_A s_B - s_A^2 \tag{7.4}$$

Analogously, the total benefits received by shepherd B are:

$$TB_B = s_B \times B = 150s_B - s_B s_A - s_B^2 \tag{7.5}$$

Shepherd A's best-response function is the value of s_A that maximizes total benefits for any value of s_B. Taking the first derivative of Equation (7.4) with respect to s_A and setting the results equal to zero, we get:

$$MB_A = \frac{\partial TB_A}{\partial s_A} = 150 - 2s_A - s_B = 0 \tag{7.6}$$

In Equation (7.6), MB_A represents the marginal benefits received by shepherd A for a given value of s_B. Solving Equation (7.6) for s_A yields Equation (7.7), which is shepherd A's best-response function.

$$s_A = 75 - \frac{s_B}{2} \tag{7.7}$$

Analogously, Equation (7.8) represents shepherd B's best-response function.

$$s_B = 75 - \frac{s_A}{2} \tag{7.8}$$

Solving Equations (7.7) and (7.8) simultaneously we obtain the Nash equilibrium number of sheep in this game, which is for both shepherds to graze 50 sheep, or a combined total of 100 sheep. This result is illustrated in Figure 7.6. It is left as an exercise for the reader to demonstrate that $s_A = s_B = 50$ can be obtained from Figure 7.6 by eliminating all strictly dominated strategies. The total benefit received by each shepherd from the Nash equilibrium strategy profile $\{50, 50\}$ is $150(50) - (50)^2 - (50)(50) = 2,500$.

The reason why this game is called a tragedy is because the Nash equilibrium strategy profile $\{50, 50\}$ results in overgrazing. To see why, let us determine the optimal number of sheep if the shepherds coordinate their grazing decisions. To do this, let $s = s_A = s_B$. The total benefit equation is:

$$TB = s \times B = s(150 - s - s) = 150s - 2s^2 \tag{7.9}$$

Taking the first derivative of Equation (7.9) with respect to s, and setting the results equal to zero, we obtain:

$$MB = \frac{\partial TB}{\partial s} = 150 - 4s = 0 \tag{7.10}$$

Solving for s, the maximum number of sheep grazed by each shepherd is 37.5. If the shepherds agree to graze 38 sheep, total benefits received by each shepherd are $150(38) - 2(38)^2 = 2,812$,

FIGURE 7.6
Nash Equilibrium for the Tragedy of the Commons

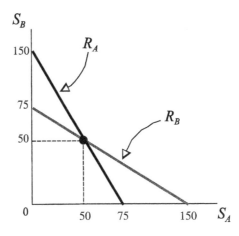

which is less than the 2,500 in benefits for each shepherd when the shepherds do not coordinate their grazing activities. In the tragedy of the commons, individual initiative does not serve the communal good. Only when the shepherds collaborate is the collective welfare maximized.

————————————— **Demonstration Problem 7.2** —————————————

Suppose that the firms in demonstration problem 7.1 merge to form a monopoly.

    a. Determine the monopolist's profit-maximizing level of output.
    b. What is the market price at the monopolist's profit-maximizing level of output? What is the monopolist's profit?
    c. Compare your answers to parts a and b above with the solution to demonstration problem 7.1. In what way are these results similar to the tragedy of the commons?

Solution

    a. The inverse of the market demand equation in demonstration problem 7.1 is:

$$P = 1,005 - 5Q \qquad\qquad (D7.2.1)$$

The monopolist's total revenue equation is:

$$TR = PQ = 1,005Q - 5Q^2 \qquad\qquad (D7.2.2)$$

Combining Equations (D7.2.2) and (D7.1.2) we obtain the monopolist's total profit, which is:

$$\pi = TR - TC = -100 + 1,000Q - 5Q^2 \qquad\qquad (D7.2.3)$$

The necessary condition for profit maximization is:

$$\frac{d\pi}{dQ} = 1,000 - 10Q = 0 \qquad\qquad (D7.2.4)$$

Solving for Q, the monopolist's profit-maximizing level of output is $Q^* = 100$. Compare this with the combined output of 133.34 in demonstration problem 7.1.

b. To obtain the market price, substitute the profit-maximizing level of output into Equation (D7.2.1):

$$P^* = 1,005 - 5(100) = \$505 \qquad (D7.2.5)$$

Substituting the profit-maximizing output level into Equation (D7.2.3) we get:

$$\pi^* = -100 + 1,000(100) - 5(100)^2 = \$49,900 \qquad (D7.2.6)$$

c. Recall that in the tragedy of the commons, the benefit-maximizing behavior of two noncooperative shepherds results in overgrazing. The shepherds' combined benefits are maximized when their efforts are coordinated. In a similar manner, the results of parts a and b above reveal that the profit-maximizing level of output for a monopolist is lower, and the market price higher, than in the two-firm case in demonstration problem 7.1. Not only that, but the monopolist's total profit is greater than the combined profits of the two firms in demonstration problem 7.1. As in the tragedy of the commons, it is in the two firms best interest to collude, behave as a profit-maximizing monopolist, and divide these profits according to the relative contributions of each firm, which in this case is a 50–50 split. What, if anything, might prevent these firms from forming a stable cartel?

Although the tragedy of the commons discussed above involves only two players, this analysis can be extended to include multiple shepherds. The fundamental problem is that private-property rights are not well defined. Overgrazing occurs because no one owns the communal grazing area. As a result, society's resources are not used efficiently, which is a form of **market failure**. Market failure occurs when the forces of supply and demand fail to generate socially efficient levels of consumption and production. In the economics literature there are four sources of market failure: Public goods, externalities, market power, and an inequitable distribution of income and wealth. Market failure is often used as a justification for government intervention in the marketplace.

> **Market failure** When the forces of supply and demand fail to generate socially efficient levels of consumption and production.

The solution to the problem of overgrazing in the tragedy of the commons is to establish well-defined private-property rights. This prescription, which is known as the Coase theorem, will be discussed in greater detail in the next section. This is precisely what occurred in fifteenth- and sixteenth-century England: Ownership of common areas was seized by the aristocracy, who levied grazing rents to maximize their incomes. Although grazing was reduced to more socially optimal levels, income was redistributed from the peasants to the landed gentry.

Unfortunately, private-property rights are not always easy to define or enforce. It is difficult, for example, to regulate the exploitation of the world's deep-sea resources or the amount of industrial and consumer pollutants spewed into the atmosphere. As a result, whales were indiscriminately hunted to near extinction, and global warming poses a serious threat to the future of mankind. In such cases, international cooperation is required, but is not always easy to achieve (see Case Study 7.1).

THE COASE THEOREM

The Coase theorem can best be understood by examining the effects of **externalities** on the efficient production of goods and services. An externality occurs when a third party is positively or negatively affected by a market transaction. In this section, we will consider the case of a negative externality in which production by one firm imposes costs on another firm in an unrelated activity. The most recognizable example of a negative externality is pollution.

> **Externality** A cost or benefit resulting from a market transaction that is imposed upon third parties.

To make the discussion more concrete, consider the following scenario. A fertilizer plant uses clean water from a nearby river in its production process, but then discharges polluted wastewater (effluents) further downstream. The amount of fertilizer produced is the result of the interaction of supply and demand in an unregulated fertilizer market. The more fertilizer produced, the more waste is dumped into the river.

Downstream from the fertilizer plant is a brewery that uses water from the river to produce beer. To meet federal and state health and safety standards, the brewery must first purify the polluted river water before it can be used. Although the brewery has nothing to do with the fertilizer market, its production costs are higher because of pollution created by the fertilizer plant. Another way of looking at this problem is to recognize that the fertilizer plant's production costs are lower than if it were required to clean up its effluents or properly dispose of its waste by-products. Figure 7.7 depicts the situation from the perspective of the brewery.

Let us begin by assuming that the fertilizer plant does not use river water in its production process. Because the fertilizer plant does not pollute the river, the downstream brewery does not incur cleanup costs. The market price of beer P_0 is determined in the competitive beer market at the intersection of the relevant demand and supply curves. The profit-maximizing brewery will produce at the output level Q_0 where the marginal private cost (MPC) equals marginal revenue (P_0). Since all costs of production are internalized, this level of output is socially optimal.

Suppose, however, that the upstream fertilizer plant uses clean water from the river, but discharges polluted water back into the river a little further downstream. This imposes cleanup costs on the brewery. The cost of cleaning up the polluted water is illustrated as the marginal externality cost (MEC) curve in Figure 7.7. If the brewery absorbs these cleanup costs, its marginal cost of production increases from MPC to MSC (marginal social cost), in which case beer production falls from Q_0 to Q_1. This level of output represents a socially nonoptimal level of beer production. In other words, "too little" beer is produced. On the other hand, the fertilizer plant lowers its marginal cost of producing fertilizer. The result is that "too much" fertilizer is produced. Pollution by the fertilizer plant results in socially inefficient levels of beer and fertilizer production. River pollution has resulted in market failure.

As in the tragedy of the commons where no one owns the communal grazing area, the private-property rights to clean water are not well defined. Individual shepherds have no incentive to graze fewer sheep. Similarly, the profit-maximizing fertilizer plant has no incentive to clean up its waste, thereby reducing output, and river pollution, to more socially optimal levels. The presence of negative externalities is a strong argument in favor of government intervention in the marketplace. In principle, there are at least two ways in which the government can intercede to promote a more socially desirable outcome. On the one hand, the government can require producers to absorb the cleanup costs by properly disposing of their waste by-products, rather than shifting these costs to third parties. Alternatively, the government can establish conditions whereby the market determines an efficient solution to the externality problem.

The solution to the pollution problem in Figure 7.7 is to get the fertilizer plant to internalize the cost of pollution. This can be accomplished in a variety of ways. The most common approach is for government to claim ownership of the river and dictate how it should be used. The government could, for example, impose an outright ban on the dumping of industrial waste into the river. This would force the fertilizer plant to, say, employ the services of a waste disposal company. Alternatively, the government could require the fertilizer plant to clean its waste before discharg-

FIGURE 7.7
Negative Externality in Production

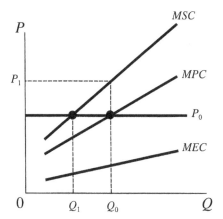

ing its waste back into the river. Either way, this will raise the marginal private cost of producing fertilizer, which will reduce fertilizer production to more socially desirable levels. Another way to accomplish this objective was proposed by Nobel laureate Ronald Coase (1960).

To understand the rationale underlying the **Coase theorem**, it will be useful to view negative externalities as the result of a "missing" market. Suppose there exists a market for clean water and the fertilizer plant has to pay a positive price for its use. If this market is efficient, the fertilizer plant's marginal private cost of production will be higher and production will decline. The reason why a market for clean water does not exist, however, is that no one "owns" the river. In the absence of well-defined property rights, a market for clean water cannot exist and a price for its use cannot be determined. This suggests that it may be possible for the government to solve the pollution problem by assigning private-property rights to the river.

> **Coase theorem** The assignment of well-defined private-property rights will result in a socially efficient allocation of productive resources and a socially optimal, market-determined level of goods and services.

To see how this might lead to a socially optimal level of production, suppose that the government grants ownership of the river to the fertilizer plant. The profit-maximizing fertilizer plant will now base its output decisions not only on the market price of fertilizer, but also on the market price of clean water. The fertilizer plant is confronted with a trade-off between the higher profits from unrestricted dumping and the loss of income from renting a clean-water river to the downstream brewery. If the marginal gain in profit from polluting is less than the loss of clean-water rental income, it will pay for the fertilizer plant to reduce its level of output, and pollution, to more socially optimal levels. In Figure 7.7, the maximum amount that the brewery is willing to pay for clean water is given by the *MEC* curve. As long as marginal revenue from selling clean water is greater than the marginal reduction in profits, it will pay for the fertilizer plant to reduce its output.

The reader may well argue that there is something fundamentally unfair about assigning private-property rights to the polluter. Just how this curious state of affairs might arise is an interesting question in its own right. It may have resulted from political rent-seeking activities by the owners of the fertilizer plant. **Political rent seeking** describes the situation in which one group attempts to gain special benefits from the government at the expense of taxpayers or some other group. This may be accomplished through the use of campaign contributions, political action committees, bribes, and so on. For now, it is important to recognize that assigning private-property rights will create a market for clean water, which results in a socially optimal level of fertilizer production and water pollution.

> **Political rent seeking** When one group attempts to gain special benefits from the government at the expense of taxpayers or some other group.

According to the Coase theorem, a socially optimal level of production does not depend on the particular assignment of private-property rights, although it will affect the distribution of social benefits. To see this, suppose that the government assigns ownership rights to the brewery. In this case, how much will the owners of the fertilizer plant be willing to pay the brewery for the right to pollute the river? The answer is that it will pay up to the point where the marginal cost of production, which includes the rental price of pollution, is equal to the market price of fertilizer.

The Coase theorem is significant precisely because it asserts that the assignment of private-property rights leads to a socially efficient level of production and, therefore, pollution. It does this by segregating issues of production efficiency and distributional equity. As we saw earlier, a solution to the tragedy of the commons resulted when the aristocracy seized control of common areas. Although overgrazing was reduced to socially efficient levels, income was redistributed from peasants to the landed gentry. This is similar to what we saw when private-property rights were assigned to the fertilizer plant. Be that as it may, regardless of who has the ownership rights to the river, the outcome is a socially efficient level of fertilizer, beer, and pollution. From a public-policy perspective, the assignment of private-property rights will affect the distribution of welfare, which may excite controversy about what constitutes a "fair" outcome.

CASE STUDY 7.1: THE GLOBAL CARBON TAX*

Scientists have estimated that for nearly twenty million years the earth's level of so-called greenhouse gasses, most notably carbon dioxide (CO_2), remained fairly constant. Beginning in the mid-1800s with the onset of the industrial revolution, however, the level of greenhouse gases began to rise at an exponential rate. Although many scientists believe that this increase is a natural phenomenon, many others believe that the increase resulted from the burning of fossil fuels, particularly coal and petroleum by-products. The problem is that this buildup may be the primary cause of what has come to be known as "global warming."

The Threat of Global Warming

The buildup of greenhouse gasses in the atmosphere may be transforming the planet into a gigantic greenhouse that traps heat and causes global temperatures to rise. It has been estimated that global warming has already reduced the depth of the winter polar ice cap by as much as 70 percent. According to these estimates, in the past fifty years 90 percent of the earth's glaciers have retreated significantly. If this trend continues, polar bears could become extinct by the end of the century—but that is not all.

As the white, reflective polar ice caps retreat, the earth's ability to reflect solar energy is reduced, which could accelerate global warming. In 2001, the Intergovernmental Panel on Climate Change (IPCC), which consisted of about 2,500 scientists from nearly a hundred countries, warned that unless greenhouse gas emissions are brought under control, average global temperatures could rise by as much as nine degrees Fahrenheit in the next century.

According to the IPCC forecasts, the melting polar icecaps could raise the sea level by as much as one meter by the year 2100, which will, for example, flood large areas of the U.S. eastern seaboard, including much of Florida. As temperatures rise, global weather patterns will become more turbulent, resulting in widespread loss of life and property damage. Changing global weather patterns could also redistribute the world's rainfall, which could lead to widespread flooding in some areas and severe droughts in others, resulting in crop failures and famine of mind-numbing proportions.

The only other similar episode in earth's history is the Permian mass extinction, which occurred about 230 million years ago. The release of greenhouse gases from volcanic eruptions and the resulting increase in global temperatures resulted in the extinction of 70 percent of animal species due to massive disruptions in the food chain. The problem is not just simply higher temperatures. A one-degree increase in average temperatures increases evaporation rates, which could result in a global shortfall of potable water.

Global Concerns

In May 1992, representatives of 172 governments convened in Rio de Janeiro to discuss global environmental problems. The United Nations Framework Conference on Environment and Development determined that the threat of global warming due to industrial emissions of greenhouse gases, most notably CO_2, methane (CH_4), nitrous oxides, water vapor and other chlorofluorocarbons, was one of the main threats confronting the human race.

The Rio Conference, also known as the Earth Summit, led to the United Nations Framework Convention on Climate Change (UNFCCC). The objective of the treaty was the "stabilization of greenhouse gas concentrations in the atmosphere at a level that would prevent dangerous anthropogenic interference with the climate system. Such a level should be achieved within a time-frame sufficient to allow ecosystems to adapt naturally to climate change, to ensure that food production is not threatened and to enable economic development to proceed in a sustainable manner" (UNFCCC 1992).

The UNFCCC called for the reduction of CO_2 and other greenhouse gas emissions to 1990 levels by the year 2000. It also called for a tax on nonrenewable energy sources, such as oil and coal. A plan put forth by the European Union called for a specific tax on CO_2 emissions—the so-called global carbon tax. It also proposed a general energy tax on the use of coal, oil, natural gas, and nuclear power.

In December 1997, an amendment to the UNFCCC was negotiated in Kyoto, Japan. The Kyoto Protocol to the UNFCCC identified mandatory targets for the reduction of greenhouse gas emissions. According to the Kyoto Protocol, Annex I (industrialized) countries are required to reduce greenhouse gas emissions an average of 5.2 percent below 1990 levels by 2012, although many scientists believe that a reduction of 50–60 percent from current levels is necessary by 2050. The Kyoto Protocol allows Annex I countries that have reduced greenhouse gas emissions below required levels to sell their remaining rights to pollute.

The Economics of Global Warming

According to Sir Nicholas Stern, the former chief economist of the World Bank, global warming represents "the biggest market failure the world has ever seen. The market hasn't worked because we haven't fixed it. Equity demands that the rich countries, who are largely responsible for this problem, do more about it" (Conway 2007).

The problem confronting policymakers is to determine the correct amount of the carbon tax. If this tax is too low, pollution will continue to be produced at levels that are not socially optimal. If the tax is too high, prices will be too high and production of socially desirable goods and services will be less than necessary to maximize social welfare. On the other hand, if the amount of the tax corresponds to the market price of pollution, a globally efficient level of pollution can be achieved.

Ronald Coase (1960) argued that a market for pollution rights is possible if private-property rights are well defined. Whoever "owns" the atmosphere can sell the right to pollute in the marketplace, which will force producers to internalize pollution costs and raise the private cost of production. A pollution market has the potential for identifying the socially optimal price of greenhouse gas emissions. A positive step in this direction took place in January 2005 when European carbon markets commenced operations. Under the Kyoto Protocol, corporations will be able to buy and sell pollution rights. The market-determined price of these pollution rights can be used to determine an optimal carbon tax.

* The author would like to thank Adam Webster of the University of Arizona for preparing this case study.

SHIFTING REACTION FUNCTIONS

Suppose that, for whatever reason, there is a change in the players' reaction functions? What effect will this have on the Nash equilibrium of games involving continuous strategies? To answer this question, let us return to the tragedy of the commons discussed earlier. Suppose that as a result of improved flock husbandry, the benefit accruing to shepherd A from grazing each sheep is:

$$B = 180 - s_A - s_B \qquad (7.11)$$

Proceeding as before, Shepherd A's reaction function becomes:

$$s_A = 90 - \frac{s_B}{2} \qquad (7.12)$$

This change is depicted as a right-shift in shepherd A's reaction function from R_A to R_A'.

Assuming that shepherd B's reaction function remains unchanged, the Nash equilibrium strategy profile for this game is $\{70, 40\}$. This new Nash equilibrium is depicted in Figure 7.8. Even though the marginal benefit to shepherd B remains unchanged, the strategic effect of improved husbandry methods adopted by shepherd A is reduced flock size. Shepherd B's best strategic response is to reduce the size of his or her flock.

Linear reaction functions of the type discussed in this chapter suggest that noncooperative static games involving continuous pure strategies have unique Nash equilibria. There is no guarantee, however, that reaction functions will be linear, in which case multiple Nash equilibria are possible. Even in these games, however, it may still be possible to identify a unique Nash equilibrium by mixing pure strategies, or because the players share a common understanding of the environment within which the game is being played (a focal point equilibrium). In the next chapter, these tools will be applied to an analysis of imperfectly competitive market structures. In fact, the concept of a reaction function for analyzing the strategic behavior of firms can be traced to the pioneering work of Augustin Cournot (1837), Joseph Bertrand (1883), and Heinrich von Stackelberg (1934).

FIGURE 7.8
**New Nash Equilibrium Resulting from a Change in Shepherd A's
Best-Response Function**

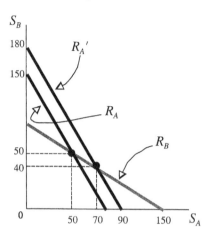

Demonstration Problem 7.3

Suppose that the total cost equations of the two firms in demonstration problem 7.1 are:

$$TC_1 = 100 - 5Q_1 \qquad (D7.3.1)$$

$$TC_2 = 100 - 10Q_2 \qquad (D7.3.2)$$

a. Calculate the best response functions for each firm. How do these reaction functions compare with the reaction functions in demonstration problem 7.1?
b. What is the new Nash equilibrium strategy profile? What are the market price and profit for each firm? How do these results compare with demonstration problem 7.1?

Solution

a. The best-response function for firm 1 remains:

$$Q_1 = 100 - 0.5Q_2 \qquad (D7.3.3)$$

The profit equation for firm 2, on the other hand, is:

$$\pi_2 = -100 + 995Q_2 - 5Q_2^2 - 5Q_1Q_2 \qquad (D7.3.4)$$

The first-order condition for profit maximization is:

$$\frac{\partial \pi_2}{\partial Q_2} = 995 - 10Q_2 - 5Q_1 = 0 \qquad (D7.3.5)$$

Solving Equation (D7.3.5) for Q_2, firm 2's best-response function is:

$$Q_2 = 99.5 - 0.5Q_1 \qquad (D7.3.6)$$

Compared with demonstration problem 7.1, firm 2's higher marginal cost of production causes its reaction to shift toward the origin.

b. Equations (D7.3.3) and (D7.3.6) may be solved simultaneously to yield the Nash equilibrium output levels $Q_1^*(Q_2^*) = 67$ and $Q_2^*(Q_1^*) = 66$. Compare this with the Nash equilibrium strategy profile in demonstration problem 7.1 of {66.67, 66.67}. The combined output of the two firms of 133 is less than the combined output in demonstration problem 7.1 of 133.34.

The market price for this Nash equilibrium strategy profile is:

$$P^* = 1{,}005 - 5(67 + 66) = \$340 \qquad\qquad (D7.3.7)$$

The market price for the Nash equilibrium strategy profile in demonstration problem 7.1 was \$338.30. The market price is predictably higher because output is lower because of firm 2's higher marginal cost of production.

The profit for each firm is:

$$\pi_1 = -100 + 1{,}005(67) - 5(67)^2 - 5(67)(66) = \$22{,}680 \qquad (D7.3.8)$$

$$\pi_2 = -100 + 995(66) - 5(66)^2 - 5(67)(66) = \$21{,}680 \qquad (D7.3.9)$$

The profit for each firm in demonstration problem 7.1 was \$22,121.12. Firm 1 did better and firm 2 did worse as a result of the increase in firm 2's marginal cost of production. The lower output and higher market price resulted in an increase in the combined profits from \$44,242.24 to \$44,360.

CHAPTER REVIEW

Static games with complete information may involve a discrete number of strategies, or a continuum of strategies. The prisoner's dilemma is an example of a noncooperative static game involving discrete pure strategies. By contrast, most strategic business and economic relationships, such as pricing and output decisions, advertising expenditures, product differentiation, research and development spending, capital investment, auction bids, bargaining positions, and so on, involve *continuous strategies.*

Nash equilibria for static games with continuous strategies can be found if all players have well-defined *reaction functions.* Also referred to as a *best-response function,* a reaction function is a formal statement of the optimal response by one player to strategies adopted by rivals.

A well-known example of a two-person, static game involving continuous strategies is the *tragedy of the commons,* which involves competition over scarce resources and the conflict between individual and communal interests. The fundamental problem in the tragedy of the commons is that private-property rights were not well defined, which results in an inefficient use of scarce resources, which reduces social welfare.

The misallocation of society's resources in the tragedy of the commons is an example of *market failure.* One possible solution to the tragedy of the commons is to establish well-defined private-property rights. The *Coase theorem* asserts that the assignment of private-property rights leads to socially efficient levels of production. It does this by separating the issues of production efficiency and distributional equity. The particular assignment of private-property rights will have income distribution implications, which may excite controversy about what constitutes a "fair" outcome.

Linear reaction functions suggest that noncooperative static games involving continuous pure strategies have unique Nash equilibria. When reaction functions are not linear, multiple Nash equilibria are possible. Even in these games, however, it may still be possible to identify a unique Nash equilibrium by mixing pure strategies, or because the players share a common understanding of the environment in which the game is being played (a focal-point equilibrium).

CHAPTER QUESTIONS

7.1   Explain the difference between static games involving discrete and continuous strategies.

7.2   Define what is meant by a player's best-response function.

7.3   In a static game involving two players with linear reaction functions, how might eliminating dominated strategies lead to a Nash equilibrium?

7.4   In a static game involving two players with continuous strategies, the simultaneous solution of linear reaction functions will always result in a unique Nash equilibrium. Do you agree with this statement?

7.5   What is the tragedy in the tragedy of the commons?

7.6   The tragedy in the tragedy of the commons is a form of market failure. Explain.

7.7   What is the Coase theorem and how might its application be applied to the tragedy of the commons? Cite a specific example.

7.8   In the tragedy of the commons we saw that improved methods of husbandry shifted shepherd A's reaction function, which resulted in a new Nash equilibrium. Can you think of any other factor that might have caused a right-shift or left-shift in shepherd A's best-response function?

CHAPTER EXERCISES

7.1   In the tragedy of the commons discussed in the text, we saw that improved methods of husbandry shifted shepherd A's reaction function, which resulted in a new Nash equilibrium.

   a.   What are the total benefits to each shepherd in a noncooperative game?
   b.   What are the total benefits to each shepherd if they agree to cooperate?

7.2   Suppose that benefits received by each shepherd from each sheep grazed is given by the equation:

$$B = 100 - s_A - s_B \qquad \text{(E7.2.1)}$$

   a.   Determine each shepherd's reaction function.
   b.   What is the Nash equilibrium strategy profile for this game?
   c.   What are the total benefits to each shepherd?
   d.   What is the strategy profile if the shepherds agree to cooperate?
   e.   Given your answer to part d, determine the total benefits to each shepherd.

7.3   Suppose that the benefits that the shepherds receive from each sheep is given by the equation:

$$B = 100 - s_A - 2s_B \qquad \text{(E7.3.1)}$$

   a.   What does the benefit function imply about the flock management by both shepherds?
   b.   Determine each shepherd's reaction function.
   c.   What is the Nash equilibrium strategy profile for this game?
   d.   What are the total benefits to each shepherd?
   e.   What is the strategy profile if the shepherds cooperate?
   f.   Given the above answers, determine the total benefits to each shepherd. How likely is cooperation between both shepherds?

7.4   An industry consisting of two firms faces the following market demand for their combined output:

$$Q_T = 10 - 0.1P \qquad \text{(E7.4.1)}$$

where $Q_T = Q_A + Q_B$. The total cost of production for each firm is:

$$TC_i = 4Q_i; \qquad i = A, B \qquad \text{(E7.4.2)}$$

a. Determine the best-response function for each firm.
b. What is the Nash equilibrium strategy profile for this game? What is the Nash equilibrium market price and profit for each firm?
c. Suppose that the two firms enter into a contract to behave jointly as a monopolist. Calculate the profit-maximizing output and market price. What is the profit for each firm?
d. How does your answer to part c compare with your answer to part b?

7.5   Suppose that the total cost functions of the two firms in chapter exercise 7.4 are:

$$TC_A = 4Q_A \qquad \text{(E7.5.1)}$$

$$TC_B = 8Q_B \qquad \text{(E7.5.2)}$$

a. What are the new reaction functions for the two firms?
b. What is the Nash equilibrium strategy profile for this game? Calculate the corresponding market price and profit for each firm.
c. How does your answer to part b compare with your answer to part b in chapter exercise 7.4?

CHAPTER

8

IMPERFECT COMPETITION

In this chapter we will:

- *Discuss the price- and output-setting behavior imperfectly competitive firms;*
- *Explore how game theory can be used to expand our understanding of strategic interaction between and among imperfectly competitive firms;*
- *Review output-setting strategies using the Cournot model;*
- *Review price-setting strategies using the Bertrand model;*
- *Investigate the Bertrand paradox;*
- *Discuss collusive price-setting behavior by imperfectly competitive firms.*

INTRODUCTION

One of a firm's most important decisions is how to price its product. A firm's ability to affect the market-determined price of its product is referred to as market power. A firm with market power is called a **price maker**. An industry comprised of a single firm has the greatest degree of market power because it controls total industry output. In general, a firm's market power is directly related to its ability to shift the market supply curve. It may also be able to exercise market power by shifting the market demand curve by changing consumer behavior, such as through advertising. A firm that does not have market power is referred to as a **price taker**. A price taker cannot significantly shift the market supply curve because its output is very small relative to the output of the entire industry.

> **Price maker** A firm with market power.
>
> **Price taker** A firm that does not have market power.

Market power is a by-product of **market structure**, which refers to the nature and degree of competition in an industry. The most significant characteristic of market structure is the number of firms in an industry. In an industry consisting of a very large number of firms, an individual firm has little or no market power. In this case, product price is not a decision variable. Managers will maximize profits by minimizing production costs. At the other extreme, an industry consisting of a single firm has the most market power. As the number of firms in an industry increases, the market power of each individual firm becomes proportionately less.

> **Market structure** The nature and degree of competition in an industry.

MARKET STRUCTURE

There are many ways to classify a firm's competitive environment. The standard textbook approach is to categorize markets in terms of certain basic characteristics that are used as benchmarks for a more detailed analysis of optimal price-setting and output-setting behavior. These characteristics include the *number and size distribution of sellers,* the *number and size distribution of buyers, product differentiation,* and the *conditions of entry into and exit from the industry.*

Number and Size Distribution of Sellers

When there are a large number of equivalently sized firms producing an identical product (perfect competition) or differentiated products (monopolistic competition), the ability of any single firm to affect the market-determined price is severely limited. On the other hand, if the industry consists of a single firm (monopoly) or a few large producers (oligopoly), market power can be considerable.

Number and Size Distribution of Buyers

When there are many small buyers of a given product, each will pay the same market-determined price. On the other hand, a single buyer or a few large buyers, such as federal or state governments, may be able to extract significant price concessions from suppliers.

Product Differentiation

Product differentiation refers to the degree to which the output of one firm differs from that of other firms in the same industry. When products are homogeneous, buyers' purchasing decisions are based mainly on the selling price. When products are differentiated, firms will charge different prices and strategic behavior is an important element in an analysis of price- and output-setting behavior. Automobiles and video game systems (Sony's PlayStation, Microsoft's Xbox, and Nintendo's Wii) are good examples of product differentiation and price dispersion.

Conditions of Entry and Exit

When it is difficult for firms to enter into an industry, incumbent firms tend to exercise a greater degree of market power and earn higher economic profits. When a firm enjoys patent protection or controls access to a vital raw material, for example, entry by rivals into the industry becomes impossible. The owner of a government patent, which is a common feature of the pharmaceutical industry, is a price-making monopolist. Industries dominated by a few large firms are typically characterized by significant capital expenditures and economies of scale. While entry by potential competitors is not impossible, it is very difficult. The automobile, airline, steel, aluminum, and petroleum-refining industries are often cited as examples.

OLIGOPOLY

The term **oligopoly** refers to an industry that is dominated by a few relatively large firms producing identical or closely related products in which entry into and exit from the industry is difficult. A special case of oligopoly is **duopoly,** which is an industry comprising just two firms. Imperfectly competitive firms fall somewhere in between perfectly competitive price takers and monopolistic price makers. The important aspect of imperfect competition is the extent to which the pricing, output, and other decisions by one firm affect, and are affected by, the decisions of rival firms.

> **Oligopoly** An industry dominated by a few large firms producing identical or closely related products in which pricing and output decisions of rivals are interdependent.

Duopoly An industry consisting of two firms producing homogeneous or differentiated products.

The pricing and output decisions of oligopolistic firms reflect the nature of competition in the industry. To be successful, each firm is required to know as much about its rivals' operations as it does about its own. Moreover, while entry into the industry is possible, it is very difficult. This allows these firms to earn positive economic profits for a significantly longer period of time than under perfect competition, where low barriers to entry suggest that incumbent firms' profits will be quickly competed away.

For oligopolies to earn economic profits in the long run, legal or economic barriers to entry must be very high. Bain (1956) argued that these barriers represent inherent advantages enjoyed by incumbent firms over latent competitors. Stigler (1968) argued that entry barriers comprise any costs not borne by incumbents that must be paid by potential rivals. Many of the barriers to entry erected by oligopolists are the same as those used by monopolists, such as the ownership of patents and copyrights. Oligopolists can also limit entry by controlling distribution outlets, such as by persuading retail chains to carry only its product. Persuasion may take the form of selective discounts, long-term supply contracts, or gifts to management. Product guarantees also serve as an effective barrier to entry. New car warranties, for example, typically require the exclusive use of authorized parts and service. Such warranties inhibit competition from producers of better and less-expensive goods and services.

The output of oligopolistic firms may be homogeneous or differentiated. The automobile industry, for example, is comprised of a relatively few large firms that produce similar, but differentiated, products. Even though the Honda Accord and the Ford Taurus are both mid-sized sedans, consumers discern clear differences between them. Firms in the steel and petroleum refining industries, on the other hand, produce uniform products. USX and Mitsubishi Steel produce high-quality cold-rolled steel. ExxonMobile and Royal Dutch Shell refine and market 87-octane gasoline.

The pricing and output behavior of each oligopoly affects, and is affected by, the pricing and output decisions of rival firms. This strategic interdependence is depicted in Figure 8.1 where DD is the market demand for the output of the entire industry, and dd is the demand for the output of each individual firm. The rationale behind Figure 8.1 is as follows. If all firms in the industry lower their price from P_1 to P_2, the market quantity demanded increases from Q_1 to Q_2. On the other hand, if a single firm reduces its price from P_1 to P_2 and the other firms do not, its sales increase from Q_1 to Q_3. This implies that over this price range, the demand curve facing the individual firm is more price elastic than the demand curve faced by the entire industry. The decision by one firm to unilaterally lower its selling price will result in a substantially larger market share, provided that its rivals do not follow suit—a dubious assumption, indeed.

MODELS OF OLIGOPOLY

The distinctive characteristic of oligopolies is the strategic relationship between and among firms in the industry. The difficulty in formulating models of strategic behavior is due to the many possible ways that managers of different firms interact. Beginning in the early nineteenth century, several models were developed to explain the strategic behavior of firms in the same or related industries. These models differ in terms of the underlying assumptions regarding the nature of a firm's decision-making process.

Sweezy ("Kinked" Demand Curve) Model

Although managers are aware of the law of demand, they are also aware that their pricing and output decisions depend on the pricing and output decisions of their rivals. Any change in prices

FIGURE 8.1

The Market Demand Curve for the Output of an Oligopolistic Industry and the Demand Curve for the Output of an Individual Oligopolist

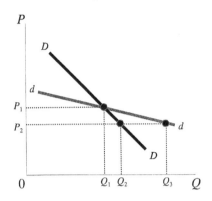

FIGURE 8.2

The "Kinked" Demand Curve (Sweezy) Model

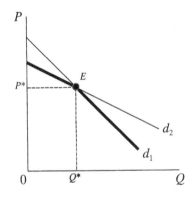

or production will provoke a response from competitors. According to Paul Sweezy (1939), this mutual interdependence will result in relatively infrequent price changes in oligopolistic industries. The **Sweezy model** argued that this price rigidity stemmed from the fact that oligopolists face a **"kinked" demand curve** for their product. This is depicted in Figure 8.2.

> **Sweezy model** A model of firm behavior that attempts to explain infrequent price changes in oligopolistic industries. The model postulates that a firm will not raise its price because this will not be matched by rivals and will result in a loss of market share. Neither will it lower its price since this will be matched by rivals, resulting in no gain in market share and a loss of revenues and profits.

According to Sweezy, the demand for the output of an oligopolist is made up of portions of two demand curves, which are labeled d_1 and d_2. This demand curve, which is characterized by a "kink" at point E, is depicted as the heavily shaded portions of d_1 and d_2 in Figure 8.2.

To understand what is going on, suppose initially that the oligopolist charges a price of P^*. If the oligopolist raises its price above P^*, it will lose market share because rivals *will not* follow suit. The loss of market share depends on the degree of product differentiation and the size of the oligopolist's loyal customer base. On the other hand, if the oligopolist attempts to increase its market share by lowering its price below P^*, this change *will* be matched by its rivals, who are unwilling to cede their own market share. In the figure, the firm will experience a small increase in sales from an increase in quantity demanded following an industry-wide decrease in prices. The gain in unit sales, however, will not compensate for lost revenues and profits because demand is less elastic for price declines than for price increases. Although the Sweezy model predates the development of modern game theory, the price P^* represents a Nash equilibrium in this price-setting game.

The "kinked" demand curve analysis has been criticized on two important points. For one thing, the analysis offers no insights into how the equilibrium price was determined in the first place. Moreover, empirical research has failed to verify the model's predictions. Stigler (1947), for example, found that firms in oligopolistic industries were just as likely to match price increases as price cuts.

─────────────── **Demonstration Problem 8.1** ───────────────

Suppose an oligopolist faces the "kinked" demand curve made up of the following equations:

$$Q_1 = 200 - 2P \qquad \text{(D8.1.1)}$$

$$Q_2 = 60 - 0.4P \qquad \text{(D8.1.2)}$$

Suppose further that the oligopolist's marginal cost is $50.

a. What are the oligopolist's price and output level?
b. Based on your answer to part a, what is the oligopolist's profit?
c. Within what range of values may marginal cost vary without affecting the market price and output level?
d. Diagram your answers to parts a, b, and c.

Solution

a. The price and output level for the oligopolist's product is determined at the "kink" formed by the intersection of the two demand curves. Solving Equations (D8.1.1) and (D8.1.2) simultaneously, we obtain the oligopolists' selling price $P^* = \$87.50$. At this price, the oligopolist produces:

$$Q^* = 200 - 2(87.50) = 60 - 0.4(87.50) = 25 \text{ units} \qquad \text{(D8.1.3)}$$

b. Since MC is constant, $MC = ATC$. By definition:

$$ATC = \frac{TC}{Q} \qquad \text{(D8.1.4)}$$

Rearranging Equation (D8.1.4) we get:

$$TC = ATC \times Q = MC \times Q = 50(25) = \$1,250 \qquad \text{(D8.1.5)}$$

Total revenue is:

$$TR = P \times Q = 87.50(25) = \$2,187.50 \qquad \text{(D8.1.6)}$$

The oligopolist's total profit is:

$$\pi = TR - TC = 2,187.50 - 1,250 = \$937.50 \qquad \text{(D8.1.7)}$$

FIGURE D8.1
Solution to Demonstration Problem 8.1

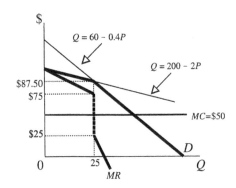

c. To determine the range of values within which marginal cost may vary without affecting the price and output level, we must first derive the firm's marginal revenue equation. Solving Equations (D8.1.1) and (D8.1.2) for price and substituting into $TR_i = PQ_i$, we obtain:

$$TR_1 = 100Q - 0.5Q^2 \qquad \text{(D8.1.8)}$$

$$TR_2 = 150Q - 2.5Q^2 \qquad \text{(D8.1.9)}$$

The corresponding marginal revenue equations are:

$$MR_1 = \frac{dTR_1}{dQ} = 100 - Q \qquad \text{(D8.1.10)}$$

$$MR_2 = \frac{dTR_2}{dQ} = 150 - 5Q \qquad \text{(D8.1.11)}$$

MR_1 is relevant for output levels between zero and 25 units. MR_2 is relevant for output levels greater than 25 units. The firm maximizes its profits where $MC = MR$. At $Q = 25$:

$$MR_1 = 100 - 25 = 75 \qquad \text{(D8.1.12)}$$

$$MR_2 = 150 - 5(25) = 25 \qquad \text{(D8.1.13)}$$

Thus, marginal cost may vary between 25 and 75 without affecting the prevailing (profit-maximizing) price and output level.

d. The above results are depicted in Figure D8.1.

___

Cournot Model

In the early-nineteenth century, French mathematician Augustin Cournot (1838) developed a model to explain how competing firms determined output levels to maximize profits. Although Cournot's model employed the techniques of standard optimization analysis, in retrospect his work represents an important contribution to literature on game theory. We will begin our discussion with a graphical description of the **Cournot model**. We will then reformulate the Cournot model as a static, output-setting game involving continuous strategies. The Cournot model is significant because it highlights the central role of strategic behavior in imperfectly competitive markets.

> **Cournot model** A static output-setting game in which firms in the same industry cannot subsequently switch strategies without great cost. In the Cournot model, prices adjust to firms' output decisions to clear the market.

The Cournot model is an output-setting game in which rival firms simultaneously determine their optimal output. An important assumption is that once output is determined, it cannot be changed, or at least not without significant economic cost. In the Cournot model, output is the only decision variable. Once the firms' output levels are determined, the market price adjusts to equate consumer demand with total industry supply. This functional relationship is summarized in Equation (8.1).

$$P = P(Q_T); \quad \frac{dP(Q_T)}{dQ_T} < 0 \tag{8.1}$$

where

$$Q_T = \sum_{i=1}^{n} Q_i \tag{8.2}$$

Our objective is to determine the Nash equilibrium output level for each firm in this one-time, static game in which player strategies are continuous. To simplify the analysis, we will assume that the industry comprises two profit-maximizing firms producing a homogeneous product. We will also assume linear market demand and constant marginal cost, which will be set equal to zero for diagrammatic convenience. These assumptions are depicted in Figure 8.3. We will analyze this static, output-setting game from firm A's perspective.

To maximize profits, firm A will produce at an output level where $MC = MR_A$. Since we have assumed that $MC = 0$, firm A will first consider producing at output level Q_A at a price of P_1. Since $MC = 0$, maximizing profits is equivalent to maximizing total revenue. Output level Q_A represents one-half of total industry supply Q' when $P = 0$. Will firm A actually produce at Q_A? In the Cournot model, only if it is the sole firm in the industry. But, in a two-firm industry, firm A must also consider the output level of firm B.

In the Cournot model, if firm A produces Q_A, the remainder of the market is available to firm B. This can be identified diagrammatically by shifting the price axis to the output level Q_A. The remaining demand curve facing firm B is given by the line segment ED. This is referred to as **residual demand**, which gives all possible price and output combinations for firm B given the level of output by firm A. In Figure 8.3, firm B's marginal revenue curve is MR_B. To maximize profits, firm B will produce where $MR_B = MC = 0$, which occurs at the output level $Q_B - Q_A$, which is $\frac{1}{2}Q_A$ or $\frac{1}{4}Q'$. If these are the output levels of both firms, total industry output will be $\frac{1}{2}Q' + \frac{1}{4}Q' = \frac{3}{4}Q'$. Do these output levels represent an optimal outcome for this game?

> **Residual demand** In the Cournot model, it is the possible price and output combinations for a firm given the output levels of rivals.

If firm B produces $\frac{1}{4}Q'$, the residual demand available to firm A is no longer Q' but Q_B, which is $\frac{3}{4}Q'$. Firm A's residual demand is given by the line segment FD in Figure 8.4. Firm A will maximize profits by producing $\frac{1}{2}$ of $\frac{3}{4}Q'$, or $\frac{3}{8}Q'$, which would result in a combined industry output of $\frac{1}{4}Q' + \frac{3}{8}Q' = \frac{5}{8}Q'$. For firm A, the output level $\frac{1}{2}Q'$ is strictly dominated by $\frac{3}{8}Q'$. This is still not the end of the story.

If firm A produces $\frac{3}{8}Q'$, firm B's residual demand is $\frac{5}{8}Q'$. Firm B will produce one-half of this, or $\frac{1}{8}Q'$. Proceeding as before, firm A's residual demand becomes $\frac{11}{16}Q'$. To maximize profits, firm A will produce one-half of $\frac{11}{16}Q'$, or $\frac{11}{32}Q'$ units of output, and so on. So, where does all this

FIGURE 8.3
Cournot Model

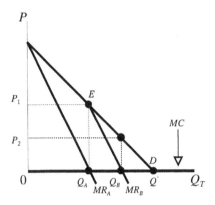

FIGURE 8.4
Determination of Market Shares in the Cournot Model

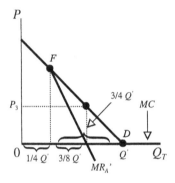

back-and-forth reasoning lead us? By eliminating strictly dominant output strategies, firm A will produce $\frac{1}{3}Q'$ at a market-determined price of P' (not shown). Because of the symmetry of the game, firm B will also produce $\frac{1}{3}Q'$. Each firm will also equally share total industry sales of $Q_T = \frac{2}{3}Q'$, where Q' is the level of output where $P = MC$.

The Cournot model can be generalized to include industries comprising more than two firms. Cournot demonstrated that for n identical firms, total industry output is:

$$Q_T = \frac{nQ'}{n+1} \tag{8.3}$$

where n is the number of firms in the industry.

Another way of finding a Nash equilibrium in the Cournot output-setting game is to identify each firm's reaction function (discussed in Chapter 7), as depicted in Figure 8.5. $Q_A{}^*(Q_B)$ is the reaction function for firm A and $Q_B{}^*(Q_A)$ is the reaction function for firm B. The Nash equilibrium for the Cournot model summarized in Table 8.1 occurs at the intersection of the firms' reaction functions, with each firm producing $\frac{1}{3}Q'$ units of output. Starting at point A in Figure 8.5, the output of firm A is $\frac{1}{2}Q'$ and the output of firm B is zero. Given firm A's output, firm B will produce $\frac{1}{4}Q'$, which occurs as point B. This illustrates the move from iteration 1 to iteration 2 in Table 8.1. If firm B produces $\frac{1}{4}Q'$, firm A's residual demand is $\frac{3}{4}Q'$, in which case firm A will maximize

FIGURE 8.5
Reaction Functions and Adjustment to a Cournot-Nash Equilibrium

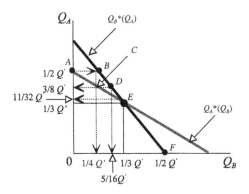

TABLE 8.1
Iterated Firm and Industry Output in the Cournot Model

Iteration	Q_A	Q_B	$Q_A + Q_B$
1	1/2 Q′	0	1/2 Q′
2	1/2 Q′	1/4 Q′	3/4 Q′
3	3/8 Q′	1/4 Q′	5/8 Q′
4	3/8 Q′	5/16 Q′	11/16 Q′
5	11/32 Q′	5/16 Q′	21/32 Q′
⋮	⋮	⋮	⋮
i	1/3 Q′	1/3 Q′	2/3 Q′

its profit by producing $\frac{3}{8}Q'$, which occurs at point C in Figure 8.5 (iteration 3 in Table 8.1). At this output, firm B's residual demand is $\frac{5}{8}Q'$. To maximize its profit, firm B will produce $\frac{5}{16}Q'$ (iteration 4), and so on. By systematically eliminating strictly dominated strategies, firm A will produce $\frac{1}{3}Q'$ units of output.

Due to the symmetry of this output-setting game, firm B begins by considering the output level $\frac{1}{2}Q'$ at point F and eliminates strictly dominated strategies until it, too, decides to produce $\frac{1}{2}Q'$ units of output. Thus, the Nash equilibrium strategy profile for this game is $\{\frac{1}{3}Q', \frac{1}{3}Q'\}$, which occurs at the intersection of the firms' reaction functions (point E).

The above analysis assumed that the demand for the output of each firm is identical, and that production occurs at zero marginal cost. Of course, neither of these assumptions is necessary. To see this, consider the following general description of the Cournot model involving two firms. Since the sum of the two firms' output equals the industry output ($Q_T = Q_1 + Q_2$), the market demand function is:

$$P = f(Q_T) = f(Q_1 + Q_2) \tag{8.4}$$

where Q_1 and Q_2 represent the outputs of firm 1 and firm 2, respectively. The total revenue functions are:

$$TR_1 = PQ_1 = Q_1 f(Q_1 + Q_2) \tag{8.5a}$$

$$TR_2 = PQ_2 = Q_2 f(Q_1 + Q_2) \tag{8.5b}$$

Each firm's profit equation is given by:

$$\pi_1 = Q_1 f(Q_1 + Q_2) - TC_1(Q_1) \tag{8.6a}$$

$$\pi_2 = Q_2 f(Q_1 + Q_2) - TC_2(Q_2) \tag{8.6b}$$

The basic behavioral postulate of the Cournot model is that each duopolist will maximize its profit given its rival's level of output. The necessary condition for firm 1 to maximize profit is:

$$\frac{\partial \pi_1}{\partial Q_1} = \frac{\partial TR_1}{\partial Q_1} - \frac{\partial TC_1}{\partial Q_1} = 0 \tag{8.7a}$$

or

$$MR_1 = MC_1 \tag{8.8a}$$

Similarly for firm 2:

$$\frac{\partial \pi_2}{\partial Q_2} = \frac{\partial TR_2}{\partial Q_2} - \frac{\partial TC_2}{\partial Q_2} = 0 \tag{8.7b}$$

or

$$MR_2 = MC_2 \tag{8.8b}$$

Assuming that the sufficient (second-order) conditions for profit maximization are satisfied, solving Equations (8.7) yield the firms' reaction functions:

$$Q_1 = Q_1^*(Q_2) \tag{8.9a}$$

$$Q_2 = Q_2^*(Q_1) \tag{8.9b}$$

Equation (8.9a) asserts that for any specified value of Q_2, the corresponding value of Q_1 maximizes π_1, and similarly for firm 2. The profit-maximizing values of Q_1 and Q_2 are depicted in Figure 8.6.

To make the above discussion more concrete, consider the following numerical example involving two firms producing a homogeneous product. The market-clearing price of the product, which is a function of total industry output in thousands of units, is given by the equation:

$$P = 25 - 0.5Q_T = 25 - 0.5(Q_1 + Q_2) \tag{8.10}$$

From Equation (8.10), the total revenue equations are:

$$TR_1 = 25Q_1 - 0.5Q_1^2 - 0.5Q_2Q_1 \tag{8.11a}$$

$$TR_2 = 25Q_2 - 0.5Q_2^2 - 0.5Q_1Q_2 \tag{8.11b}$$

Each firm's total cost of production is given by the equation:

$$TC_i = 10Q_i; \qquad i = 1, 2 \tag{8.12}$$

Combining Equations (8.11) and (8.12), total economic profit for each firm ($\pi_i = TR_i - TC_i$) is given by the equations:

$$\pi_1 = 15Q_1 - 0.5Q_1^2 - 0.5Q_2Q_1 \tag{8.13a}$$

FIGURE 8.6
Cournot-Nash Equilibrium

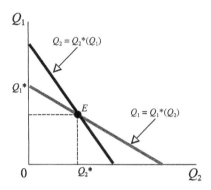

$$\pi_2 = 15Q_2 - 0.5Q_2^2 - 0.5Q_1Q_2 \tag{8.13b}$$

Taking the first partial derivative of Equations (8.13) with respect to each firm's output, and setting the resulting equation equal to zero is the necessary (first-order) conditions for profit maximization.

$$\frac{\partial \pi_1}{\partial Q_1} = 15 - Q_1 - 0.5Q_2 = 0 \tag{8.14a}$$

$$\frac{\partial \pi_2}{\partial Q_2} = 15 - Q_2 - 0.5Q_1 = 0 \tag{8.14b}$$

Solving Equations (8.14) yields each firm's reaction function.

$$Q_1 = 15 - 0.5Q_2 \tag{8.15a}$$

$$Q_2 = 15 - 0.5Q_1 \tag{8.15b}$$

Since both firms have identical total revenue and cost functions, simultaneously solving Equations (8.15) results in the profit-maximizing output levels $Q_1{}^*(Q_2{}^*) = Q_2{}^*(Q_1{}^*) = 10$, or 10,000 units. The Nash equilibrium strategy profile for this output-setting game is $\{10, 10\}$. Consider carefully the notation used in this solution. The profit-maximizing level of output for each firm is a function of the profit-maximizing level of its rival. This is because the reaction functions represent a complete description of the profit-maximizing firms' output strategies. Substituting these optimal values into Equation (8.10) results in a market-clearing price of $P^* = \$15$. Substituting these output levels into Equations (8.13) we obtain each firm's maximum profit of $\pi_1{}^* = \pi_2{}^* = 50$, or $50,000.

─────────────── **Demonstration Problem 8.2** ───────────────

Consider an industry comprising two firms producing a homogeneous product. The market demand and firms' cost functions are:

$$P = 200 - 2(Q_1 + Q_2) \tag{D8.2.1}$$

$$TC_1 = 4Q_1 \tag{D8.2.2}$$

$$TC_2 = 4Q_2 \qquad \text{(D8.2.3)}$$

a. Calculate each firm's reaction function.
b. Determine the market-clearing price, and output and economic profit for each firm.

Solution

a. The total revenue function for firms 1 and 2 are:

$$TR_1 = 200Q_1 - 2Q_1^2 - 2Q_1Q_2 \qquad \text{(D8.2.4a)}$$

$$TR_2 = 200Q_2 - 2Q_1Q_2 + 2Q_2^2 \qquad \text{(D8.2.4b)}$$

The firms' total economic profit equations are:

$$\pi_1 = 200Q_1 - 2Q_1^2 - 2Q_1Q_2 - 4Q_1 = 196Q_1 - 2Q_1^2 - 2Q_1Q_2 \qquad \text{(D8.2.5a)}$$

$$\pi_2 = 200Q_2 - 2Q_1Q_2 - 2Q_2^2 - 4Q_2 = 196Q_2 - 2Q_1Q_2 - 2Q_2^2 \qquad \text{(D8.2.5b)}$$

The necessary conditions for profit maximization are:

$$\frac{\partial \pi_1}{\partial Q_1} = 196 - 4Q_1 - 2Q_2 = 0 \qquad \text{(D8.2.6a)}$$

$$\frac{\partial \pi_2}{\partial Q_2} = 196 - 2Q_1 - 4Q_2 = 0 \qquad \text{(D8.2.6b)}$$

Solving Equations (D8.3.6) we obtain the firms' reaction functions:

$$Q_1 = 49 - 0.5Q_2 \qquad \text{(D8.2.7a)}$$

$$Q_2 = 49 - 0.5Q_1 \qquad \text{(D8.2.7b)}$$

b. Solving simultaneously Equations (D8.3.7) we get $Q_1{}^*(Q_2{}^*) = Q_2{}^*(Q_1{}^*) = 32.67$. Total industry output is $Q_1{}^* + Q_2{}^* = 65.34$. Substituting these results into Equations (D8.2.5) the profit of each firm is:

$$\pi_1{}^* = 196(32.67) - 2(32.67)^2 - 2(32.67)(32.67) = \$2,134 \qquad \text{(D8.2.8a)}$$

$$\pi_2{}^* = 196(32.67) - 2(32.67)^2 - 2(32.67)(32.67) = \$2,134 \qquad \text{(D8.2.8b)}$$

The market-clearing price is:

$$P^* = 200 - 2(32.67 + 32.67) = \$69.32 \qquad \text{(D8.2.9)}$$

Cournot-Nash Equilibrium

It is important to underscore the fact that Cournot assumes that each firm simultaneously determines its level of output without knowing the output chosen by its rivals. Moreover, once output decisions are made, they cannot be changed, or at least not without great cost. These assumptions transform the Cournot model into a noncooperative, static game in continuous strategies. In this setting, Cournot argued that firms would choose an output level corresponding to the intersection of their respective reaction functions. In the above example, the output for each firm is 10,000 units, with each firm earning a maximum profit of $50,000. This solution is depicted in Figure 8.7.

FIGURE 8.7
Cournot-Nash Equilibrium

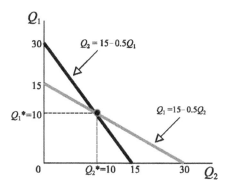

In the game depicted in Figure 8.7, the strategy profile {$10, 10$} is a Nash equilibrium if, and only if, each firm's strategy is to maximize its own profit given the profit-maximizing output strategy of its rival. This occurs if the output combination ($Q_1^*(Q_2^*), Q_2^*(Q_1^*)$) simultaneously lies on the reaction function of both firms, which is at the intersection of the firms' linear reaction functions. Since the solution profile {$Q_1^*(Q_2^*), Q_2^*(Q_1^*)$} is a Nash equilibrium for the Cournot output-setting game, we will refer to this as a **Cournot-Nash equilibrium**.

Cournot-Nash equilibrium A Nash equilibrium for a Cournot output-setting game.

A Cournot-Nash equilibrium is optimal only if the players are rational and adopt the strategy corresponding to the intersection of the players' reaction functions. Is it possible to rationalize some other strategy profile? To answer this question, we will simplify the Cournot output-setting game by first eliminating all dominated strategies. Once this is done, we will reexamine the simplified, output-setting game for additional dominated strategies and remove those as well. We will repeat this process until all dominated strategies have been eliminated. In the game depicted in Figure 8.7, which is derived from the reaction functions in Equations (8.15), this iterative process should result in the unique Cournot-Nash equilibrium strategy profile {$10, 10$} that was identified above using standard optimization analysis.

Since both firms earn zero economic profits by producing nothing, only positive levels of output will be considered. If there are any positive output levels that result in an economic loss, these dominated strategies will be eliminated. From Equation (8.10), neither firm will produce an output level greater than, or equal to, 50,000 units since this results in a zero price. What is more, there is no output level by firm 2 that will induce firm 1 to produce more than 15,000 units. The reason for this is that firm 1 will maximize profits by producing 15,000 units, but only if firm 2 produces nothing. In this example, 15,000 units is the profit-maximizing level of output for a monopolist. Thus, any output level greater than 15,000 constitutes a dominated strategy for firm 1 and must be eliminated. Because of the symmetry of this output-setting game, the same must also be true for firm 2. These dominated strategies are depicted by the heavily shaded lines above 15,000 units along both quantity axes in Figure 8.8.

After eliminating the dominated strategies in Figure 8.9, we then examine the simplified game in Figure 8.9 for additional dominated strategies. This simplified game is bounded from above by the dashed lines in Figure 8.8 at the 15,000 units of output. From Equation (8.15a), if firm 1 produces less than 15,000 units, firm 2 will always produce more than 7,500 units. Similarly, from Equation (8.15b), if firm 2 produces less than 15,000 units, firm 1 will also produce more than

FIGURE 8.8
First Iterated Elimination of Dominated Strategies

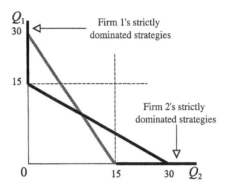

FIGURE 8.9
Second Iterated Elimination of Dominated Strategies

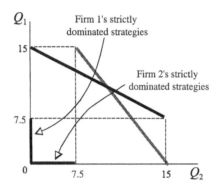

7,500 units. Thus, output levels less than 7,500 units, which are depicted as the heavily shaded lines in Figure 8.9, are dominated strategies and should also be eliminated.

After eliminating the dominated strategies in Figure 8.9, we are left with the simplified output-setting game in Figure 8.10. In the first iteration, the range of output levels not dominated is [0, 15]. In the second iteration, the range of nondominated output levels is narrowed to [7.5, 15]. In the third iteration, this range is narrowed further to [7.5, 11.25]. In the fourth iteration, the range narrows again [9.375, 11.25]. Proceeding in the same manner, in the fifth iteration (not shown), the range of output narrows further to [9.375, 10.3125]. Continuing in this manner, the range of nondominated output strategies shrinks to the limiting Cournot-Nash equilibrium strategy profile {*10, 10*}.

Dynamic Interpretation of the Cournot-Nash Equilibrium

In the previous paragraphs we assumed that each firm's output decision was determined simultaneously. The Cournot-Nash equilibrium was determined by eliminating strictly dominated strategies. It is possible, however, to interpret the Cournot model as a sequential move game. In fact, this was the interpretation first proposed by Cournot.

We begin by assuming that firm 1 chooses an output level in period 1. Firm 2 responds to firm 1's move by choosing an optimal level of output in period 2. Firm 1 responds to firm 2's choice

FIGURE 8.10
Third and Fourth Iterated Elimination of Dominated Strategies

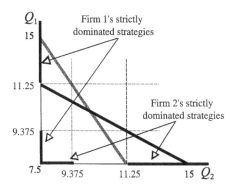

of output in period 3 by altering its optimal output level, and so on. Figure 8.5 can be used to illustrate this sequential process. The interesting thing about this dynamic process of move and countermove is that our choice of a starting period is irrelevant. The game will always converge to a Cournot-Nash equilibrium.

CASE STUDY 8.1: COURNOT-NASH EQUILIBRIUM IN THE CORN WET MILLING INDUSTRY

An interesting, real-world example of the Cournot output-setting model was offered by Porter and Spence (1982). Firms in the corn wet milling industry extract cornstarch and corn syrup from corn. Until the 1960s, the corn syrup industry might be described as a stable oligopoly. This changed when several new firms entered into the industry, including two of the world's largest processors of agricultural products: Archer Daniels Midland Company and Cargill. The immediate effect of the increase in capacity and competition was a significant decline in corn syrup prices. By the early 1970s, however, prices and output were back on the upswing.

In 1972, the production of high fructose corn syrup (HFCS) became commercially viable. HFCS can be used as a substitute for more expensive sugar in the production of many products, including soft drinks. Porter and Spence analyzed the process whereby firms in the corn wet milling industry expanded production capacity to accommodate the new HFCS technology. These authors simulated the competitive behavior of the industry's 11 largest producers. The simulations assumed that the firms' expansion decisions were based on industry capacity and price expectations for cornstarch, corn syrup, and HFCS. These assumptions are consistent with the Cournot output-setting model in which the production decision of each firm depends on the output decisions of rivals, and that prices adjust to industry output to clear the market.

The objective of the Porter and Spence study was to determine whether predicted HFCS capacity expansion in the corn wet milling industry were consistent with actual capacity expansion. If so, this would provide compelling evidence of the existence of a Cournot-Nash equilibrium. Table C8.1 summarizes the results of this study. Porter and Spence found statistically significant evidence of a Cournot-Nash equilibrium in the corn wet milling industry in the years 1973 and 1974. The results are less compelling for the years 1975 and

TABLE C8.1
**Actual and Predicted HFCS Capacity in the Corn Wet Milling Industry
(billions of pounds)**

	1973	1974	1975	1976	Post-1976	Total
Actual	0.6	1.0	1.4	2.2	4	9.2
Predicted	0.6	1.5	3.5	3.5	0	9.1

1976, although this may have reflected difficulties in accurately timing the firms' capacity expansion. In 1976, the industry had 4 billion pounds of HFCS capacity under construction, which came on line in the post-1976 period. For the entire period covered by their study, Porter and Spence predicted equilibrium capacity of 9.1 billion pounds. Actual industry capacity was 9.2 billion pounds. Overall, the Porter and Spence study appears to provide persuasive evidence that the behavior of firms in the corn wet milling industry is consistent with the assumptions of the Cournot output-setting model.

Bertrand Model: Homogeneous Products

If you are uncomfortable with the underlying assumptions of the Cournot model, you are not alone. One of the first economists to criticize the Cournot model was the nineteenth-century French mathematician and economist Joseph Bertrand (1883). Recall that the Cournot model assumes that firms simultaneously set output levels and that price adjusts to clear the market. By contrast, Bertrand argued that firms in the real world compete on the basis of price (not output), and that consumers respond by deciding how much to purchase from each firm. Unlike the Cournot model, in which firms decide how much to produce without knowing beforehand the amounts produced by rivals, the **Bertrand model** assumes that firms set their prices without knowing the prices charged by rivals. While Cournot assumes that the market price is a function of total industry output, Bertrand assumes that output and sales are a function of the prices charged.

> **Bertrand model** A static price-setting game in which firms in the same industry cannot subsequently switch strategies without great cost. In the Bertrand model, industry output adjusts to firms' pricing decisions.

The predictions of the Bertrand model depend on whether firms in the same industry produce a homogeneous or a differentiated product. Our discussion of the Bertrand model begins by assuming that firms in a duopolistic industry face a linear demand for a homogeneous product. We will also assume that both firms are identical and produce at constant marginal cost. Since the firms' output are perfect substitutes, consumers will buy from the firm charging the lowest price. More specifically, if $P_i < P_j$, the demand for firm i's product is $Q_i = Q_i(P_i)$ and $Q_j = 0$. If both firms charge the same price ($P_i = P_j = P$), the firms will split the market, that is, $Q_i = Q_j = \frac{1}{2}Q_T$.

Bertrand-Nash Equilibrium for Homogeneous Products

What is the optimal pricing strategy for each firm? Suppose that firm 1 expects firm 2 to set its price above the monopoly price P_m? Firm 1's best response is to charge the monopoly price. Firm 1 will sell to the entire market and earn monopoly profits. If firm 1 expects firm 2 to charge a price that is below the monopoly price but greater than marginal cost (MC), firm 1 will set its price just below that of firm 2. While firm 1 will not earn monopoly profits, it will earn positive economic profits since $P > MC$, while firm 2 will sell nothing and earn zero profits. Finally, if firm 2 sets its price below marginal cost, firm 1's optimal strategy is to set $P = MC$. Firm 1 will produce and sell nothing, while 2 earns an economic loss. Firm 1's reaction (best-response) function $P_1 = P_1^*(P_2)$ is depicted in Figure 8.11. For $P_2 > MC$, $P_1 = MC$. For $P_m > P_2 > MC$, firm 1 will set its price just below P_2. Finally, for $P_2 > P_m$, $P_1 = P_m$.

If firm 2 has the same marginal cost as firm 1, its reaction function $P_2 = P_2^*(P_1)$ is identical to that of firm 1. These reaction functions, which are symmetrical with respect to the 45-degree line, are depicted in Figure 8.12. The **Bertrand-Nash equilibrium** for this game occurs at the intersection of the two reaction functions at point E, where $P_1^*(P_2^*) = P_2^*(P_1^*) = MC$.

Is it possible for some other price $P' > MC$ to constitute a Bertrand-Nash equilibrium when both firms produce identical products? At this price, the two firms will split the market and earn positive economic profits of $0.5[Q'(P')(P' - MC)]$. This price, however, does not constitute a Nash equilibrium because either firm will be able to almost double its profit by charging a slightly lower price and capturing the entire market. From this we can conclude that the only possible equilibrium price is $P_1^*(P_2^*) = P_2^*(P_1^*) = MC$. All other prices are dominated strategies.

> **Bertrand-Nash equilibrium** A Nash equilibrium for a Bertrand price-setting game.

The Bertrand Paradox

The remarkable thing about price competition in industries consisting of two firms with the same marginal cost producing a homogeneous product is that the Nash equilibrium for this price-setting game is $P = MC$. This is precisely what we would expect to observe in industries comprising a large number of firms producing a homogeneous product with low entry and exit barriers. According to the Bertrand model, it takes just two firms to guarantee a perfectly competitive outcome. This remarkable prediction is referred to as the **Bertrand paradox**.

> **Bertrand paradox** A Nash equilibrium in which duopolies that produce a homogeneous product and have symmetric marginal cost charge a price equal to marginal cost and earn zero economic profit. This is the same outcome as in perfect competition.

The predictions of the Bertrand model are paradoxical because this outcome is not observed in reality. In the real world, firms in the same industry tend to charge a price greater than marginal cost, and as the number of firms in the industry declines, prices rise. By contrast, the Bertrand model predicts that firms charge a price equal to marginal cost, regardless of the number of firms in the industry, and earn zero economic profits. So how can we explain the discrepancy between what goes on in the real world and the predictions of the Bertrand model? There are at least three reasons why the predictions of the Bertrand model are not observed in the real world, including *capacity constraints, differentiated products,* and *collusion.*

Capacity Constraints

The Bertrand paradox depends crucially on the assumption that competing firms have the production capacity to satisfy total market demand at any price. If this condition is not satisfied, the Bertrand-Nash equilibrium is for each firm to charge the same price that is greater than marginal cost, in which case both firms earn positive economic profits.

To see how capacity constraints resolve the Bertrand paradox, we will assume that the output of each firm is capacity constrained at Q_i^0 below some price. As before, we will continue to maintain all of the same assumptions in the previous section. If both firms initially charge the same price,

FIGURE 8.11
Firm 1's Reaction Function

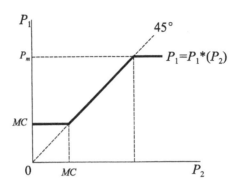

FIGURE 8.12
Bertrand-Nash Equilibrium for Homogeneous Products

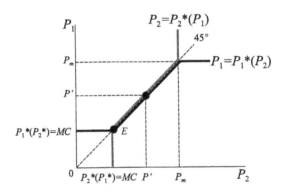

they will share the market. If firm 1 lowers its price by just a little, its sales will double provided that it has enough capacity to satisfy total market demand, while sales of firm 2 fall to zero. On the other hand, if firm 1 cannot supply enough output to satisfy total market demand, sales of firm 2 will not fall to zero. We will now argue that the Bertrand-Nash equilibrium for this game is for both firms to charge the same price, which is greater than marginal cost. When both firms are capacity constrained, each firm will earn a positive economic profit, otherwise exit the industry.

Let $P = P(Q)$ be the inverse of the market demand function. If both firms produce at capacity, the market price is $P' = P(Q_1^0 + Q_2^0)$. Suppose that firm 1 charges $P_1 = P'$. Can firm 2 do better by charging a different price? One possibility is for firm 2 to charge $P_2 < P_1 = P'$. This pricing strategy does not help firm 2 because it is already producing at capacity. In this case, firm 2 will suffer lower profits because marginal cost is constant.

Suppose, instead, that firm 2 charges $P_2 > P_1 = P'$. Unlike the Bertrand model in which both firms have unlimited capacity, firm 1's sales will not drop to zero. The reason for this is that firm 1 is producing at capacity and is unable to satisfy the increase in demand. This pricing strategy does not help firm 2. Recall that profit-maximizing behavior requires that a firm produce up to the output level where $MR = MC > 0$. This implies that demand is price elastic. Raising price, therefore, results in a decline in total profits because of the drop in total revenues at constant marginal cost. Thus, the optimal price for firm 1 is $P_1 = P'$. By analogous reasoning, the optimal price for firm

2 must be $P_2 = P'$. Thus, the Bertrand-Nash equilibrium for the price-setting game in which both firms are capacity constrained is $P_1 = P_2 = P' > MC$.

Bertrand Model: Differentiated Products

We have so far considered the predictions of the Bertrand price-setting model in which competing firms produce a homogeneous product. This form of market structure, however, is not often observed in the real world. Most industries are comprised of firms that produce close, not perfect, substitutes. Examples of industries comprising firms producing differentiated products abound, including clothing, soft drinks, beer, cosmetics, computer hardware, computer software, automobiles, petroleum refineries, fast-food franchises, and so on. McDonald's, Wendy's, Carl's Jr., White Castle, Burger King, Checkers, Sonic, and others all sell hamburgers, but the provender offered by fast-food restaurants are hardly identical. Not only do their products differ in terms of ingredients, taste, and appearance, the restaurants differ in style, motif, customer service, packaging, and so on.

Differences between and among products may be real or imagined. Regular (87-octane) gasoline has a precise chemical composition. Yet, many consumers willingly pay higher prices charged by ExxonMobile or Texaco because they believe their product is superior to those of less-well-known vendors. Firms in industries producing differentiated products often commit substantial sums to advertising, promotion, product placement, and packaging to reinforce real and perceived product differences. The goal is to create customer loyalty and reinforce brand-name identification, which makes a firm's product more price inelastic and increases the firm's market power.

One way to model strategic interaction among firms producing differentiated products is to specify separate demand equations for the output of each firm in the industry. In the two-firm case, these demand functions may be written:

$$Q_1 = f_1(P_1, P_2) \tag{8.16a}$$

$$Q_2 = f_2(P_1, P_2) \tag{8.16b}$$

Unlike the case of homogeneous products discussed above, with differentiated products neither firm is able to capture the entire market by undercutting its rival's price because some customers have a preference for the product of one firm over another. Even if one firm were to charge nothing, the other firm would still find it profitable to charge a positive price for its product because of customer or brand-name loyalty. In this setting, a firm that raises its price will experience a decline in the quantity demanded for its product, while the demand for the output of its rival will increase as some customers defect to the relatively cheaper good. This raises the second firm's marginal revenue, which will make it profitable to raise prices. To simplify the analysis, we will assume that both firms have the same total cost equation, which exhibits constant marginal cost. The firms' total revenue and total cost equations are given by Equations (8.17) and (8.18), respectively.

$$TR_1 = P_1 f_1(P_1, P_2) \tag{8.17a}$$

$$TR_2 = P_2 f_2(P_1, P_2) \tag{8.17b}$$

$$TC_1(P_1, P_2) = TC_2(P_1, P_2) \tag{8.18}$$

The firms' total profit equations are:

$$\pi_1 = P_1 f_1(P_1, P_2) - TC_1(P_1, P_2) \tag{8.19a}$$

$$\pi_2 = P_2 f_2(P_1, P_2) - TC_2(P_1, P_2) \tag{8.19b}$$

Bertrand-Nash Equilibrium for Differentiated Products

In the Bertrand model, the objective of each firm in the industry is to maximize Equations (8.19) with respect to its own selling price. Each firm assumes that its rival's price is invariant with respect to its own pricing decision. The necessary conditions for profit maximization are summarized in Equations (8.20).

$$\frac{\partial \pi_1}{\partial P_1} = \frac{\partial TR}{\partial P_1} - \frac{\partial TC_1}{\partial P_1} = 0 \qquad (8.20a)$$

$$\frac{\partial \pi_2}{\partial P_2} = \frac{\partial TR}{\partial P_2} - \frac{\partial TC_2}{\partial P_2} = 0 \qquad (8.20b)$$

The Bertrand-Nash equilibrium is a price that simultaneously maximizes the profit of both firms. The reaction functions are found by solving Equations (8.20) for each firm's price. These reaction functions, which are summarized by Equations (8.21), express the profit-maximizing price charged by each firm as a function of the price charged by its rival.

$$P_1 = P_1^*(P_2) \qquad (8.21a)$$

$$P_2 = P_2^*(P_1) \qquad (8.21b)$$

The reaction functions for firms 1 and 2 are depicted in Figure 8.13, respectively. Contrary to the reaction functions depicted in Figure 8.12, the differences in the shapes of these reaction functions reflect the differentiated nature of the firms' products. Unlike the reaction functions in the Cournot model, these reaction functions have different slopes. To understand why, suppose that firm 1 charges a zero price for its product. It would be in firm 2's best interest to charge at least $P_2^{min} = MC_2$. Thus, any price below P_2^{min} represents a dominated strategy because it results in negative economic profits. The same also holds true for firm 1. Any price below P_1^{min} also represents a dominated strategy. Now, if firm 2 charges a price above P_2^{min}, some of its customers will defect to firm 1. Firm 1 will respond by increasing its own price. Remember that each point along the firms' reaction function represents a profit-maximizing price given the price charged by the other firm. In the Bertrand model with differentiated products, equilibrium occurs at the intersection of the firms' reaction functions. This occurs at point E in Figure 8.13. Unlike the Bertrand model depicted in Figure 8.12, the equilibrium prices are greater than marginal cost, which implies that both firms will earn positive economic profit.

The Bertrand-Nash equilibrium strategy profile for the game depicted in Figure 8.13 is $\{P_1^*(P_2^*), P_2^*(P_1^*)\}$. The proof of this is straightforward. This strategy profile is a Nash equilibrium if, and only if, each firm's pricing strategy maximizes its profit given the profit-maximizing price charged by its rival. This is possible if, and only if, the coordinates $(P_1^*(P_2^*), P_2^*(P_1^*))$ simultaneously lie on both linear reaction functions. Since a rational player would never choose a dominated strategy, the Bertrand price-setting game can be simplified by eliminating all dominated strategies. After this is done, we then examine the simplified game for additional dominated strategies and remove them as well. This process is repeated until all dominated strategies have been eliminated.

Recall from above that firm 2 would never charge a price less than P_2^{min}. Any price strategy below P_2^{min} must be eliminated. It is also easy to see why any price above P_2^{max} must also be eliminated. This is the price that firm 2 would charge if it were a profit-maximizing monopolist. Any higher price would result in lower economic profits, and must be eliminated as a dominated strategy. These dominated strategies are depicted in Figure 8.14 by the heavily shaded lines along

FIGURE 8.13

Bertrand-Nash Equilibrium Occurs at the Intersection of the Firms' Reaction Functions

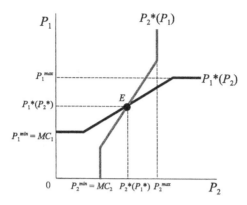

FIGURE 8.14

First and Second Iterated Elimination of Dominated Strategies

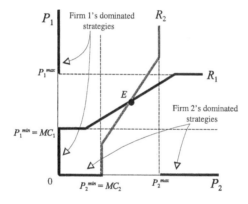

the price axes. Removing these strictly dominated strategies is the first iteration in the process of simplifying the price-setting game. By analogous reasoning, prices below P_1^{min} or above P_1^{max} represent dominated strategies for firm 1. Eliminating these dominated strategies is the second iteration in simplifying the Bertrand price-setting game. The new, simplified price-setting game is depicted in Figure 8.15. In this simplified game, prices below $P'_1{}^{min}$ and above $P'_1{}^{max}$ now represent strictly dominated strategies for firm 1.

In the reduced price-setting game in Figure 8.15, prices below $P'_2{}^{min}$ represent dominated strategies. The reason for this is that $P'_2{}^{min}$ is the minimum price that firm 2 would charge if it only sold its output to customers who would never buy from firm 1, even if firm 1 were giving away its product for free. Thus, prices below $P'_2{}^{min}$ are dominated strategies. The same also holds true for firm 1. Any price less than P_1^{min} represents a dominated strategy for firm 1. Prices above P, say, P_1^* and P_2^*, are also dominated strategies because it would pay for a firm to undercut its rival by some small amount to attract price-sensitive buyers. These dominated strategies are eliminated in the third and fourth iteration of the price-setting game. This process of iteratively eliminating dominated strategies continues until we are left with the Bertrand-Nash equilibrium solution profile $\{P_1^*(P_2^*), P_2^*(P_1^*)\}$.

FIGURE 8.15
Third and Fourth Iterated Elimination of Dominated Strategies

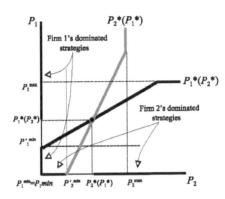

Demonstration Problem 8.3

Suppose that two firms producing a differentiated product are Bertrand price-setting competitors. The demand equations for the output of these profit-maximizing firms are:

$$Q_1 = 48 - 0.5P_1 + 0.25P_2 \tag{D8.3.1a}$$

$$Q_2 = 48 - 0.5P_2 + 0.25P_1 \tag{D8.3.1b}$$

The firms' total cost equations are:

$$TC_1 = 4Q_1 \tag{D8.3.2a}$$

$$TC_2 = 4Q_2 \tag{D8.3.2b}$$

a. Determine the reaction functions for each firm.
b. What is the Bertrand-Nash equilibrium strategy profile for this game?
c. Calculate the profits for each firm.

Solution

a. The total revenue equations for firms 1 and 2 are:

$$TR_1 = P_1Q_1 = 48P_1 - 0.5P_1^2 + 0.25P_1P_2 \tag{D8.3.3a}$$

$$TR_2 = P_2Q_2 = 48P_2 - 0.5P_2^2 + 0.25P_1P_2 \tag{D8.3.3b}$$

The total cost equations are:

$$TC_1 = 4Q_1 = 192 - 2P_1 + P_2 \tag{D8.3.4a}$$

$$TC_2 = 4Q_2 = 192 - 2P_2 + P_1 \tag{D8.3.4b}$$

The firms' profit equations are:

$$\pi_1 = TR_1 - TC_1 = -192 + 50P_1 - 0.5P_1^2 + 0.25P_1P_2 - P_2 \tag{D8.3.5a}$$

$$\pi_2 = TR_2 - TC_2 = -200 + 50P_2 - 0.5P_2^2 + 0.25P_1P_2 - P_1 \tag{D8.3.5b}$$

The necessary conditions for profit maximization are:

$$\frac{\partial \pi_1}{\partial P_1} = 50 - P_1 + 0.25P_2 = 0 \tag{D8.3.6a}$$

$$\frac{\partial \pi_2}{\partial P_2} = 50 - P_2 + 0.25P_1 = 0 \tag{D8.3.6b}$$

Solving Equations (D8.3.6) for each firm's price we get:

$$P_1 = 50 + 0.25P_2 \tag{D8.3.7a}$$

$$P_2 = 50 + 0.25P_1 \tag{D8.3.7b}$$

Solving the reaction functions simultaneously, the Bertrand-Nash equilibrium strategy profile for this game is {*$80, $80*}.

c. Substituting the Bertrand-Nash equilibrium prices into Equations (D8.3.5) we obtain:

$$\pi_1^* = \pi_2^* = -192 + 50(80) - 0.5(80)^2 + 0.25(80)(80) = \$2{,}208 \tag{D8.3.8}$$

The duopoly models discussed thus far have been criticized for the simplicity of their underlying assumptions. As E. H. Chamberlin (1933, 46) noted: "When a move by one seller evidently forces the other to make a countermove, he is very stupidly refusing to look further than his nose if he proceeds on the assumption that it will not." Chamberlin was quick to realize that mutual interdependence would lead imperfectly competitive firms to explicit or tacit alliances, in which they would charge monopoly prices and split the profits. In Chapter 4 we introduced the concept of trigger strategies in connection with the conditions necessary to bind firms to a collusive agreement. In the next section, we will apply these concepts to the Bertrand price-setting game to provide another possible resolution of the Bertrand paradox.

Collusion

To keep matters simple, we will assume that an industry consists of two Bertrand price-setting competitors that produce a homogeneous product. We have already seen that the simple Bertrand model predicts that both firms will engage in marginal cost pricing and earn zero economic profit. On the other hand, the firms could agree to form a cartel to coordinate their pricing strategies. We will model this game as a finitely repeated, static game in pure strategies with an uncertain end (see Chapter 5). This game, which is depicted in Figure 8.16, involves two players, firm A and firm B, and three pricing strategies, P^m, $P^m - \varepsilon$, and P^c.

If the firms behave as Bertrand competitors, they will set prices equal to marginal cost, that is, $P^c = MC$, and the payoff to each will be zero economic profits. On the other hand, if the firms agree to form a cartel and charge a monopoly price P^m, the industry as a whole will earn monopoly profits (π^m) and the payoff to each firm will be $\frac{1}{2}\pi^m$. Suppose, on the other hand, that firm B remains faithful to the agreement while firm A defects by charging a lower price $P^m - \varepsilon$, where ε is an arbitrarily very small number. By slightly undercutting firm B, firm A will capture the entire market while firm B sells nothing. In this case, firm B will produce and sell nothing, and will incur losses equal to its sunk costs ($-F$). By contrast, firm A will capture the entire market and earn $\pi^m - \nu$, which is only slightly less than π^m because of the slightly lower price charged. If firm A defects, firm B will retaliate by reverting to Bertrand price-setting competition in all subsequent stages of this game. In this case, the payoffs to both firms will be zero economic profits.

FIGURE 8.16

Collusion in the Bertrand Price-Setting Game

Firm B

		P^m	$P^m - \varepsilon$	P^c
	P^m	$(\frac{1}{2}\pi^m, \frac{1}{2}\pi^m)$	$(-F, \pi^m-\nu)$	$(-F, 0)$
Firm A $P^m - \varepsilon$		$(\pi^m-\nu, -F)$	$(\frac{1}{2}\pi^m-\nu, \frac{1}{2}\pi^m-\nu)$	$(-F, 0)$
	P^c	$(0, -F)$	$(0, -F)$	$(0, 0)$

Payoffs: (Firm A, Firm B)

The present value of the expected stream of payoffs for either firm from violating the agreement is $\pi^m - \nu$. The reason for this is that if one firm remains faithful and the other firm defects, the defector will earn $\pi^m - \nu$ in the first stage, but will earn zero profits in all subsequent stages as the players revert to Bertrand price-setting behavior. It is left as an exercise for the reader (see Chapter 5) to show that the present value of the expected stream of future payoffs by cooperating is $[\frac{1}{2}\pi^m (1 + i)/(i + \theta)]$, where i is the interest rate at which future payoffs can be reinvested (the discount rate) and θ is the probability that the game will come to an end in the next stage. For collusion to be viable, it must be the case that:

$$\frac{1}{2}\pi^m \frac{1+i}{i+\theta} > \pi^m - \nu \qquad (8.22)$$

Under the assumption of Bertrand price-setting competition, firms engage in marginal cost pricing. When games are infinitely repeated, or finitely repeated with an uncertain end, it would appear that there are real benefits for firms that engage in price fixing since this produces large and positive economic profits. In fact, collusion may constitute a Nash equilibrium provided that the discount rate and the probability that the game will come to an end in the next stage are sufficiently low. Even if the discount rate is sufficiently low, collusion may not be feasible if there is a high probability that the game will come to an end, such as might occur with an imminent failure of one of the firms.

Another reason why cartels might not endure is that collusion may not, in fact, constitute a Nash equilibrium. Although collusion may appear to be stable when the present value of the cooperation is greater than the present value of defection, the very concept may be unrealistic. If deviation from monopoly pricing is punished with a price war, the firm violating a collusive agreement has an incentive to reestablish the cartel at some future date. If the rivals agree, which seems reasonable since it results in a higher payoff, the threat of retaliation by means of a price war may not be credible. This underscores the importance of the underlying assumptions when attempting to predict the stability of collusive arrangements.

CASE STUDY 8.2: THE OPEC PRICE-FIXING CARTEL

Perhaps the best example of a price-fixing cartel is the Organization of Petroleum Exporting Countries (OPEC). OPEC was organized in 1960 by Saudi Arabia, Venezuela, Kuwait, Iraq, and Iran in response to efforts by a cartel of U.S. oil refiners, led by Standard Oil of New

Jersey, to reduce the price paid for imported crude oil. Until 1972 when OPEC imposed a boycott on crude oil exports as a geopolitical weapon against the U.S. in response to the crisis in the Middle East, the cartel had little effect on global oil prices. It was not until the early 1980s that OPEC adopted an official policy of manipulating world oil prices by restricting output for purely economic reasons.

The ability of OPEC to regulate world oil prices depends critically on the cartel's share of the world's oil supplies, but more importantly on the willingness of each member nation to abide by fixed production quotas. In 1982, OPEC established an 18-million-barrel-per-day production ceiling, which was down from 31 million barrels per day three years earlier, to maintain market prices at $34 per barrel. Except for Saudi Arabia, each member of the cartel was assigned an individual production quota. As the world's largest producer of crude oil, Saudi Arabia assumed the role of "swing producer." It would adjust its output as necessary to maintain world oil prices at their targeted level.

Maintaining cartel discipline has been difficult in the face of divergent national interests. During the 1980–1985 Iran–Iraq War, for example, both countries attempted to raise revenues to finance the conflict by increasing output. Despite Saudi Arabian efforts to maintain pro-duction level, the resulting oil glut caused world prices to plummet. Additional downward pressure on world oil prices occurred when OPEC's share of total output declined following the discovery of significant crude oil deposits in the North Sea by the British National Oil Company (BNOC). When BNOC slashed the price of North Sea oil to $3 per barrel in 1983, OPEC responded by reducing output by 3 percent and cutting the price by 15 percent. In addition to the development of North Sea oil reserves, the discovery and development of substantial reserves in the Gulf of Mexico and elsewhere has significantly reduced OPEC's share of the world's total supply of oil. In 2007, the 12-member cartel accounted for about 40 percent of global production, with oil prices hovering around $70 per barrel. This relatively high price of oil not only reflects efforts by OPEC to regulate the global supply of oil, but the very high rates of economic growth in China and India.

There have been many attempts to organize international commodities cartels over the years, including those for copper, tin, coffee, tea, and cocoa. A few of these attempts have had short-term success, such as in the bauxite and uranium industries. Even fewer cartels have had long-term success. Besides OPEC, perhaps the best-known example is the DeBeers diamond cartel (see Case Study 8.3). As a general rule, however, the success of cartels has been fleeting, primarily because of divergent member interests, and an inability to enforce agreements or punish violators.

CASE STUDY 8.3: BLOOD DIAMONDS

A classic example of a cartel is the diamond industry. The acknowledged global industry leader is DeBeers Consolidated Mines, Ltd., which has the diamond-mining monopoly in South Africa and interests in several other countries as well. Although the company's global market share is relatively small, especially since the discovery of diamonds in Russia, the company exercises considerable market power. The reason for this is DeBeers' London-based marketing subsidiary, the Central Selling Organization (CSO).

The role of the CSO is to serve as liaison between diamond mines, cutters, and polishers. More than 80 percent of the world's uncut diamonds go through CSO, only a fraction of which come from DeBeers' mines. CSO attempts to keep diamond prices high by regulating supply by adding to its inventory during periods of slack demand and reducing inventories when demand is robust. The high profit margins earned by DeBeers are a temptation to other mining companies to bypass CSO entirely by selling directly to diamond cutters. Two factors keep them from doing so. To begin with, not only do many diamond producers believe that the cartel is in the industry's best interest, but there is no incentive for any individual member to defect. As well as keeping prices high, DeBeers also accounts for a significant portion of the industry's total advertising expenditures. Other companies in the industry are able to "free ride" on DeBeers' price and advertising activities.

Another reason why individual mining companies do not defect is the fear of retaliation. In 1981, President Mobutu of Zaire announced that his country would no longer sell its diamonds through CSO. Instead, it signed contracts with London-based and Antwerp-based diamond brokers. Two months later, a flood of industrial diamonds into the market caused prices to fall by 40 percent. Although the source of the increased supply is uncertain, many believe that the culprit was DeBeers. Although this move was costly, DeBeers retaliation against maverick Zaire was successful. In 1983, the country requested renewal of its contract with CSO, which was granted on terms less favorable than the original contract.

Unfortunately, the DeBeers diamond cartel may have had unintended and tragic consequences. Spurred by high prices, blood diamonds, also known as conflict diamonds, conflict stones, and war diamonds, have been used by numerous rebel groups to fuel brutal civil wars in Angola, the Congo, Liberia, and Sierra Leone, which have resulted in the death, displacement, and dismemberment of millions of people. Blood diamonds have been used by groups such as al-Qaeda to finance terrorism and money laundering operations. Blood diamonds smuggled out of the rebel-held northern Côte d'Ivoire and eastern Congo are being used to fund organized crime at home and abroad.

In 1998 the United Nations launched a campaign against the purchase and sale of blood diamonds. This effort has placed the global diamond industry, which was previously shrouded in secrecy, under intense international scrutiny. In May 2000, major diamond trading and producing countries, representatives of the diamond industry, and special interest groups met in Kimberley, South Africa, to consider ways to tackle the conflict diamond problem. The meeting, which was hosted by the South African government, was the start of an important and often contentious three-year negotiating process to establish the Kimberley Process, an international diamond certification scheme. According to the World Diamond Council, by 1999 the illicit diamond trade had been reduced to just over 3 percent of global diamond production. By 2004, this figure was reported to have fallen to about 1 percent.

CHAPTER REVIEW

One of a firm's most important decisions is how to price its product. A firm's ability to affect the market-determined price of its product is called *market power.* A firm with market power is referred to as a *price maker.* Market power is a by-product of *market structure,* which refers to the nature and degree of competition in an industry.

The most significant characteristic of market structure is the number of firms in an industry. A firm operating in an industry consisting of a very large number of firms has little or no market power. As the number of firms in an industry declines, the market power of each individual firm becomes proportionately greater.

The term *oligopoly* refers to a few relatively large firms producing identical or closely related products in industries characterized by significant barriers to entry and exit. A special case of an oligopoly is *duopoly,* which is an industry comprising just two firms. In oligopolistic markets, the pricing and output decisions of firms are mutually interdependent, that is, output and other decisions by one firm affect, and are affected by, the decisions of rival firms.

One of the earliest models of strategic behavior of firms in imperfectly competitive industries is the *Sweezy* ("*kinked*" *demand curve*) *model.* According to this theory, firms will match price cuts by rivals to maintain market share, but will not match price increases to capture market share. Economists have questioned the "kinked" demand curve analysis because it provides no insights into how equilibrium prices are initially determined. Moreover, empirical research has generally failed to verify predictions of the model.

The *Cournot model* is an output-setting game in which rival firms simultaneously determine optimal levels of output. An important assumption is that once output is determined, it cannot be changed, or at least not without significant economic cost. In the Cournot model, output is the only decision variable. Once the firms' output levels are determined, the market price adjusts to equate consumer demand with total industry supply. In the Cournot model, the profit-maximizing level of output for each firm is a function of the profit-maximizing levels of output of rival firms. A *Cournot-Nash equilibrium* exists when firms simultaneously produce their profit-maximizing level of output.

Unlike the Cournot model, in which rival firms decide how much to produce without knowing beforehand the amounts produced by rivals, the *Bertrand model* assumes that firms set their prices without knowing the prices charged by rivals. While Cournot assumes that the market price is a function of total industry output, Bertrand assumes that output and sales are a function of the prices charged. A *Bertrand-Nash equilibrium* exists when firms simultaneously charge a profit-maximizing price for their product. In models in which all firms are identical, each firm will charge the same price, which is also equal to marginal cost.

The Bertrand price-setting model predicts that it takes just two firms to guarantee an imperfectly competitive outcome in which each firm earns zero economic profits. This prediction is referred to as the *Bertrand paradox* because this outcome is not generally observed in reality. In the real world, firms in the same industry tend to charge a price greater than marginal cost, and as the number of firms in the industry declines, prices rise. By contrast, the Bertrand model predicts that firms charge a price equal to marginal cost, regardless of the number of firms in the industry, and earn zero economic profits. There are at least three reasons why the predictions of the Bertrand model are not observed in the real world, including *capacity constraints, differentiated products,* and *collusion.*

CHAPTER QUESTIONS

8.1   Imperfectly competitive markets are characterized by interdependence in pricing and output decisions. Explain.

8.2   Oligopolies are characterized by a "few" firms in the industry. What is meant by a "few" firms, and when does a "few" become "too many?"

8.3   Product differentiation is an essential characteristic of oligopolistic market structures. Do you agree? Explain.

8.4   The "kinked" demand curve model suffers from the same weakness as the Cournot and Bertrand models in that it fails to consider the interdependence of pricing and output decisions of rival firms in oligopolistic industries. Do you agree? Explain.

8.5   The prisoner's dilemma is an example of a one-time, two-player, static game in pure strate-gies. If the players are allowed to cooperate, a Nash equilibrium is no longer possible. Do you agree with this statement? If not, then why not?

8.6   What is a reaction (best-response) function?

8.7   What is a Cournot-Nash equilibrium? Be specific.

8.8   What was Bertrand's criticism of the Cournot output-setting model? Do you agree with this criticism?

8.9   What is a Bertrand-Nash equilibrium? Be specific.

8.10   According to the Bertrand model, price competition results in each firm earning zero eco-nomic profit. What assumptions are necessary for this outcome?

8.11   What is the Bertrand paradox? What modifications can be made to the assumptions of the Bertrand price-setting model to resolve the paradox?

8.12   What important assumption about the demand for a firm's product is necessary in the Ber-trand price-setting model for differentiated products?

8.13   What is a cartel? In what way is an analysis of a cartel similar to an analysis of a monopoly?

8.14   Explain fully how collusive price-fixing agreements among firms may be used to resolve the Bertrand paradox. Be sure to discuss the role of discount rates and the probability that the price-setting game will come to an end.

8.15   In what way does the application of game theory to explain interdependent behavior among firms represent an improvement over earlier models of imperfect competition?

CHAPTER EXERCISES

8.1   Suppose that an oligopolist charges a price of $500 and sells 200 units of output per day. If the oligopolist increases its price above $500, quantity demanded will decline by 4 units for every $1 increase in price. On the other hand, if the oligopolist lowers price below $500, quantity demanded will increase by only 1 unit for every $1 decrease in price. If the marginal cost of production is constant, within what range may marginal cost vary without the profit-maximizing oligopolist altering price or output level?

8.2   Magnum Opus (MO) is an oligopolistic firm that faces a "kinked" demand curve for its product. If MO charges a price above the prevailing market price, the demand curve for its product is:

$$Q_1 = 40 - 2P \qquad \text{(E8.2.1a)}$$

On the other hand, if MO charges a price below the prevailing market price, it faces the demand equation:

$$Q_2 = 12 - 0.4P \qquad \text{(E8.2.1b)}$$

a. What is the market-clearing price for MO's product?
b. At the prevailing market price, what is MO's total output?
c. What is MO's marginal revenue function?

    d. If MO is a profit maximizer, at the prevailing market price what is the possible range of values for marginal cost?

8.3   Suppose that the demand function for an industry's output is $P = 55 - Q$. Suppose that the industry consists of two profit-maximizing firms with constant average total and marginal cost equal to \$5 per unit. Both firms are Cournot competitors.

    a. Determine the reaction function for each firm.
    b. What is the Cournot-Nash equilibrium for this game?
    c. What is the output and profit for each firm?

8.4   Consider the following market demand equation for the output of two firms that are Cournot competitors.

$$P = 100 - 5(Q_1 + Q_2) \qquad \text{(E8.4.1)}$$

The total cost equations for the two firms are:

$$TC_1 = 5Q_1 \qquad \text{(E8.4.2a)}$$

$$TC_2 = 5Q_2 \qquad \text{(E8.4.2b)}$$

    a. Determine each firm's reaction function.
    b. What is the Cournot-Nash equilibrium for this game?
    c. Given your answer to part b, what is each firm's output and economic profit?

8.5   Suppose that the inverse market demand equation for the homogeneous output of a two-firm industry is:

$$P = A - (Q_1 + Q_2) \qquad \text{(E8.5.1)}$$

The firms' total cost equations are:

$$TC_1 = B \qquad \text{(E8.5.2a)}$$

$$TC_2 = C \qquad \text{(E8.5.2b)}$$

where A, B, and C are positive constants. What is the profit-maximizing level of output for each firm?

8.6   Suppose that the demand equations for the product of two profit-maximizing firms in a duopolistic industry are:

$$Q_1 = 50 - 5P_1 + 2.5P_2 \qquad \text{(E8.6.1a)}$$

$$Q_2 = 20 - 2.5P_2 + 5P_1 \qquad \text{(E8.6.1b)}$$

The firms' total cost functions are:

$$TC_1 = 25Q_1 \qquad \text{(E8.6.2a)}$$

$$TC_2 = 50Q_2 \qquad \text{(E8.6.2b)}$$

Suppose that these firms are Bertrand competitors.

    a. What is each firm's reaction function?
    b. What is the Bertrand-Nash equilibrium for this game?
    c. What is the profit-maximizing output and profit for each firm?

8.7   This is an example of the Cournot model. Suppose that an industry consisting of two firms produces a homogeneous product. The market demand and individual firm cost equations are:

$$P = 200 - 2(Q_1 + Q_2) \tag{E8.7.1}$$

$$TC_1 = 2Q_1 \tag{E8.7.2a}$$

$$TC_2 = 4Q_2^2 \tag{E8.7.2b}$$

a. Calculate each firm's reaction function.
b. What is the Cournot-Nash equilibrium for this game?
c. Calculate the profit for each firm.

8.8   Suppose that an industry consists of two output-setting firms that produce a homogeneous product. Each must decide how much to produce without knowing how much its rival will produce. The industry demand equation is:

$$P = 145 - 5(Q_1 + Q_2) \tag{E8.8.1}$$

where Q_1 and Q_2 represent the outputs of firm 1 and firm 2, respectively. The total cost equations of the two firms are:

$$TC_1 = 3Q_1 \tag{E8.8.2a}$$

$$TC_2 = 5Q_2 \tag{E8.8.2b}$$

If neither firm can alter its level of output once determined:

a. Calculate each firm's reaction function.
b. Calculate the equilibrium price and profits for each firm. Assume that each duopolist is a profit maximizer.

8.9   Suppose that there are two firms producing an identical product for sale in the same market. The market demand for the product is given by the equation:

$$Q = 1,200 - P \tag{E8.9.1}$$

The marginal cost of product by both firms is $MC = 0$.

a. Firm 1 has 200 units of capacity and firm 2 has 300 units of capacity. What is the Bertrand-Nash equilibrium for this game?
b. Given your answer to part a, calculate each firm's total profit.

8.10   This is an example of the Bertrand model involving two firms producing differentiated products. Suppose that the demand functions for two profit-maximizing firms in a duopolistic industry are

$$Q_1 = 20 - 2P_1 + P_2 \tag{E8.10.1a}$$

$$Q_2 = 20 - 2P_2 + P_1 \tag{E8.10.1b}$$

Suppose, further, that the firms' total cost functions are

$$TC_1 = 10Q_1 \tag{E8.10.2a}$$

$$TC_2 = 10Q_2 \tag{E8.10.2b}$$

where P_1 and P_2 represent the prices charged by each firm producing Q_1 and Q_2 units of output.

a. What is the equilibrium price charged by both firms?
b. What is the equilibrium output for each firm?
c. What are the profits for each firm?

8.11 Suppose that the demand equations for the differentiated products of two profit-maximizing firms are:

$$Q_1 = 250 - 5P_1 + 2P_2 \qquad \text{(E8.11.1a)}$$

$$Q_2 = 250 - 5P_2 + 2P_1 \qquad \text{(E8.11.1b)}$$

Suppose, further, that the firms' total cost functions are

$$TC_1 = 25 + 5Q_1 \qquad \text{(E8.11.2a)}$$

$$TC_2 = 25 + 5Q_2 \qquad \text{(E8.11.2b)}$$

where P_1 and P_2 represent the prices charged by each firm producing Q_1 and Q_2 units of output. Both firms are Bertrand competitors.

a. What is the equilibrium price charged by both firms?
b. What is the equilibrium output of each firm?
c. What are the profits of each firm?

CHAPTER

9

PERFECT COMPETITION AND MONOPOLY

In this chapter we will:

- *Review the standard models of price and output determination under perfect competition and monopoly;*
- *Explore how game theory can provide insights into perfect competition and monopoly;*
- *Discuss the conditions for perfect competition and monopoly under Cournot output-setting behavior;*
- *Review some of the origins of monopoly power;*
- *Introduce the idea of contestable monopolies.*

INTRODUCTION

In economics we assume that decision makers attempt to maximize some objective function subject to one or more binding constraints. This assumption, which is referred to as bounded rationality, asserts that firms endeavor to maximize profit, or some other equally rational objective, such as maximizing market share, subject to a fixed operating budget, resources prices, market structure, and so on. In this chapter we will examine the market extremes of perfect competition and monopoly. In each case, we will begin with a review of the standard models of price and output setting behavior in which an analysis of strategic behavior plays little or no role. This will be followed with a brief discussion of how the principles of game theory may be used to enhance our understanding of these market structures.

STANDARD MODEL OF PERFECT COMPETITION

In the standard model, the term perfect competition is somewhat misleading since strategic interaction between and among firms in the same industry does not take place in the conventional sense. Managers do not explicitly consider the pricing and output decisions of rival firms. This follows from the strict application of conditions for perfectly competitive markets in which the output of any individual firm is very small relative to total industry supply. Each perfectly competitive firm is said to be a price taker. The market price is parametric to the output decision-making process. In other words, perfectly competitive firms are not Bertrand price-setting competitors. On the other hand, our earlier discussion of Cournot output-setting behavior can shed light on our understanding of perfectly competitive market structures.

The term **perfect competition** is used to describe an industry consisting of a large number of equivalently sized firms, each producing an identical good or service. Because the relative contribution of each firm to total industry output is very small, changes in output will not significantly shift the industry supply curve. Thus, perfectly competitive firms do not have the ability to affect the market-determined price.

FIGURE 9.1

Short-Run, Perfectly Competitive Equilibrium in Which the Individual Firm Earns a Positive Economic Profit

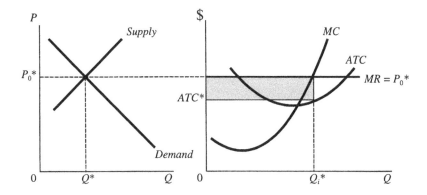

Perfect competition A market structure consisting of a large number of utility-maximizing buyers and profit-maximizing sellers of a homogeneous good or service in which factors of production are perfectly mobile, buyers and sellers have perfect information, and entry into and exit from the industry is very easy.

Another important characteristic of perfect competition is that firms produce a homogeneous product. The purchasing decisions of buyers, therefore, are based entirely on the selling price of the good or service. As a result, individual firms are unable to raise their prices above the market-determined price since they will be unable to attract buyers. Conversely, price cutting is counterproductive since perfectly competitive firms can sell all of their output at the higher, market-clearing price.

Perfect competition is also characterized by easy entry into, or exit from, the industry. This characteristic allows investors to easily shift resources between industries to exploit above-normal profit opportunities. Alternatively, below-normal or negative economic profits provide an incentive for investors to exit the industry and shift resources into industries where profits are higher.

Equilibrium Price and Output

In the standard model, the collective output decisions of individual firms affect the market-determined price of the industry's homogeneous good or service. The equilibrium price and quantity are determined at the intersection of the market demand and supply curves. On the other hand, an individual firm is powerless to affect the market price of its product. Revenues, costs, and profits are a function of output alone. This is depicted in Figure 9.1.

──────────────────── **Demonstration Problem 9.1** ────────────────────

Suppose that a perfectly competitive industry consists of 1,000 identical firms producing an identical product. The market demand (Q_D) and supply (Q_S) equations are:

$$Q_D = 170,000,000 - 10,000,000P \qquad \text{(D9.1.1)}$$

$$Q_S = 70,000,000 + 15,000,000P \qquad \text{(D9.1.2)}$$

  a. Calculate the market equilibrium price and quantity.
  b. Given your answer to part a, what is the output of an individual firm?
  c. Suppose that one of the firms exits the industry. What effect will this have on the equilibrium market price and quantity?

Solution

  a. Solving Equations (D9.1.1) and (D9.1.2) simultaneously, we obtain an equilibrium price of $P^* = \$4$. Substituting this price back into Equations (D9.1.1) or (D9.1.2) we obtain the equilibrium level of output $Q^* = 130,000,000$.
  b. Since there are 1,000 identical firms in the industry, the output of any individual firm Q_i is:

$$Q_i = \frac{Q^*}{1,000} = \frac{130,000,000}{1,000} = 130,000 \tag{D9.1.3}$$

  c. The supply equation of any individual firm in the industry is:

$$Q_i = \frac{Q^s}{1,000} = 70,000 + 15,000P \tag{D9.1.4}$$

Subtracting the output of the individual firm from market supply we obtain:

$$Q^s - Q_i = 69,930,000 + 14,985,000P \tag{D9.1.5}$$

Solving Equations (D9.1.1) and (D9.1.5) we obtain the new equilibrium price and quantity of $P^* = \$4.0052$ and $Q^* = 129,948,000$. This problem illustrates the virtual inability of an individual firm in a perfectly competitive industry characterized by a large number of firms to affect the equilibrium market price of a product by changing its level of output, which in this case amounts to reducing output from 130,000 units to zero.

Short-Run Profit-Maximizing Price and Output

It is easy to demonstrate that the necessary (first-order) condition for profit maximization in the short run is to produce up to the output level where the addition to total cost associated with producing one more unit of output (marginal cost) is equal to the addition to total revenue from its sale (marginal revenue).[1] This necessary profit-maximizing condition is summarized in Equation (9.1). Equation (9.1) is sometimes referred to as the golden rule of profit maximization since it applies to all firms regardless of market structure.

$$MR = MC \tag{9.1}$$

Since a perfectly competitive firm is too small to affect the market price, its only decision variable is its level of output. A perfectly competitive firm is a price taker. The firm's marginal revenue is equal to the market-determined selling price ($MR = P_0$).[2] The necessary condition for profit maximization for a perfectly competitive firm may be rewritten as:

$$P_0 = MC \tag{9.2}$$

This condition is depicted on the right-hand side of Figure 9.1.

The law of diminishing marginal product to a variable input implies that beyond some point, the firm's marginal cost curve increases at an increasing rate. If the firm experiences increasing returns in the early stages of production, this will generate a U-shaped average total cost curve. Figure 9.1 illustrates that a perfectly competitive firm maximizes its profits in the short run by producing at the output level Q_i^* at which $P_0^* = MC$. The firm's economic profit is shown as area of the shaded rectangle. This is because $\pi(Q) = TR(Q) - TC(Q)$, where $TR = P_0 \times Q$ and $TC = ATC \times Q$. In general, a firm earns a positive economic profit (above-normal rate of return) whenever price is greater than average total cost.

───────────────── **Demonstration Problem 9.2** ─────────────────

Consider the following total monthly cost equation of a perfectly competitive firm:

$$TC = 1,000 + 2Q + 0.01Q^2 \qquad\qquad (D9.2.1)$$

The market price of the product is $10. What is the firm's monthly profit-maximizing level of output? What is the firm's monthly profit?

Solution

Profit is maximized at the output level where $MR = MC$. The firm's marginal cost equation is:

$$\frac{dTC}{dQ} = MC = 2 + 0.02Q \qquad\qquad (D9.2.2)$$

For a perfectly competitive firm $P = MR$. Thus, profit-maximizing level of output is determined by setting $P^* = MC$, that is:

$$10 = 2 + 0.02Q \qquad\qquad (D9.2.3)$$

Solving Equation (D9.2.3), the profit-maximizing level of output is $Q^* = 400$. Total economic profit is:

$$\pi = TR - TC = \$10(400) - (1,000 + 2(400) + 0.01(400)^2) \qquad (D9.2.4)$$

Long-Run Profit-Maximizing Price and Output

The situation depicted in Figure 9.1 represents profit maximization for a perfectly competitive firm in the short run. Positive economic profits will attract new firms and productive resources into the industry, which causes the market supply curve to shift to the right, causing the market equilibrium price to fall. This is depicted in Figure 9.2. Likewise, the existence of negative economic profit provides an incentive for firms and productive resources to exit the industry. This will result in a leftward shift of the market supply curve and an increase in the market equilibrium price (not shown in Figure 9.2).

In Figure 9.2, a right-shift of the market supply curve results in a decline in the market price from P_0^* to P_0' and an increase in the equilibrium output from Q^* to Q', which are the break-even price and output for the individual perfectly competitive firm. In this case, the firm earns zero economic profit.[3] There is no longer an incentive for firms to enter the industry. The industry depicted in Figure 9.2 is in long-run competitive equilibrium.

FIGURE 9.2

**Long-Run Perfectly Competitive Equilibrium in Which the Individual Firm
Earns Zero Economic Profit**

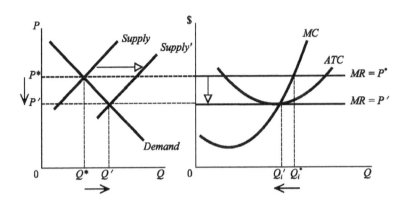

--- **Demonstration Problem 9.3** ---

A perfectly competitive industry consists of 300 firms with identical cost structures. The respective market demand (Q_D) and market supply (Q_S) equations for the good produced by this industry are:

$$Q_D = 3{,}000 - 60P \tag{D9.3.1}$$

$$Q_S = 500 + 40P \tag{D9.3.2}$$

   a. What are the profit-maximizing price and output for each firm?
   b. Assume that each firm is in long-run competitive equilibrium. Determine each firm's total revenue, total economic cost, and total economic profit.

Solution

   a. Firms in a perfectly competitive industry are characterized as "price takers." The profit-maximizing condition for firms in a perfectly competitive industry is $P = MC$. The market equilibrium price $P^* = \$25$ is found by setting $Q_D = Q_S$. The market-equilibrium output $Q^* = 1{,}500$ is found by substituting this result into Equation (D9.3.1) or (D9.3.2). Since there are 300 firms in the industry, each firm supplies $Q_i = 1500/300 = 5$ units.
   b. The total revenue of each firm in the industry is:

$$TR = P^*Q_i = 25(5) = \$125 \tag{D9.3.3}$$

In the long run, each firm earns zero economic profit. Since economic profit is defined as the difference between total revenue and total economic cost, the total economic cost of each firm is $125.

--- **Demonstration Problem 9.4** ---

A perfectly competitive firm faces the following total variable cost equation:

$$TVC = 150Q - 20Q^2 + Q^3 \tag{D9.4.1}$$

Below what price should this firm shut down?

Solution

The shut-down level of output corresponds to minimum average variable cost. Average variable cost is given by the equation:

$$AVC = \frac{TVC}{Q} = 150 - 20Q + Q^2 \qquad (D9.4.2)$$

To determine the output level that minimizes the firm's average variable cost, take the first derivative of Equation (D9.4.2) with respect to output and set the result to zero.

$$\frac{dAVC}{dQ} = -20 + 2Q = 0 \qquad (D9.4.3)$$

Solving Equation (D9.4.3), the shut-down level of output is $Q_{SD} = 10$ units. The profit-maximizing firm produces where $P = MC$. To obtain the shut-down price, substitute Q_{SD} into the marginal cost function.

$$P = MC = \frac{dTVC}{dQ} = 150 - 40Q + 3Q^2 = 150 - 40(10) + 3(10)^2 = \$50 \qquad (D9.4.4)$$

From Equation (D9.4.4), if the price falls below $50 per unit, the firm should shut down.

Demonstration Problem 9.5

Trim and Fit Inc. (TF) is a small company in the highly competitive food supplements industry. The market determined price of a 100-tablet vial of TF's most successful food supplement, *Forever Young,* is $10. TF's total cost (*TC*) function is:

$$TC = 100 + 2Q + 0.01Q^2 \qquad (D9.5.1)$$

a. What is TF's profit-maximizing level of output? At this output level, what is TF's total profit? Is TF in short-run or long-run competitive equilibrium? Explain.
b. What is TF's break-even level of output?

Solution

a. TF's total profit equation is:

$$\pi = PQ - TC = 10Q - (100 + 2Q + 0.01Q^2) \qquad (D9.5.2)$$

The necessary condition for profit maximization is:

$$\frac{d\pi}{dQ} = 8 - 0.02Q = 0 \qquad (D9.5.3)$$

Solving Equation (D9.5.3) for Q, the profit-maximizing level of output is 400 units of output. Substituting this result into Equation (D9.5.2), TF's maximum profit is:

$$\pi^* = -100 + 8(400) - 0.01(400)^2 = \$1,500 \qquad (D9.5.4)$$

TF is in short-run competitive equilibrium since it is earning positive economic profit. This will attract new investment. This will shift the market supply curve to the right, which causes the market equilibrium price to fall.

b. In the long run, perfectly competitive industry firms earn zero economic profits, in which case there is no incentive for new firms to enter or exit the industry. The firm will break even at the output level where the market price is equal to minimum average total cost. ATC is given by the equation:

$$ATC = \frac{TC}{Q} = \frac{100 + 2Q + 0.01Q^2}{Q} = 100Q^{-1} + 2 + 0.01Q \qquad (D9.5.5)$$

Minimizing Equation (D9.5.5) with respect to output we obtain:

$$\frac{dATC}{dQ} = -100Q^{-2} + 0.01 = 0 \qquad (D9.5.6)$$

Solving Equation (D9.5.6) we find that the break-even level of output is 100 units. Substituting this result into Equation (D9.5.5) we get:

$$P_{be} = ATC_{min} = -100Q^{-1} + 2 + 0.01Q = \frac{100}{100} + 2 + 0.01(100) = 4 \qquad (D9.5.7)$$

For this break-even price and output to constitute a long-run equilibrium, the firm's total economic profit must be zero. Substituting this result into Equation (D9.5.2) we obtain:

$$\pi = 4(100) - [100 + 2(100) + 0.01(100)^2] = 0 \qquad (D9.5.8)$$

PERFECT COMPETITION AND GAME THEORY

The preceding paragraphs summarize the classic textbook treatment of price and output determination in perfectly competitive markets. In this model, the market consists of a large number of buyers whose objective is to maximize total utility per period of time subject to a fixed household budget. On the supply side, the market consists of a large number of equivalently sized firms with the objective of maximizing profits subject to a fixed operating budget. In the standard model, the interaction between buyers and sellers boils down to a collection of noncooperative games in which buyers and sellers (the players) make decisions (adopt strategies) to maximize their respective objective functions (the payoffs).

Although the relationship between the buyers and sellers is clearly rivalrous, a central tenet of perfect competition is that the actions (moves) of an individual player have no effect on the behavior of rivals. Individual consumers and producers in perfectly competitive markets cannot unilaterally affect the market price of a good or service. Perfectly competitive firms do not, for example, engage in price wars. Thus, players do not think strategically in this noncooperative game. This would seem to suggest that game theory has little to offer in advancing our understanding of price and output determination in perfectly competitive markets. Nothing, however, could be further from the truth.

Game theorists have been able to demonstrate the existence of equilibria in perfectly competitive markets, although the manner in which this was done might appear to be somewhat contrived. To demonstrate the existence of equilibria in perfectly competitive markets, game theorists introduced the fictitious player Nature into this noncooperative game. Nature is not an economic agent in the usual sense that it attempts to maximize some objective function subject to binding constraints. Rather, the objective of benevolent Nature is to minimize market disequilibrium. The strategies adopted by buyers and sellers in this noncooperative game have no effect on the behavior of rivals. As a consequence, these strategies play no role in Nature's decisions.

Although the introduction of benevolent Nature enables game theorists to prove the existence of equilibria in perfectly competitive markets, this somewhat creative approach might be somewhat less than satisfying. On the other hand, it has been demonstrated that perfect competition represents a special case of a more conventional approach to explaining competitive behavior between and among firms in an industry. This more general approach will be discussed in the next section.

PERFECT COMPETITION AND THE COURNOT MODEL

In the previous chapter we introduced the Cournot output-setting model to advance our understanding of strategic behavior in a two-firm industry. As it turns out, perfect competition may be viewed as a special case of this model of strategic behavior. In this section, we will demonstrate that as the number of firms in an industry becomes very large, the market share of each becomes very small. In the extreme, the Cournot-Nash equilibrium collapses to the standard model of perfect competition. To see how, consider again the Cournot-Nash equilibrium depicted in Figure 8.7, which is replicated in Figure 9.3.

The difference between Figures 8.7 and 9.3 is the addition of line segment $0EN$, which satisfies the equation $Q_1 = Q_2$. Along this line segment, each firm produces the same output. Point E identifies firm 1's optimal level of output given the output level chosen by firm 2. Now, consider what happens as the number of firms in this industry increases. Recall that in the Cournot duopoly model, the market price is a function of the combined output of both firms in the industry. Consider the profit function of firm 1 in a multifirm industry:

$$\pi_1 = Q_1 f(Q_T) - TC_1(Q_1) \tag{9.3}$$

Equation (9.3) asserts that firm 1's total profit depends on total industry output (Q_T). Firm 1's reaction function may be written as:

$$Q_1 = Q_1^*(Q_{T-1}) \tag{9.4}$$

where Q_{T-1} represents the total output of the remaining $T-1$ firms in the industry, or:

$$Q_{T-1} = \sum_{i=2}^{T} Q_i \tag{9.5}$$

To keep the discussion simple, we will continue to assume that every firm in the industry produces the same level of output. Figure 9.4 replicates Figure 9.3 but replaces firm 2's reaction function with the reaction function for the remaining $T-1$ firms. The vertical and horizontal axes in Figure 9.4 measure output of firm 1 and remaining $T-1$ firms, respectively. The intersection of the reaction functions identifies the Cournot-Nash equilibrium output levels in this industry. In general, the equation of the linear isoquant from the origin is:

$$Q_1 = \frac{1}{(T-1)} Q_{T-1} \tag{9.6}$$

Alternatively, the Cournot-Nash equilibrium may be determined from the intersection of firm 1's reaction function and Equation (9.6).

FIGURE 9.3
Cournot-Nash Equilibrium for a Two-Firm Industry

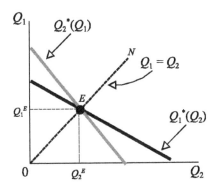

FIGURE 9.4
Cournot-Nash Equilibrium for a T-Firm Industry

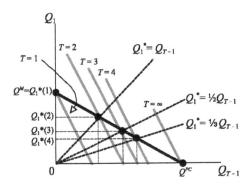

It should be clear from Figure 9.4 that as the number of firms in the industry increases, the isoquant described by Equation (9.6) becomes flatter. As T becomes very large, the share of output going to firm 1, and every other firm in the industry, becomes very small. In the limit, as the number of firms approaches infinity, the output of each individual firm approaches zero.

Although the output of each individual firm declines and the number of firms in the industry increases, total industry output increases. The limiting value is the perfectly competitive level of output (Q^{PC}) in Figure 9.4. At the other extreme, when there is only one firm in the industry, the limiting value occurs where firm 1 produces Q^M units of output.

———————————————— **Demonstration Problem 9.6** ————————————————

The demand for the output of a multifirm industry producing a homogeneous product is given by the equation:

$$Q_T = 100 - 5P \tag{D9.6.1}$$

where Q_T represents total industry output. Suppose that the total cost of production by any individual firm in the industry is:

$$TC_i = 5Q_i \tag{D9.6.2}$$

a. Define the total number of firms in the industry as $Q_T = Q_1 + Q_{T-1}$, where Q_1 is the output of firm 1 and Q_{T-1} is the output of the remaining $T-1$ firms in the industry. If all profit-maximizing firms in the industry are the same size, determine the reaction functions of firm 1 and the remaining $T-1$ firms in this industry.

b. Determine the Nash-Cournot equilibrium level of output for each firm in an industry consisting of 2, 3, 5, and 10 equally sized firms.

Solution

a. The inverse of Equation (D9.6.1) is:

$$P = 20 - 0.2Q_T = 20 - 0.2(Q_1 + Q_{T-1}) \tag{D9.6.3}$$

The total revenue equation for firm 1 is:

$$TR_1 = PQ_1 = 20Q_1 - 0.2Q_1^2 - 0.2Q_1Q_{T-1} \tag{D9.6.4}$$

The total profit equation for firm 1 is:

$$\pi_1 = TR_1 - TC_1 = 15Q_1 - 0.2Q_1^2 - 0.2Q_1Q_{T-1} \tag{D9.6.5}$$

The necessary condition for profit maximization is:

$$\frac{\partial \pi_1}{\partial Q_1} = 15 - 0.4Q_1 - 0.2Q_{T-1} = 0 \tag{D9.6.6}$$

To obtain firm 1's reaction function, solve Equation (D9.6.6) for Q_1.

$$Q_1 = 37.5 - 0.5Q_{T-1} \tag{D9.6.7}$$

In an analogous manner, the reaction function for the remaining $T-1$ firms is:

$$Q_{T-1} = 37.5 - 0.5Q_1 \tag{D9.6.8}$$

b. Solving Equation (9.6) for Q_{T-1} we get:

$$Q_{T-1} = (T-1)Q_1 \tag{D9.6.9}$$

Substituting Equation (D9.6.9) into Equation (D9.6.7) and solving for Q_1 we obtain:

$$Q_1 = \left[\frac{1}{0.5(1+T)}\right] 37.5 \tag{D9.6.10}$$

Equations (D9.6.9) and (D9.6.10) can be used to determine the Nash-Cournot equilibrium level of output for firm 1, which is equal to the output of every other firm in the industry, and the remaining $T-1$ firms in the industry. It is left as an exercise for the student to show that when there are $T = 2$ firms in the industry, each firm will produce 25 units of output. When there are $T = 3$ firms in the industry, firm 1 will produce 18.75 units of output. The output of the remaining firms in the industry from Equation (D9.6.9) is $2 \times 18.75 = 37.5$, for total industry output of $18.75 + 37.5 = 56.25$. The output of each firm declines while industry output increases. When $T = 5$, the output of firm 1 is 12.5 units and industry output is 62.5 units. Finally, for $T = 10$, the output of firm 1 is 6.8 units and the output of the industry is 68.2 units.

HOW MANY FIRMS ARE NECESSARY FOR PERFECT COMPETITION?

We saw in the previous section that the Cournot model converges to perfect competition as the number of firms in the industry becomes very large. Government regulators are frequently concerned about the allocative inefficiency associated with the exercise of market power. Allocative inefficiency occurs when firms are able to charge prices that are above their marginal cost of production. Under perfect competition, allocative efficiency is maximized because firms without market power set price equal to marginal cost. This raises the following question: How many firms does it take for an industry to approximate the conditions of perfect competition?

Assuming that firms are Cournot competitors, $4/(T + 1)^2$ is a measure of allocative inefficiency as a fraction of the allocative inefficiency that would prevail under a monopoly. Table 9.1 summarizes the Cournot-Nash equilibrium allocative inefficiency for T identical firms. With seven firms in the industry, for example, allocative inefficiency under Cournot competition is only 6.25 percent of monopoly. When there are 10 firms in the industry, allocative inefficiency falls to 3.31 percent. With 15 identical firms, the industry approximates zero allocative inefficiency. In short, it does not take very many firms in an industry producing a homogeneous good to approximate the conditions prevailing under perfect competition.

STANDARD MODEL OF MONOPOLY

We saw in the case of perfect competition that entry and exit of firms to and from an industry tends to drive economic profit to zero. But, what if entry into the industry is not possible? How are the conclusions of the preceding section altered when a single firm constitutes the entire industry? In this section we will review the standard model of monopoly, which in terms of the number of firms in an industry is the polar opposite of perfect competition.

A **monopoly** is an industry consisting of a single firm that produces a standard good for which there are no substitutes. Unlike the perfectly competitive firm, the output of the monopolist is synonymous with industry supply. As a result, the market price is no longer parametric. Changes in total output will raise or lower the market-clearing price. For this reason, a monopolist is a *price maker*.

> **Monopoly** An industry consisting of a single producer of a good for which there are no substitutes, and where entry into the industry is impossible.

The standard textbook treatment of a monopoly is depicted in Figure 9.5. The monopolist faces the downward-sloping market demand curve. We have assumed that the monopolist faces U-shaped marginal and average total cost curves. The profit-maximizing monopolist will produce where marginal cost equals marginal revenue. This occurs at point E at the output level Q_m. The monopoly price (P_m) is determined along the market demand curve. At this price-quantity combination, the economic profit earned by the monopolist is the area of the shaded rectangle AP_mBC.

If entry into the industry is relatively easy, the existence of economic profits will attract investment into the industry. In this case, the industry will cease to be a monopoly and the demand for the output of any individual firm becomes increasingly elastic (flatter). As this process continues, the demand curve for the output of an individual firm eventually becomes infinitely elastic (horizontal). If entry is not possible, the monopolist's privileged position and economic profits are secure. Other things being equal, the profit-maximizing price and output for the monopolist are the same in both the short and the long run. Restricted entry guarantees that the monopolist's economic profits will not be competed away.

TABLE 9.1
**Cournot-Nash Equilibrium Allocative Inefficiency as a Fraction of
Allocative Inefficiency Under Monopoly**

T	Inefficiency (percent)
1	1 (100)
2	4/9 (44.44)
3	1/4 (25)
4	4/25 (16)
5	1/9 (11.11)
6	4/49 (8.16)
7	1/16 (6.25)
⋮	⋮
10	4/121 (3.31)
⋮	⋮
15	1/64 (1.56)

FIGURE 9.5
A Monopolist's Profit-Maximizing Price and Output

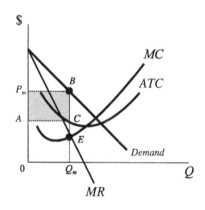

Demonstration Problem 9.7

Suppose that the market demand curve for the output of a monopolist is:

$$Q = 3,000 - 60P \tag{D9.7.1}$$

The monopolist's total cost equation is:

$$TC = 100 + 5Q + \frac{Q^2}{480} \tag{D9.7.2}$$

a. What are the profit-maximizing price and output?
b. Given your answer to part a, what is the firm's economic profit? Is this a short- or
long-run equilibrium?

Solution

a. The inverse of the market demand equation is:

$$P = 50 - \frac{Q}{60} \qquad (D9.7.3)$$

The total revenue ($P \times Q$) equation is:

$$TR = \left(50 - \frac{Q}{60}\right)Q = 50Q - \frac{Q^2}{60} \qquad (D9.7.4)$$

The marginal revenue equation is:

$$MR = \frac{dTR}{dQ} = 50 - \frac{Q}{30} \qquad (D9.7.5)$$

The monopolist's marginal cost equation is:

$$MC = \frac{dTC}{dQ} = 5 + \frac{Q}{240} \qquad (D9.7.6)$$

Total profit is maximized at the output level where $MR = MC$. Equating Equations (D9.7.5) and (D9.7.6) and solving for Q we get the profit-maximizing output of 1,200 units. Substituting this result back into Equation (D9.7.3) we obtain:

$$P^* = 50 - \frac{1,200}{60} = 50 - 20 = \$30 \qquad (D9.7.7)$$

b. The monopolist's profit at $P^* = \$30$ and $Q^* = 1,200$ is:

$$\pi^* = TR - TC = 30(1,200) - \left[100 + 5(1,200) + \frac{(1,200)^2}{480}\right] = \$26,900 \qquad (D9.7.8)$$

If the entry of new firms into the industry is impossible, the monopolist's profit is a long-run equilibrium. If entry is relatively easy, this is a short-run outcome since new firms will be attracted into the industry by the lure of monopoly profits.

SOURCES OF MONOPOLY POWER

The standard model of monopoly assumes that entry by other firms into the industry is impossible. *Ceteris paribus*, a monopolist will earn economic profit in the long run. But, how does a firm become a monopolist? A common reason is that the firm is the exclusive beneficiary of a **government franchise**. In the fifteenth and sixteenth centuries, for example, governments of some maritime countries granted trading monopolies in exchange for a share of the profits.[4] A notable example of a government-sanctioned monopoly is the U.S. Postal Service, which has the sole authority to provide local mail delivery service. Government franchises are often associated with public utilities, such as Consolidated Edison in New York City.

Government franchise A publicly authorized monopoly.

Government franchised monopolies are sometimes justified because there exists significant economies of scale. **Economies of scale** exist when a firm's per unit cost of production declines with a proportional increase in the use of all factors of production. In the case of public utilities, for example, it is more cost efficient for a single firm to generate and distribute electricity. Such firms are referred to as **natural monopolies**. In principle, these cost savings are passed along to the consumer in the form of lower prices. In exchange for this franchise, public utilities agree to price regulation. Fairness is another reason frequently cited in defense of government regulation. In many states, local telephone service is subject to regulation to ensure that low-income consumers have access to affordable service. Profits earned by telephone companies from its business customers are used to subsidize private household service, which is often priced below marginal cost.

> **Economies of scale** A decline in a firm's per unit cost of production following a proportional increase in the use of all factors of production.

> **Natural monopoly** When a firm that can satisfy total market demand at lower per unit cost than an industry consisting of two or more firms.

A **patent** is another legal barrier to the entry by new firms into an industry. A patent is the exclusive right to market a product or process by its inventor.[5] In the United States, patent protection is granted by the U.S. Congress for a period of twenty years. The rationale underlying patent protection is that it provides an incentive for product research, development, invention, and innovation. Without patent protection, investors are less likely to incur the financial risks and development costs associated with bringing a new product to market. On the other hand, patent protection discourages competition, which could result in product innovation, the development of more efficient production techniques, and lower prices. For these reasons, patents are not granted in perpetuity.

> **Patent** An exclusive right granted to an inventor by government to a product or a process.

Arguments for and against patents have recently taken center-stage in the debate over the escalating cost of health care in the United States. A frequently cited culprit is the high price of prescription drugs. Pharmaceutical companies have been granted thousands of patents for a wide range of new prescription medicines. The high price of prescription medicines places a financial burden on low-income individuals, particularly the elderly, who rely upon government assistance. Proponents argue that high prices are necessary to compensate pharmaceutical companies for billions of dollars invested in research and development. Some of these prescription medicines are never brought to the market, or fail to receive approval from the U.S. Food and Drug Administration.

Monopolists can also attempt to protect their exclusive market position by filing lawsuits against potential competitors claiming patent or copyright infringement. Start-up companies typically need to get their products to market as quickly as possible to generate cash flow. Regardless of the merits of the lawsuit, such cash-poor companies are financially unprepared to weather such legal challenges. In the end, these companies may be forced out of business, or may even be acquired by the monopolist.

Another barrier to entry is exclusive control of an essential factor of production. Until the 1940s, for example, the Aluminum Company of America (Alcoa) owned or controlled nearly 100 percent of the world's bauxite deposits. Because aluminum is refined bauxite, Alcoa emerged as the sole producer and distributor. A firm that controls the entire supply of a vital factor of production will control production of the downstream product for the entire industry.

MONOPOLY AND GAME THEORY

It is difficult to see how game theory, which is the study of strategic behavior, can be of much help in analyzing the pricing and output behavior of a monopolist since there is only one player. Like any other profit-maximizing firm, a monopolist will produce at an output level and charge a price such that $MR = MC$. Unlike the behavior of firms in imperfectly competitive markets, the monopolist need not worry about the response of rivals since there aren't any. But, is this always true? This would certainly seem to be the case in situations where entry by new firms is absolutely impossible, as would be the case with government franchised monopolies. Suppose, however, that the firm's monopoly position is based on something that is unique about the firm's product or its production technology, such as enjoying significant economies of scale. In this case, the threat of competition is always present. Such an industry has been referred to as a *contestable monopoly*. So, how might a monopolist thwart such a latent challenge to its primacy in the industry? Does it mean that a monopolist might choose not to behave like a monopolist?

CONTESTABLE MONOPOLY

A monopolist's ability to earn economic profits in the long run is related to entry conditions. When it is impossible for new firms to enter the industry, economic profits are likely to endure. When entry is possible, economic profits will be competed away with an increase in industry output and a fall in the market price. Yet, the requirement that potential competitors actually enter the industry for product prices to fall has been criticized as overly restrictive. Baumol, Panzar, and Willig (1982) have argued that it is not necessary for new firms to actually enter a market for lower prices and economic profits to result. They argue that it is only necessary that the threat of entry by potential competitors exist for this to occur. This situation is referred to as a **contestable market**. These authors also argue that government policies that seek to artificially increase the number of firms in an industry may be counterproductive since they will increase production costs and reduce consumer welfare.

> **Contestable market** An industry with an unlimited number of potential competitors producing a homogeneous product using identical production technologies, have no significant sunk costs, and are Bertrand competitors.

Several conditions must be satisfied for a market to be considered contestable. For one thing, there must exist an unlimited number of potential, profit-maximizing competitors that produce a homogeneous product using the same production technology. Moreover, entry into the industry does not involve substantial **sunk costs**, which are costs that are not recoverable once incurred. Finally, when a firm sets its price, it ignores the prices charged by rivals. In other words, a contestable market is a variation of Bertrand price competition.

> **Sunk cost** A cost that is not recoverable once incurred.

A **contestable monopoly** is a contestable market consisting of a single firm. The reader may recall that the Bertrand paradox predicts that regardless of the number of firms in an industry, price competition will lead firms to engage in marginal cost pricing and earn zero economic profits. For similar reasons, if the market is contestable, Bertrand price competition predicts that the monopolist will produce a level of output sufficient to deter entry by potential competitors. In other words, a monopolist in a contestable market will set a price that results in zero economic profit.

> **Contestable monopoly** A contestable market consisting of a single firm.

To make the discussion more concrete, suppose there are two firms, Alpha Company and Beta Company, which produce an identical product. Alpha Company is the incumbent monopolist and Beta Company is the latent challenger. Each firm is capable of satisfying the entire market, sunk costs are negligible, and both firms use the same production technology. Alpha company's total cost equation is:

$$TC = 70 + 100Q \tag{9.7}$$

Alpha Company's marginal cost and average total cost equations are:

$$MC = \frac{dTC}{dQ} = 100 \tag{9.8}$$

$$ATC = \frac{70}{Q} + 100 \tag{9.9}$$

The market demand equation is:

$$Q = 18 - 0.1P \tag{9.10}$$

The inverse of the market demand equation is:

$$P = 180 - 10Q \tag{9.11}$$

The total and marginal revenue equations are:

$$TR = PQ = 180Q - 10Q^2 \tag{9.12}$$

$$MR = \frac{dTR}{dQ} = 100 - 20Q \tag{9.13}$$

This example is depicted in Figure 9.6. The profit-maximizing level of output in which entry is impossible is determined at the intersection of marginal revenue and marginal cost curves. This profit-maximizing price and output level are $140 and 4,000 units, respectively. In the above example, Alpha Company is a natural monopoly in that per unit cost of production declines throughout the entire range of output. To avoid earning an economic loss, the lowest price that the monopolist can charge occurs where the average total cost curve intersects the demand curve, which occurs at 7,000 units and a price of $110. From society's perspective, this is a second-best solution. The long-run perfectly competitive price and output level occurs at the intersection of the marginal cost and demand curves, which occurs at a price of $100 and an 8,000 units of output. This price/ quantity combination is a first-best level of output since it results in zero deadweight loss.

Assume that Alpha Company and Beta Company are playing a Bertrand price-setting game. Both companies simultaneously announce a price for their products. This price cannot be easily changed once the announcement has been made. In the above example, Alpha Company knows that if it announces a price that is greater than average total cost, Beta Company will capture the entire market by charging a slightly lower price. This leads to a version of the Bertrand paradox. In spite of its monopoly position, Alpha Company has no choice but to set price equal to average total cost to deter entry by Beta Company.

The role of sunk cost is critical to our analysis of a contestable monopoly. As we saw, if Alpha Company charges a price that is greater than average total cost, Beta Company will capture the entire market by charging a slightly lower price. However, if Beta Company incurs substantial sunk costs by entering the market it will not be able to contest the market without incurring economic losses. The greater this barrier to entry, the higher the price and the greater Alpha Company's

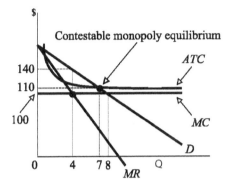

FIGURE 9.6
Contestable Monopoly

economic profit. In fact, if sunk costs are sufficiently high, Alpha Company may be able to engage in monopoly pricing. The presence of sunk costs provides another explanation for why the predictions of the simple Bertrand model are not observed in the real world.

CHAPTER REVIEW

In this chapter we examined the market extremes of perfect competition and monopoly. *Perfect competition* is a term used to describe a market structure characterized by a large number of more or less equally sized firms producing an identical product in which entry into and exit from the industry is easy. A *monopoly* consists of a single firm that produces a good for which there are no close substitutes and entry by new firms into the industry is impossible.

Perfectly competitive firms are called *price takers* because they do not have the ability to affect the market price by changing their level of output. By contrast, the output of the monopolist is the industry supply. Since changing output will shift the market supply curve and change the market price, a monopolist is said to be a *price maker.*

All profit-maximizing firms produce at an output level where marginal revenue equals marginal cost ($MR = MC$). For a perfectly competitive firm, marginal revenue equals the market-determined price ($MR = P_0$). The profit-maximizing condition for a perfectly competitive firm may be rewritten as $P_0 = MC$. When $P_0 > ATC$, perfectly competitive firms earn economic profits and new investment will be attracted into the industry. When $P_0 < ATC$, firms earn economic losses and exit the industry. Long-run perfectly competitive equilibrium occurs when $P_0 = MC = minimum\ ATC$. In this case, each firm earns zero economic profit and the industry is in long-run competitive equilibrium. From society's perspective this is a first-best solution since allocative efficiency is maximized.

Game theorists have demonstrated the existence of equilibrium in perfectly competitive markets by introducing a fictitious player called Nature into a noncooperative output-setting game. The objective of Nature is to minimize disequilibrium in a perfectly competitive market.

Perfect competition may be viewed as a special case of the Cournot model. As the number of equally sized firms in an industry becomes very large, the market share of each firm approaches zero. In this case, the Cournot-Nash equilibrium collapses to the standard model of perfect competition. At the other extreme, when there is only one firm in the industry, the Cournot-Nash equilibrium collapses into the standard model of monopoly.

A profit-maximizing monopolist will produce at an output level where $MR = MC$. Since $P > MR$, the monopolist earns positive economic profit. Since $MC < ATC$, there is allocative inefficiency. If a single firm earns monopoly profits, its unique position in the market could be threatened by the lure of positive economic profits. This situation is referred to as a *contestable monopoly*. Under certain conditions, if firms engage in Bertrand price competition, a monopolist will produce a level of output sufficient to deter entry by potential competitors. The monopolist will set a price that results in zero economic profit.

The role of *sunk cost* is crucial to the analysis of a contestable monopoly. A sunk cost is not recoverable once incurred. If potential competitors incur significant sunk costs by entering the market, the monopolist will be able to charge a high price and earn positive economic profits. If sunk costs are substantial, the incumbent may even be able to charge monopoly prices. The presence of significant sunk costs provides another explanation for why the predictions of the simple Bertrand model are not observed in the real world.

CHAPTER QUESTIONS

9.1 Price competition is characteristic of perfectly competitive firms. Do you agree with this statement? Explain.

9.2 In the standard model, perfectly competitive firms are referred to as price takers. What does this imply for the price and output decisions of profit-maximizing firms?

9.3 Under what circumstances should a perfectly competitive firm, which is earning an economic loss at the profit-maximizing level of output, shut down? Diagram your answer.

9.4 A profit-maximizing firm produces at an output level where the price is less than the average total cost. Under what conditions will this firm continue to operate?

9.5 Suppose that a profit-maximizing, perfectly competitive firm produces where price is less than the average total cost. Under what conditions will this firm shut down? Explain.

9.6 A perfectly competitive firm in short-run equilibrium must also be in long-run competitive equilibrium. Do you agree? Explain.

9.7 A perfectly competitive firm in long-run equilibrium must also be in short-run competitive equilibrium. Do you agree? Explain.

9.8 In the standard model, a perfectly competitive firm in long-run equilibrium produces at minimum per unit cost. Do you agree? Explain.

9.9 A perfectly competitive firm will operate in the short run as long as total revenues cover all of the firm's total variable costs and some of its fixed costs. Explain.

9.10 The perfectly competitive firm's marginal cost curve is also its supply curve. Do you agree with this statement? If not, why not?

9.11 When price is greater than average variable cost, a perfectly competitive firm can expect the market price to fall. Do you agree with this statement? Explain.

9.12 Perfectly competitive firms are price takers because they do not have market power. As a result, game theory provides few insight into perfect competition. Do you agree? Explain.

9.13 How can we use the Cournot model to explain the existence of perfect competition?

9.14 To maximize total revenue the monopolist must charge the highest price possible. Do you agree? Explain.

9.15 A monopolist does not have a supply curve. Explain.

9.16 When is maximizing total revenues the same thing as maximizing total profits?

9.17 Why do governments grant patents and copyrights?

9.18 Monopolies are industries consisting of a single firm. Since there is only one player, game theory is of little value in advancing our understanding of monopolistic market structures. Do you agree? Explain.

9.19 Monopoly and perfect competition are limiting cases of the Cournot model. Explain.

9.20 Explain how government regulators can apply the Cournot output-setting model to assess the level of market power and allocative inefficiency.

9.21 What is a contestable monopoly? Does this represent a special case of monopolistic market structures, or is it a more general case in which the tools of game theory can be used to explain pricing and output behavior?

9.22 What are the game theoretic implications of contestable monopolies?

9.23 What do contestable markets imply about government regulation to promote allocative efficiency?

9.24 What are the implications of significant sunk costs for the Bertrand price-setting game for contestable monopolies?

CHAPTER EXERCISES

9.1 A firm faces the following total cost equation for its product:

$$TC = 500 + 5Q + 0.025Q^2 \tag{E9.1.1}$$

The firm can sell its product for \$10 per unit of output.

a. What is the profit-maximizing output level?
b. Verify that the firm's profit corresponding to this level of output represents a maximum.

9.2 The total cost (TC) and demand equations for a monopolist are:

$$TC = 100 + 5Q^2 \tag{E9.2.1}$$

$$P = 200 - 5Q \tag{E9.2.2}$$

a. What is the profit-maximizing quantity?
b. What is the profit-maximizing price?

9.3 Bucolic Farms supplies milk to B&Q Foodstores. Bucolic has estimated the following total cost function for its product.

$$TC = 100 + 12Q + 0.06Q^2 \tag{E9.3.1}$$

where Q is measured in 100 gallons of milk.

a. Determine the ATC, AVC, MC, and TFC.
b. What are Bucolic's shut-down and break-even price and output levels?
c. Suppose that there are 5,000 nearly identical milk producers in this industry. What is the market supply curve?
d. Suppose that the market demand function is:

$$Q_D = 660,000 - 16,333.33P \tag{E9.3.2}$$

What are the market equilibrium price and quantity?

   e. Determine Bucolic's profit.

   f. Assuming that there is no change in demand or costs, how many milk producers will remain in the industry in the long run?

9.4   A monopoly faces the following demand and total cost equations for its product.

$$Q = 30 - \frac{P}{3} \tag{E9.4.1}$$

$$TC = 100 - 5Q + Q^2 \tag{E9.4.2}$$

   a. What are the firm's short-run, profit-maximizing price and output level?

   b. What is the firm's economic profit?

9.5   The demand equation for a product sold by a monopolist is:

$$Q = 25 - 0.5P \tag{E9.5.1}$$

The total cost equation of the firm is:

$$TC = 225 + 5Q + 0.25Q^2 \tag{E9.5.2}$$

   a. Calculate the profit-maximizing price and quantity?

   b. What is the firm's profit?

9.6   The market equation for a product sold by a monopolist is:

$$Q = 100 - 4P \tag{E9.6.1}$$

The total cost equation of the firm is:

$$TC = 500 + 10Q + 0.5Q^2 \tag{E9.6.2}$$

   a. What are the profit-maximizing price and quantity?

   b. What is the firm's maximum profit?

9.7   In a multifirm industry, derive the general reaction function for firm 1 and the reaction function for the remaining $T - 1$ firms.

9.8   The demand for the output of a multifirm industry is given by the equation:

$$Q_T = 250 - 10P \tag{E9.8.1}$$

where Q_T represents the output of the entire industry. The total cost equation of any individual firm in the industry is:

$$TC_i = 10 + 4Q_i \tag{E9.8.2}$$

   a. Determine the reaction functions of firm 1 and the remaining $T - 1$ firms in this industry.

   b. Determine the Nash-Cournot equilibrium level of output for each firm in an industry consisting of 2, 3, 5, 10, and equally sized firms.

ENDNOTES

   1. The firm's profit function is:

$$\pi(Q) = TR(Q) - TC(Q) \tag{F9.1.1}$$

where π is total profit, TR is total revenue, which is defined as the selling price of the product times the number of units sold ($TR = P \times Q$), and total cost (TC), which is an increasing function of output. To find the level of output that maximizes total profit, take the first derivative of Equation (F9.1.1) with respect to Q and set the result equal to zero.

$$\frac{d\pi(Q)}{dQ} = \frac{dTR(Q)}{dQ} - \frac{dTC(Q)}{dQ} = 0 \tag{F9.1.2}$$

Equation (F9.1.2) may be rewritten as:

$$M\pi = MR - MC = 0 \tag{F9.1.3}$$

where $M\pi$ is marginal profit. Rearranging Equation (F9.1.3) we obtain Equation (9.1).

2. Total revenue for a perfectly competitive firm is:

$$TR = P_0 \times Q \tag{F9.2.1}$$

where P_0 is a constant price. Thus, the marginal revenue of this firm is:

$$MR = \frac{dTR}{dQ} = \frac{d(P_0 \times Q)}{dQ} = P_0 \tag{F9.2.2}$$

3. A firm that earns zero economic profits earns a positive accounting profit that is equal to a normal rate of return on its investment.

4. For example, Queen Isabella of Spain granted Christopher Columbus a monopoly in exchange for a share of any gold found in the New World. In 1602, the States-General of the Netherlands granted a spice trading monopoly to the Dutch East India Company. The Hudson Bay Company, which was established in 1670, was granted a general trading monopoly by the English monarchy in the area around Hudson Bay in Canada.

5. In the United States, patents are granted under Article I, Section 8, of the U.S. Constitution, which gives the Congress the authority to "promote the progress of science and the useful arts, by securing for limited times to authors and inventors the exclusive right to their respective writings and discoveries."

CHAPTER

10 STRATEGIC TRADE POLICY

In this chapter we will:

- *Extend our analysis of imperfect competition to include trade policy as a static game in pure discrete and continuous strategies;*
- *Demonstrate how the Cournot output-setting model can be used to explain the rationale underlying the use export subsidies;*
- *Use game theory to demonstrate the gains from international trade;*
- *Demonstrate how consumer and producer surplus may be used to measure changes in national welfare from strategic policy using the one-dollar, one-vote rule.*

INTRODUCTION

In this chapter, we will extend our analysis of imperfect competition to include strategic trade policy, which refers to measures taken by governments to influence the decisions of consumers and producers engaged in international trade. Our fundamental understanding of the origins and benefits of international trade can be traced to the classic works of David Ricardo (1807), Eli Heckscher (1919), and Bertil Ohlin (1933). An analysis of strategic trade policy using the modern tools of game theory, however, is of a much more recent vintage.

> **Strategic trade policy** Measures adopted by government to influence the decisions of consumers and producers engaged in international trade.

An analysis of international oligopolies and the role of strategic trade policy was one of the first applications of modern game theory. Prior to the 1980s, the standard textbook treatment of **commercial policy** tended to assume that international markets were perfectly competitive. Although distortions arising from imperfect competition led to discussions of second-best trade policies, strategic interaction between and among firms engaged in international trade was largely ignored (see, for example, Bhagwati, Ramaswami, and Srinivasan [1969]).

> **Commercial policy** Policies adopted by a government to influence the quantity and composition of a country's international trade.

The use of game theory to analyze strategic behavior was a major leap forward in our understanding of commercial policy, intraindustry trade, economies of scale, research and development, and technology transfer. One of the first attempts to apply oligopoly theory to international trade (Brander 1981) dealt with intraindustry trade in homogeneous products. Two papers that are often cited as seminal contributions to the theory of strategic trade policy are Spencer and Brander (1983, 1985). Both papers used the Cournot output-setting model to analyze the behavior of domestic and foreign firms that compete in third-country markets. The model of strategic trade policy presented in this chapter is based on the analysis presented in the Brander and Spencer (1985) paper. Other important contributions to the theory of strategic trade policy include Dixit (1979, 1984), Brander and Spencer (1981), Krugman (1984), and Eaton and Grossman (1986).

International trade policy instruments such as import tariffs, import quotas, export subsidies, countervailing duties, domestic content requirements, antidumping laws, product safety standards, trade regulations, and so on, redirect the flow of goods and services across national boundaries. Their continued use contradicts our basic understanding of the virtues of free international trade. Despite this, the use of commercial trade policy to influence the direction and composition of international trade is widespread and pervasive. While this chapter does not rule out the possibility that the formulation of trade policy may be based on somewhat less than noble intentions, the analysis presented in this chapter provides insights into the necessary emergence of such multilateral organizations as the General Agreement of Tariffs and Trade (GATT) and its successor, the World Trade Organization (WTO). It analyzes their role in the systematic reduction of trade barriers, which have dramatically improved global living standards.

CASE STUDY 10.1: BOEING VERSUS AIRBUS

On October 12, 2004, the U.S. government filed a complaint with the World Trade Organization (WTO) claiming that the European Union (EU) had subsidized Airbus production of commercial aircraft. Later that same day, the EU countered with a similar complaint against the U.S. government. This was yet another battle in an ongoing war dating back to the 1960s when a consortium of European governments decided to support its fledgling civilian aircraft industry in order to compete with such American companies as Boeing, McDonnell Douglas, and Lockheed.

The development of Airbus went largely unnoticed until the 1980s when it managed to capture about 20 percent of the global market. Following complaints from Boeing that the success of Airbus was the direct result of government subsidies, Washington entered into discussions with the EU in 1992. These negotiations resulted in a bilateral agreement to limit direct and indirect government support for the production of new airplane development (launch aid). Despite these limitations, EU subsidies enabled Airbus to grab an ever-increasing share of the global market. By 2004, Airbus claimed more than 50 percent of all new aircraft deliveries.

Increasingly unhappy with Airbus's growing global market share, the U.S. Trade Representative initiated discussions with the EU to revise the 1992 agreement. Despite these efforts, no progress was made, and in September of that year Washington announced that it was abrogating the agreement. It subsequently filed a complaint with the WTO under its general subsidy rules. Washington's complaint focused on $17 billion in low-interest-rate loans with repayment contingent on future Airbus sales and $6.5 billion in development and production subsidies for its A380 super jumbo jet. The EU countered by claiming that Boeing had received $23 billion in research and development contracts from NASA and the U.S. Department of Defense. Airbus claimed that these contracts had applications to civilian aircraft production. The EU also argued that Boeing had received billions of dollars in federal, state, and local tax breaks.

Despite its lead over Boeing, in 2005 Airbus announced its intention to seek government loans for the development of a new model, the A350, which was designed to rival Boeing's B787 "Dreamliner." The U.S. government petitioned the WTO to rule on this new development. The EU responded with its own complaint against the United States. Together, these cases were the largest ever to go before the WTO. While both companies probably violated WTO rules, it was unclear what the WTO can do about it. In a worst-case scenario, the dispute will lead to bilateral sanctions that could put a serious dent in transatlantic trade, thereby harming both European and U.S. economies. It is clearly in both parties' best interest to settle their disputes.

STRATEGIC TRADE POLICY WITH DISCRETE PURE STRATEGIES

In this section we will consider the implications of strategic trade policy when modeled as a pure-strategy static game involving two companies that dominate the global market for a particular product. The game depicted in Figure 10.1 was inspired by the real-world competition existing between the world's largest producers of commercial aircraft—the Boeing Company of the United States and Airbus S.A.S, which is owned by a consortium of European stakeholders from Great Britain, France, Germany, and Spain (see Case Study 10.1). We will refer to this as the jumbo jet game.

Let us begin by supposing that Boeing and Airbus are considering investing $10 billion to develop a new, technologically advanced, commercial, jumbo jet airliner. While both companies have enough production capacity to satisfy total market demand, each must capture a significant share of the global market to justify these up-front development costs. If only one firm enters the market, it will earn $150 billion in operating profits. If both firms enter the market, competition for sales will drive operating profits to zero, in which case both companies will lose their initial $10 billion investment. The game depicted in Figure 10.1 has two Nash equilibrium strategy profiles: {*Don't produce, Produce*} and {*Produce, Don't produce*}. The payoffs are highlighted in boldface.

Producing the new commercial airliner is clearly a risky venture for both companies. Neither company is willing to commit $10 billion in development costs without some reasonable expectation that its rival will stay out of the market. Since its rival's intentions are unknown, the threat by either company to produce the new jumbo jet may result in neither firm producing the new jet aircraft. Can each company's government do anything to break this logjam?

Suppose that the U.S. government offers Boeing a $15 billion research and development subsidy under the pretense of promoting, say, job growth or developing synergies in high-technology industries. The revised normal form of this game is depicted in Figure 10.2. Boeing now has a strictly dominant strategy to produce the new jumbo jet. The strategy profile {*Produce, Don't produce*} represents a unique Nash equilibrium in which Boeing earns $165 billion in operating profits, while Airbus stays out of the market and earns nothing. These payoffs are highlighted in boldface.

The research and development subsidy would appear to be a no-lose proposition. Boeing earns $165 billion in operating profits while Airbus is no worse off. The U.S. government presumably achieves its objective of promoting job growth and creating synergies in high-technology production. The world's consumers gain because they have access to a new, and presumably less expensive and more efficient, mode of transportation.[1] But, not so fast. Since this is a static game, suppose that the European Union (EU) decides to subsidize Airbus by the same amount. This new situation is depicted in Figure 10.3.

When both companies receive $15 billion in research and development subsidies, the Nash equilibrium strategy profile for this game is {*Produce, Produce*}. While neither company loses, U.S. and EU taxpayers may be made worse since they pay the amount of the subsidy. But, is the world better off because both companies compete in the jumbo jet market? Clearly the world's air

FIGURE 10.1
Jumbo Jet Game

		Airbus	
		Produce	Don't produce
Boeing	Produce	(−10, −10)	**(150, 0)**
	Don't produce	**(0, 150)**	(0, 0)

Payoffs: (Boeing, Airbus)

FIGURE 10.2
Jumbo Jet Game in Which Boeing Receives a Research and Development Subsidy

		Airbus	
		Produce	Don't produce
Boeing	Produce	(5, −10)	**(165, 0)**
	Don't produce	(0, 150)	(0, 0)

Payoffs: (Boeing, Airbus)

FIGURE 10.3
Jumbo Jet Game in Which Both Players Receive a Research and Development Subsidy

		Airbus	
		Produce	Don't produce
Boeing	Produce	**(5, 5)**	(165, 0)
	Don't produce	(0, 165)	(0, 0)

Payoffs: (Boeing, Airbus)

travelers gain because they now have access to a new and better mode of transportation at lower fares due to increased competition. Whether the United States and EU are net winners, or losers, depends on where these consumers are located. If most of these air travelers are located in the United States, the EU may be a net loser, and *vice versa*. On the other hand, if most of these air travelers are located outside the United States and EU, both may be losers.

The analysis is complicated when there are more than two companies vying for sales in the global jumbo-jet market. Figure 10.4 depicts the no-subsidy case in which a third company, which we will call Air Nippon, enters into the mix. In this case, we have assumed that when three companies enter the market, each will lose $15 billion. This larger loss results from $10 billion in sunk research and development expenditures, and an additional loss of $5 billion in operating losses due to lower aircraft prices due to more intense global competition. In this game there are three Nash equilibrium strategy profiles: {*Don't produce, Don't produce, Produce*}, {*Don't produce, Produce, Don't produce*}, and {*Produce, Don't produce, Don't produce*}. Production of a new jumbo jet will only take place if a company can be certain that its two rivals will not produce. As before, since production entails a great deal of risk, and because a rival's intentions are unknown, the threat by either company to produce may result in no production whatsoever.

FIGURE 10.4
Three-Player Jumbo Jet Game without Government Subsidies

Air Nippon

		Produce		Don't produce	
		Airbus		**Airbus**	
		Produce	Don't produce	Produce	Don't produce
Boeing	Produce	(−15, −15, −15)	(−10, 0, −10)	(−10, −10, 0)	**(150, 0, 0)**
	Don't produce	(0, −10, −10)	**(0, 0, 150)**	**(0, 150, 0)**	(0, 0, 0)

Payoffs: (Boeing, Airbus, Air Nippon)

FIGURE 10.5
Three-Player Jumbo Jet Game with Government Subsidies

Air Nippon

		Produce		Don't produce	
		Airbus		**Airbus**	
		Produce	Don't produce	Produce	Don't produce
Boeing	Produce	**(0, 0, 0)**	(5, 0, 5)	(5, 5, 0)	(165, 0, 0)
	Don't produce	(0, 5, 5)	(0, 0, 165)	(0, 165, 0)	(0, 0, 0)

Payoffs: (Boeing, Airbus, Air Nippon)

Let us now assume that the government of each producer provides a $15 billion research and development subsidy. This new situation is depicted in Figure 10.5. The outcome is similar to that depicted in Figure 10.3. When each firm receives a subsidy, the Nash equilibrium strategy profile is for all three firms to develop and produce a new jumbo jet, although each earns zero economic profit.

The results presented in Figures 10.3 and 10.5 underscore several important points about the effectiveness of a strategic trade policy. To begin with, success is in the eye of the beholder. If the subsidy is sufficiently large, firms benefit from increased production and greater economic profits. Workers in the aerospace industry benefit directly from the jobs that increased production create. International air travelers benefit from more technologically advanced modes of transportation. They may also benefit from lower airfares resulting from increased competition. Lower air transportation costs may also stimulate global commerce and a rise in output and incomes. On the other hand, taxpayers may be made worse off because they foot the bill for the government subsidy. This is especially true for taxpayers who are not employed in the aerospace industry or do not travel by air. Despite all this, the above analysis of strategic trade policy with discrete pure strategies illustrates how the government may be able to use export subsidies to redirect the flow of international trade. In the next section, we will expand our analysis by considering strategic trade policy with continuous pure strategies.

——————————————— **Demonstration Problem 10.1** ———————————————

Suppose there are three identical firms considering producing a homogeneous product for sale in the global marketplace. Figure D10.1.1 summarizes the payoffs in billions of dollars.

FIGURE D10.1.1

Three-Player, Static Game in Demonstration Problem 10.1a

Firm C

		Produce		Don't produce	
		Firm B		**Firm B**	
		Produce	Don't produce	Produce	Don't produce
Firm A	Produce	(−1, −1, −1)	**(1, 0, 1)**	**(1, 1, 0)**	(10, 0, 0)
	Don't produce	**(0, 1, 1)**	(0, 0, 10)	(0, 10, 0)	(0, 0, 0)

Payoffs: (Firm A, Firm B, Firm C)

FIGURE D10.1.2

Three-Player, Static Game in Demonstration Problem 10.1b

Firm C

		Produce		Don't produce	
		Firm B		**Firm B**	
		Produce	Don't produce	Produce	Don't produce
Firm A	Produce	(0, −1, −1)	**(2, 0, 1)**	**(2, 1, 0)**	(11, 0, 0)
	Don't produce	(0, 1, 1)	(0, 0, 10)	(0, 10, 0)	(0, 0, 0)

Payoffs: (Firm A, Firm B, Firm C)

a. If larger payoffs are preferred, what is the Nash equilibrium strategy profile for this game?
b. Suppose that country *A* offers $1 billion to firm *A* as a production subsidy. If larger payoffs are preferred, what is the Nash equilibrium strategy profile for this game?
c. Suppose that countries *A* and *B* offer firms *A* and *B* $1 billion in production subsidies. If larger payoffs are preferred, what is the Nash equilibrium strategy profile for this game?
d. Suppose that countries *A*, *B*, and *C* offer firms *A*, *B*, and *C* $1 billion in production subsidies. If larger payoffs are preferred, what is the Nash equilibrium strategy profile for this game?

Solution

a. The game depicted in Figure D10.1.1 has three Nash equilibria: {*Don't produce, Produce, Produce*}, {*Produce, Don't produce, Produce*}, and {*Produce, Produce, Don't produce*}. The payoffs for these strategy profiles are highlighted in boldface.
b. The revised version of this game when firm *A*'s production is subsidized is depicted in Figure D10.1.2. The Nash equilibrium strategy profiles for this game are {*Produce, Don't produce, Produce*} and {*Produce, Produce, Don't produce*}. The payoffs for these strategy profiles are highlighted in boldface.
c. The revised version of this game when the production by firms *A* and *B* are subsidized is depicted in Figure D10.1.3. The unique Nash equilibrium strategy profile for this game is {*Produce, Produce, Don't produce*}. The payoffs for this strategy profile are highlighted in boldface.

FIGURE D10.1.3

Three-Player, Static Game in Demonstration Problem 10.1c

Firm C

		Produce		Don't produce	
		Firm B		**Firm B**	
		Produce	Don't produce	Produce	Don't produce
Firm A	Produce	(0, 0, −1)	(2, 0, 1)	**(2, 2, 0)**	(11, 0, 0)
	Don't produce	(0, 2, 1)	(0, 0, 10)	(0, 11, 0)	(0, 0, 0)

Payoffs: (Firm A, Firm B, Firm C)

FIGURE D10.1.4

Three-Player, Static Game in Demonstration Problem 10.1d

Firm C

		Produce		Don't produce	
		Firm B		**Firm B**	
		Produce	Don't produce	Produce	Don't produce
Firm A	Produce	**(0, 0, 0)**	(2, 0, 2)	(2, 2, 0)	(11, 0, 0)
	Don't produce	(0, 2, 2)	(0, 0, 11)	(0, 11, 0)	(0, 0, 0)

Payoffs: (Firm A, Firm B, Firm C)

d. The revised version of this game where the production of all three firms is subsidized is depicted in Figure D10.1.4. This game has the unique Nash equilibrium strategy profile {*Produce, Produce, Produce*}, although each firm earns zero economic profit on production of the new product.

STRATEGIC TRADE POLICY WITH CONTINUOUS PURE STRATEGIES

We will begin our discussion of strategic trade policy with continuous pure strategies by first considering the behavior of two profit-maximizing monopolies producing identical goods for sale only in their respective home markets, perhaps because of extremely high trade barriers and transportation costs. **Autarky** describes the situation in which a country does not engage in international trade. We will assume that both firms use an identical production technology that exhibits **constant returns to scale**. By constant returns to scale we mean that an equal proportional increase in the use of all factors of production results in an equivalent proportional increase in total output.[2]

> **Autarky** When a country does not engage in international trade.
>
> **Constant returns to scale** When a proportional increase in the use of all inputs results in the same proportional increase in total output.

To keep the discussion simple, we will assume that neither monopoly incurs fixed cost, and that the marginal cost of production is constant at $1 per unit. Total output is in millions of units and prices are in thousands of dollars. Finally, we will assume that the domestic market in each country is identical and is described by the following linear demand equations:

$$Q_{US}^{US} = 25 - 5P_{US} \tag{10.1a}$$

$$Q_J^J = 25 - 5P_J \tag{10.1b}$$

In Equations (10.1), P_{US} and P_J represent unit prices in the United States and Japan, respectively. The superscript on output denotes the nationality of the producer. The subscript on output denotes the market in which the company sells its product. Equation (10.2b), for example, indicates the demand for the Japanese company output in the Japanese market. The inverse of Equations (10.1) are:

$$P_{US} = 5 - \frac{1}{5}Q_{US}^{US} \tag{10.2a}$$

$$P_J = 5 - \frac{1}{5}Q_J^J \tag{10.2b}$$

The objective of each firm is to maximize their respective autarkic profit functions:

$$\pi_{US}^{US} = \left(31 - \frac{1}{5}Q_{US}^{US}\right)Q_{US}^{US} - 2Q_{US}^{US} \tag{10.3a}$$

$$\pi_J^J = \left(5 - \frac{1}{5}Q_J^J\right)Q_J^J - Q_J^J \tag{10.3b}$$

To determine the autarkic profit-maximizing level of output for each firm, take the first derivative of each company's profit function with respect to total output and set the resulting equation equal to zero. For the U.S. company, this becomes:

$$\frac{d\pi^{US}}{dQ_{US}^{US}} = \left(5 - \frac{1}{5}Q_{US}^{US}\right) - \frac{1}{5}Q_{US}^{US} - 1 = 0 \tag{10.4}$$

Solving Equation (10.4), the U.S. company's profit-maximizing level of output in its home market is 10 million units. It is left as an exercise for the student to demonstrate that the second-order (sufficient) condition for profit-maximization is satisfied. By substituting this output level into Equation (10.2a), the market clearing price is $P_{US} = \$3$ per unit. Substituting the profit-maximizing output level into Equation (10.3a), the U.S. company's maximum profit is $20 billion. Because of the symmetry of the example, these same results also hold for the Japanese company.

NATIONAL WELFARE

In the remainder of this chapter we will attempt to assess the effectiveness of strategic trade policy by examining how it affects national welfare. National welfare is defined as the sum of consumer and producer surpluses. A net increase in consumer and producer surplus is generally taken to mean that an economy is made better off by a change in the prevailing trade regime. Before proceeding to a discussion of the effectiveness of strategic trade policy, we will begin by reviewing the concepts of producer and consumer surplus.

Producer surplus is defined as the difference between revenues earned from the production and sale of a given quantity of output less the minimum amount that the firm must receive to produce it. For a profit-maximizing firm, this amount is equal to the sum of the marginal cost of producing each unit of output less total fixed cost. In the above example, we assumed that the monopolist incurred zero total fixed cost. Thus, producer surplus in each country under autarky is equal to the firm's total economic profit of $20 billion.

> **Producer surplus** The difference between the total revenues earned from the production and sale of a given quantity of output and the minimum amount that the firm would accept to produce that output.

What is the total net benefit to the consumer from the no-trade scenario? **Consumer surplus** is defined as the difference between the amount that consumers are prepared to pay for a given quantity of output and their actual expenditures. Diagrammatically, consumer surplus is the area under the market demand curve up to the equilibrium level of output (Q^*) minus total consumer expenditures (P^*Q^*). In the above example, consumer surplus (CS) is:[3]

$$CS = \int_0^Q P(Q)\,dQ - P(Q)Q = \int_0^Q \left(5 - \frac{1}{5}Q\right)dQ - \left(5 - \frac{1}{5}Q\right)Q = \frac{1}{10}Q^2 \qquad (10.5)$$

> **Consumer surplus** The difference between what a consumer is willing to pay for a given quantity of a good or service and the amount that he or she actually pays. Diagrammatically, consumer surplus is the area below a downward-sloping demand curve but above the selling price.

Since the inverse of the demand equation is linear of the form $P(Q) = a - bQ$, consumer surplus can be easily calculated as:

$$CS = \frac{1}{2}\left[(a - bQ)\right]Q = \frac{1}{2}\left(\frac{1}{5}Q^2\right) = \frac{1}{10}Q^2 \qquad (10.6)$$

In the above example, consumer surplus in the United States (and Japan) under autarky is depicted as the area of the shaded triangle in Figure 10.6. Substituting the profit-maximizing output level of 10 million units into Equation (10.5) or (10.6), consumer surplus at a price of $3 is equal to $10 billion. In the above example, national welfare for both the United States and Japan is equal to $CS + PS = \$10 + \$20 = \$30$ billion.

Combining consumer and producer surpluses to assess net national welfare effects can be tricky because it is not possible objectively to quantify the social impact of market changes without imposing value judgments on the groups affected. Although economists generally assume diminishing marginal utility of money, it is not possible to assert whether a dollar received by one person has greater or lesser psychic value than a dollar lost by another person. Unfortunately, there is no theorem of economic behavior that can resolve this dilemma.

To overcome this difficulty, economists often adopt the imperfect convention of assuming that a dollar transferred from one group to another has equivalent value. This is referred to as the **one-dollar, one-vote rule**. Adopting this rule implies a willingness to judge trade issues based on the estimated monetary values of consumer and producer surplus, without regard to the distribution of national well-being. This is not to say that distributional issues are unimportant. One way to resolve issues of unfair distribution of national welfare is to compel the "winners" to compensate the "losers." How this is accomplished is problematic, although in democratic societies this is normally resolved at the ballot box.

> **One-dollar, one-vote rule** The arbitrary value judgment that a dollar transferred to one group is equivalent in value to a dollar transferred from another group.

FIGURE 10.6
Consumer Surplus Under Autarky

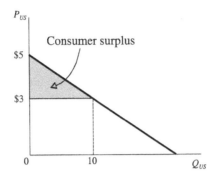

INTRAINDUSTRY TRADE

The principle of comparative advantage and the Heckscher-Ohlin theorem are at the heart of our formal understanding of the benefits of international trade. Neoclassical trade theory, which is based on the assumptions of perfect competition and constant returns to scale, asserts that a country will tend to produce and export goods and services that use its relatively abundant factors of production intensively and import products that use its relatively scarce factors intensively. Heckscher-Ohlin predicts how trade will affect the incomes of those who own the resources used in production. Despite these predictions, the data reveal something quite different. Most international trade is carried out by a relatively small number of highly industrialized countries with similar resource endowments, production technology, and consumer preferences. What is more, much of this trade involves the export and import of goods and services produced by companies in the same industry. The name given to this is **intraindustry trade**.

> **Intraindustry trade** When a country exports and imports the same, or very similar, goods and services.

An explanation for this apparent contradiction may be found in the underlying assumptions of neoclassical trade theory. For one thing, international trade is often carried out by imperfectly competitive companies, which allows us to consider trade issues involving strategic behavior. An extreme version of imperfect competition in international trade involves natural monopolies, which are characterized by increasing, not constant, returns to scale. To understand the contribution of game theory to our understanding of international trade and commercial policy, we will consider the situation in which a domestic and foreign monopoly, each using the same production technology, are engaged in the production and cross-border sales of an identical product. We will assume that both countries have identical resource endowments and consumer preferences. Neoclassical trade theory predicts that under perfect competition no trade will take place between the two countries, largely because of transportation costs and similar trade impediments. As we will see, each firm will find it profitable to sell in its home market and in the market of its foreign rival. We will also demonstrate that it may be possible for the government of either country to unilaterally improve its own country's national welfare by subsidizing exports of the domestic producer. By changing the equilibrium price and market shares in the foreign market, the export subsidy enhances the profits of the domestic producer at the expense of its foreign rival.

IMPERFECT COMPETITION AND INTERNATIONAL TRADE

To highlight the role of imperfect competition in international trade, we will assume that there are two countries with identical resource endowments. In each country there is a single firm that produces a good for which there are no close substitutes. The two firms' goods are identical. Both firms use the same production technology, which exhibit constant returns to scale. In the absence of international trade (autarky), each firm behaves as a monopolist. With the opening of intraindustry trade each company behaves strategically since each must consider the actions of its rival in its home and foreign markets. The total profit earned by each firm is equal to the sum of the profits earned in its home market and profits earned in the home market of its competitor. As before, we will assume zero total fixed cost and that the marginal cost of production is $1 per unit. Finally, it costs each firm $1 per unit to transport its product to the foreign market. The domestic and foreign profit equations of the U.S. company are:

$$\pi_{US}^{US} = P_{US}\left(Q_{US}^{US} + Q_{US}^{J}\right)Q_{US}^{US} - Q_{US}^{US} \tag{10.7a}$$

$$\pi_{J}^{US} = P_{J}\left(Q_{J}^{J} + Q_{J}^{US}\right)Q_{J}^{US} - 2Q_{J}^{US} \tag{10.7b}$$

Combining Equations (10.7a) and (10.7b) we obtain Equation (10.8a), which is the total profit equation of the U.S. company engaged in international trade. Proceeding in a similar manner, Equation (10.8b) is the total profit equation of the Japanese company.

$$\pi^{US} = \pi_{US}^{US} + \pi_{J}^{US} = P_{US}\left(Q_{US}^{US} + Q_{US}^{J}\right)Q_{US}^{US} + P_{J}\left(Q_{J}^{US} + Q_{J}^{J}\right)Q_{J}^{US} - Q_{US}^{US} - 2Q_{J}^{US} \tag{10.8a}$$

$$\pi^{J} = \pi_{J}^{J} + \pi_{US}^{J} = P_{J}\left(Q_{J}^{J} + Q_{J}^{US}\right)Q_{J}^{J} + P_{US}\left(Q_{US}^{US} + Q_{US}^{J}\right)Q_{US}^{J} - Q_{J}^{J} - 2Q_{US}^{J} \tag{10.8b}$$

The domestic demand equations in each country are:

$$P_{US} = 5 - \frac{1}{5}\left(Q_{US}^{US} + Q_{US}^{J}\right) \tag{10.9a}$$

$$P_{J} = 5 - \frac{1}{5}\left(Q_{J}^{J} + Q_{J}^{US}\right) \tag{10.9b}$$

Substituting Equations (10.9) into Equations (10.8), the profit equations become:

$$\pi^{US} = \left[5 - \frac{1}{5}\left(Q_{US}^{US} + Q_{US}^{J}\right)\right]Q_{US}^{US} + \left[5 - \frac{1}{5}\left(Q_{J}^{J} + Q_{J}^{US}\right)\right]Q_{J}^{US} - Q_{US}^{US} - 2Q_{J}^{US} \tag{10.10a}$$

$$\pi^{J} = \left[5 - \frac{1}{5}\left(Q_{J}^{J} + Q_{J}^{US}\right)\right]Q_{J}^{J} + \left[5 - \frac{1}{5}\left(Q_{US}^{US} + Q_{US}^{J}\right)\right]Q_{US}^{J} - Q_{J}^{J} - 2Q_{US}^{J} \tag{10.10b}$$

If we assume that both companies play a noncooperative, pure-strategy static game in which domestic sales and exports are determined simultaneously, and that prices adjust so as to "clear" the market, we have a Cournot output-setting game in continuous strategies. This game is complicated by the fact that we are playing two Cournot output-setting games, one for the home market and one for the foreign market. The U.S. company, for example, must not only choose how much to produce and sell at home, but also how much to export. The same is also true for the Japanese company.

FIGURE 10.7

The U.S. and Japanese Best-Response Functions in the Japanese Market

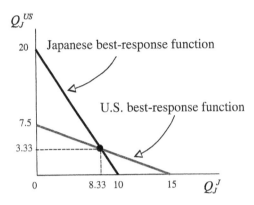

Let us first concentrate on the Cournot output-setting game that is being played in the Japanese market. Maximizing Equation (10.10a) with respect to Q_J^{US} and rearranging the terms we obtain Equation (10.11a), which is the best-response (reaction) function of the U.S. company in the Japanese market. In an analogous manner, Equation (10.11b) represents the best-response function of the Japanese company in its home market.

$$Q_J^{US} = 7.5 - \frac{1}{2}Q_J^J \tag{10.11a}$$

$$Q_J^J = 10 - \frac{1}{2}Q_J^{US} \tag{10.11b}$$

By solving Equations (10.11) simultaneously we obtain the Cournot-Nash equilibrium for this output-setting game, which is depicted in Figure 10.7. In this game, the U.S. company sells 3.33 million units and the Japanese company sells 8.33 million units in the Japanese market. Total market sales have increased from 10 million units under autarky to 11.67 million units under free trade. By symmetry, the outcome of the game being played in the U.S. market is identical, except that company unit sales are reversed.

Substituting these equilibrium output levels into Equations (10.10), producer surplus of each company from the opening of trade falls by $3.87 billion to $16.13 billion, which consists of $13.91 billion earned in the home market and $2.22 billion earned in the foreign market. At one level, this is not surprising since each company was extracting monopoly profits under autarky, but is now engaged in imperfect competition (duopoly) in the international market. On the other hand, the increased output, and lower price, in both markets results in an increase in consumer surplus. Clearly, consumers in both countries are better off from intraindustry trade.

With free trade, domestic sales in both markets increased from 10 million units to 11.67 million units and the market price falls from $3 to $2.67 per unit. As a result, consumer surplus in both countries rises from $10 billion to $13.60 billion, or an increase of $3.60 billion. This increase in the U.S. market is illustrated by the shaded area in Figure 10.8. The picture is identical for the Japanese market (not shown).

In keeping with the predictions of standard international trade theory, using the one-dollar, one-vote rule it would appear that both countries have been made worse off as a result of intraindustry trade. While domestic consumers gained from bilateral, intraindustry trade by $3.60 billion, pro-

FIGURE 10.8

Change in U.S. Consumer Surplus Resulting from an Opening of Intraindustry Trade

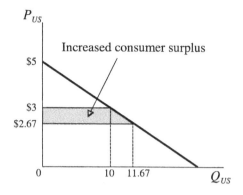

ducers were made worse off by $3.87 billion. The net change in national welfare in both countries was –$0.27 billion. Why, then, should these companies sell their product overseas?

TO TRADE OR NOT TO TRADE?

If producers are made worse off, why engage in international trade? Game theory provides insights into the strategic decision-making process. Suppose that the U.S. company adopts a *trade* strategy, while the Japanese company adopts a *no-trade* strategy. What effect will this have on the profits of each company? With the strategy profile {*Trade, No trade*}, the U.S. and Japanese profit functions are:

$$\pi^{US} = \left[5 - \frac{1}{5}Q_{US}^{US}\right]Q_{US}^{US} + \left[5 - \frac{1}{5}\left(Q_J^J + Q_J^{US}\right)\right]Q_J^{US} - Q_{US}^{US} - 2Q_J^{US} \tag{10.12a}$$

$$\pi_J^J = \left[5 - \frac{1}{5}\left(Q_J^J + Q_J^{US}\right)\right]Q_J^J - Q_J^J \tag{10.12b}$$

The best-response functions in the Japanese market are given by Equations (10.11). As before, total output is $Q_J^J + Q_J^{US} = 8.33 + 3.33 = 11.67$ million units. Total output in the U.S. market is $Q_{US}^{US} + Q_{US}^J = 10.0 + 0 = 10$ million units. Substituting these results into Equation (10.12) the profits for the strategy profile {*Trade, No trade*} are $22.23 billion and $13.93 billion for the U.S. and Japanese companies, respectively. Analogously, the profits for the strategy profile {*No trade, Trade*} are ($13.93, $22.23). The payoffs for the strategy profiles in this game are summarized in Figure 10.9.

Domestic producers have been made worse off from the opening of intraindustry trade, but what about domestic consumers? Figure 10.10 summarizes consumer surplus in both countries from the four possible strategy profiles. Clearly, consumers in both countries benefit from intraindustry trade. The Nash equilibrium strategy profile {*Trade, Trade*} in Figure 10.9 results in a $3.60 billion increase in consumer surplus. How has the national welfare of both countries been affected by the opening of trade? The national welfare payoffs, which are equal to the sum of consumer and producer surpluses, are summarized in Figure 10.11. Before trade, national welfare in both countries is $30 billion. Using the one-dollar, one-vote rule, both countries are worse off after trade since national welfare in both countries falls by $0.27 billion. The net loss to producers from intraindustry trade is greater than the net gain to consumers from lower prices.

FIGURE 10.9

Producer Surplus and the Prisoner's Dilemma in International Trade

		Japan	
		No trade	*Trade*
United States	*No trade*	($20, $20)	($13.93, $22.23)
	Trade	($22.23, $13.93)	**($16.13, $16.13)**

Payoffs: (U.S. company, Japanese company)

FIGURE 10.10

Consumer Surplus from Alternative Trade Strategy Profiles

		Japan	
		No trade	*Trade*
United States	*No trade*	($10, $10)	($13.62, $10)
	Trade	($10, $13.62)	**($13.62, $13.62)**

Payoffs: (U.S. consumers, Japanese consumers)

FIGURE 10.11

National Welfare from Alternative Trade Strategy Profiles

		Japan	
		No trade	*Trade*
United States	*No trade*	($30, $30)	($27.53, $32.23)
	Trade	($32.23, $27.53)	**($29.78, $29.78)**

Payoffs: (United States, Japan)

Although it is taken as axiomatic in orthodox modern trade theory that countries benefit from the opening of trade, this is certainly not the case in this example. One reason is that standard theories of international trade assume perfectly competitive domestic and international markets. This assumption implies that long-run economic profits before and after the opening of international trade are zero. Thus, an increase in consumer surplus implies an increase in national welfare. In our example, output is determined by monopolists in the no-trade scenario. With the opening of intraindustry trade, the market is dominated by two large producers with market power to extract economic profits. Another problem with interpreting these results has to do with measurement problems inherent in comparing the net gains and losses of the different groups using the one-dollar, one-vote rule. There is no way to know for sure whether the gains to consumers have greater or lesser weight than the losses to producers, especially in games characterized by imperfect competition.

In the game depicted in Figure 10.9, both companies have a strictly dominant *trade* strategy. The two companies are caught in a prisoner's dilemma because the Nash equilibrium strategy profile {*Trade, Trade*} results in a lower payoff for both companies ($16.13 billion) than under the no-trade scenario ($20 billion). As we will examine in the next section, it is certainly plausible that both companies would seek intervention by their respective governments to save them from themselves. Although the textbook objective of trade policy is to maximize national welfare, there is nothing that rules out the possibility that trade officials might use their authority to reward special interest groups, perhaps as a payback for political contributions. One such abuse involves an attempt to tilt competitive advantage in the direction of the domestic producer by providing export subsidies.

─────────────── **Demonstration Problem 10.2** ───────────────

Suppose that the inverse demand equations for a homogeneous good produced by a single firm in the U.S. and Japanese markets:

$$P_{US} = 10 - \frac{1}{2}\left(Q_{US}^{US} + Q_{US}^{J}\right) \tag{D10.2.1a}$$

$$P_{J} = 5 - \frac{1}{5}\left(Q_{J}^{J} + Q_{J}^{US}\right) \tag{D10.2.1b}$$

Neither company incurs fixed costs and output is in thousands of units. Total variable cost for both companies is $1 per unit of output. To sell in the Japanese market, the U.S. company incurs $4 per unit in transportation costs. To sell in the U.S. market, the Japanese company incurs $2 per unit in transportation costs.

a. What is the total profit equation for each company?
b. What is the best-response function of each company in the U.S. market? What is the best-response function of each company in the Japanese market?
c. What is the Cournot-Nash equilibrium in the U.S. market? In the Japanese market?
d. Given your answers to part c, calculate each company's profit. What is the most likely explanation for your results?

Solution

a. $$\pi^{US} = \left[10 - \frac{1}{2}\left(Q_{US}^{US} + Q_{US}^{J}\right)\right]Q_{US}^{US} + \left[5 - \frac{1}{5}\left(Q_{J}^{J} + Q_{J}^{US}\right)\right]Q_{J}^{US} - Q_{US}^{US} - 4Q_{J}^{US} \tag{D10.2.2a}$$

$$\pi^{J} = \left[10 - \frac{1}{2}\left(Q_{US}^{US} + Q_{US}^{J}\right)\right]Q_{US}^{J} + \left[5 - \frac{1}{5}\left(Q_{J}^{J} + Q_{J}^{US}\right)\right]Q_{J}^{J} - Q_{J}^{J} - 2Q_{US}^{J} \tag{D10.2.2b}$$

b. Maximizing the profit equation of each company with respect to sales in the U.S. market results in the following best-response functions:

$$Q_{US}^{US} = 9 - \frac{1}{2}Q_{US}^{J} \tag{D10.2.3a}$$

$$Q_{US}^{J} = 8 - \frac{1}{2}Q_{US}^{US} \tag{D10.2.3b}$$

Similarly, the best-response functions in the Japanese markets are:

$$Q_{J}^{US} = 2.5 - \frac{1}{2}Q_{J}^{J} \tag{D10.2.4a}$$

$$Q_{J}^{J} = 10 - \frac{1}{2}Q_{J}^{US} \tag{D10.2.4b}$$

c. Solving Equations (D10.2.3) simultaneously, the U.S. company will sell 6.67 thousand units and the Japanese company will sell 4.67 thousand units. Substituting Equation (D10.2.4b) into Equation (D10.2.4a), the output of the U.S. firm in Japan is − 3.33 thousand units. Since negative output is not feasible, the U.S. firm will produce nothing for the Japanese market. Substituting this result into Equation (D10.2.4b), the Japanese company will produce 10 thousand units of output for the Japanese market.

d. Substituting the results from part c into Equation (D10.2.2), the profits of the U.S. and Japanese firms are $22.21 and $30.88 million, respectively. The most likely explanation for these results is that high transportation costs for the U.S. company make it unprofitable to sell in the Japanese market. The Japanese firm is able to compete in the U.S. market, while maintaining its monopoly status in its home market.

EXPORT SUBSIDIES

Suppose initially that the U.S. government subsidizes exports to Japan, but that the Japanese government does not subsidize exports to the United States. An export subsidy constitutes a payment made to a firm or industry to promote foreign sales of domestically produced goods and services. Export subsidies come in a variety of forms, such as providing low-interest loans to exporters or their foreign customers, charging below-market prices for productive inputs, or by providing tax relief based on the value of goods exported.

> **Export subsidy** A payment made by government to a firm or industry to promote overseas sales.

Suppose in our example that the U.S. government subsidizes U.S. company sales in the Japanese market by $0.5Q_J^{US}$. Because of this subsidy, the profit earned by the U.S. company in the Japanese market is now:

$$\pi_J^{US} = \left[5 - \frac{1}{5}\left(Q_J^{US} + Q_J^J\right)\right]Q_J^{US} - 2Q_J^{US} + 0.5Q_J^{US} \qquad (10.13)$$

Why this particular subsidy? The size of the subsidy depends on what U.S. trade officials are trying to accomplish. Suppose that the objective is to maximize exporter profits in the foreign market. In general, if the subsidy S_{US} is applied to each unit sold, the profit function is:

$$\pi_J^{US} = \left[5 - \frac{1}{5}\left(Q_J^{US} + Q_J^J\right)\right]Q_J^{US} - 2Q_J^{US} + S_{US}Q_J^{US} \qquad (10.14)$$

To determine the "optimal" subsidy, maximize Equation (10.14) with respect to Q_J^{US}. The U.S. company's best-response function is:

$$Q_J^{US} = 7.5 - \frac{1}{2}Q_J^J + \frac{5}{2}S_{US} \qquad (10.15)$$

The best-response function of the Japanese company in its home market is still Equation (10.11b). Solving these best-response functions simultaneously we get:

$$Q_J^{US} = \frac{10}{3} + \frac{10}{3}S_{US} \qquad (10.16a)$$

$$Q_J^J = \frac{25}{3} - \frac{5}{3} S_{US} \qquad (10.16b)$$

The purpose of the subsidy is to maximize the profit of the U.S. company in the Japanese market. Substituting Equations (10.16) into Equation (10.7b) we obtain:

$$\pi_J^{US} = \frac{20}{9} + \frac{10}{9} S_{US} - \frac{10}{9} S_{US}^2 \qquad (10.17)$$

Maximizing Equation (10.17) we get:

$$\frac{d\pi_J^{US}}{dS_{US}} = \frac{10}{9} - \frac{20}{9} S_{US} = 0 \qquad (10.18)$$

It is left as an exercise for the reader (see Appendix to Chapter 1) to show that the sufficient conditions for profit maximization are satisfied. Solving Equation (10.18), the optimal subsidy is S_{US}, = \$0.50. While the total profit of the Japanese company continues to be characterized by Equation (10.10b), the total profit equation of the U.S. company is now:

$$\pi^{US} = \left[5 - \frac{1}{5}\left(Q_{US}^{US} + Q_J^{US}\right)\right]Q_{US}^{US} + \left[5 - \frac{1}{5}\left(Q_J^J + Q_J^{US}\right)\right]Q_J^{US} - Q_{US}^{US} - 1.5Q_J^{US} \qquad (10.19)$$

It is left as an exercise for the reader to verify that the companies' best-response functions in the Japanese market are:

$$Q_J^J = 10 - \frac{1}{2} Q_J^{US} \qquad (10.20a)$$

$$Q_J^{US} = 8.75 - \frac{1}{2} Q_J^J \qquad (10.20b)$$

The Nash equilibrium for this game in the Japanese market is for the U.S. company to produce 5 million units and for the Japanese company to produce 7.5 million units. The market-clearing price is \$2.5 per unit. Profits of the U.S. and Japanese companies in the Japanese market are \$5 billion and \$11.25 billion, respectively. Total U.S. company profit with the subsidy increased from \$16.11 billion to \$18.89 billion, while total profit of the Japanese company has fallen to \$13.47 billion.

How were consumers in both countries affected by the U.S. subsidy? To answer this question, it is important to remember that domestic consumers must be taxed in order to pay for the subsidy. To keep matters simple, we will assume that the reduction in consumer income has no effect on domestic demand for the product, but reduces consumer surplus by the amount of the tax. The lower price and increased sales in the Japanese market resulted in an increase in consumer surplus from \$13.59 billion to \$15.63 billion. Net consumer surplus in the U.S. market is $CS_{US} - S_{US} =$ \$11.09 billion. As a result of the U.S. export subsidy, national welfare in the United States rose from \$29.70 billion to \$29.98 billion, while national welfare in Japan has fallen from \$29.70 billion to \$29.10 billion.

RECIPROCITY

Suppose that the Japanese company, not pleased with this outcome, petitions its own government for an "optimal" export subsidy. If the request is granted, the total profit equation of the Japanese company becomes:

FIGURE 10.12
Company Profits with and Without Export Subsidies

Japan

		No subsidy	Subsidy
United States	No subsidy	($16.11, $16.11)	($13.47, $18.89)
	Subsidy	($18.89, $13.47)	**($16.25, $16.25)**

Payoffs: (United States, Japan)

$$\pi^J = \left[5 - \frac{1}{5}\left(Q_J^J + Q_J^{US} \right) \right] Q_J^J + \left[5 - \frac{1}{5}\left(Q_{US}^{US} + Q_{US}^J \right) \right] Q_{US}^J - Q_J^J - 1.5 Q_{US}^J \tag{10.21}$$

From Equations (10.19) and (10.21), the best-response functions of the two companies in the U.S. market are:

$$Q_{US}^{US} = 10 - \frac{1}{2} Q_{US}^J \tag{10.22a}$$

$$Q_{US}^J = 8.75 - \frac{1}{2} Q_{US}^{US} \tag{10.22b}$$

Solving the best-response functions simultaneously, the profit-maximizing levels of output in the Japanese and U.S. markets are $Q_J^J = Q_{US}^{US} = 7.5$ and $Q_J^{US} = Q_{US}^J = 5$. Substituting these results into Equations (10.20) and (10.21), the Cournot-Nash equilibrium strategy profile for this game is {*Subsidy, Subsidy*} is depicted in Figure 10.12. Each company earns a profit of $16.25 billion. Producers in both countries have been made better off as a result of the export subsidies, but what about the consumers?

Recalling that the export subsidies must be borne by consumers in the form of taxes, net consumer surplus in both countries is $15.63 − $2.5 = $13.13 billion. Under the no-trade scenario, consumer surplus is $10 billion. Under the trade scenario without subsidies, consumer surplus is $13.60 billion. Although consumers in both markets are still better off, they have been harmed by the export subsidy while the two companies have been enriched.

The effects of intraindustry trade on producer and consumer surpluses are summarized in Figures 10.12 through 10.14. Whether or not an export subsidy should be granted constitutes a game between the U.S. and Japanese governments. The choice of strategy depends on each government's objective. Suppose that the objective is to maximize producer surplus. This game is depicted in Figure 10.12.

If neither government grants an export subsidy, both companies will earn profits of $13.89 billion from their domestic sales and $2.22 billion from exports, for total profits of $16.11 billion. If only one government grants the "optimal" export subsidy, that country's company will earn $13.89 billion from its domestic sales and $5 billion from its Japanese sales, for total profits of $18.89 billion. On the other hand, the profits on domestic sales of the company that does not enjoy the benefit of an export subsidy will fall to $11.25 billion and its profits on sales in the foreign market will fall to $2.22 billion. Total profits fall from $16.11 billion to $13.47 billion. Finally, if both governments simultaneously provide "optimal" export subsidies, each company will earn $11.25 billion from domestic sales and $5 billion from foreign sales, for total sales of $16.25 billion. In this game, each government has a strictly dominant *subsidy* strategy. The Cournot-Nash equilibrium for this game is {*Subsidy, Subsidy*}. These payoffs in Figure 10.12 are highlighted in boldface.

FIGURE 10.13
Consumer Surplus Net of Taxes to Pay for Export Subsidies

		Japan	
		No subsidy	Subsidy
United States	No subsidy	**($13.59, $13.59)**	($15.63, $11.09)
	Subsidy	($11.09, $15.63)	($13.13, $13.13)

Payoffs: (United States, Japan)

The payoffs in Figure 10.12 do not represent the total benefits to each country from each strategy profile because they do not consider the effects of the export subsidies on consumer surplus. The beneficiaries of the export subsidy are the domestic company in the form of higher profits and foreign consumers who pay a lower price. The loser is the domestic consumer who pays for these benefits in the form of higher taxes. Figure 10.13 summarizes the payoffs in the form of consumer surplus net of export subsidies from each strategy profile.

The best outcome for consumers comes about under the free trade scenario. When only one government grants an export subsidy, the net payoff to domestic consumers falls by $2.5 billion and increases by $2.03 billion in the foreign market. Finally, when both governments subsidize exports, the net payoff to consumers in both countries falls by $0.47 billion. In this game, if the objective of government is to maximize consumer surplus, the Cournot-Nash equilibrium strategy profile for this game is {*No subsidy, No subsidy*}. These payoffs in Figure 10.13 are highlighted in boldface.

Comparing the Cournot-Nash equilibrium strategy profiles in Figures 10.12 and 10.13, it is interesting to note that the gain to producers from the export subsidies is outweighed by the losses of domestic consumers. If governments are only interested in the welfare of domestic producers, the Cournot-Nash equilibrium strategy profile is {*Subsidy, Subsidy*}. If governments are primarily concerned with the welfare of consumers, the Cournot-Nash equilibrium strategy profile is {*No subsidy, No subsidy*}. Suppose, however, that the objective of each government is to maximize net national welfare, which is defined as the sum of the payoffs in Figures 10.12 and 10.13. This game is depicted in Figure 10.14.

What is remarkable about the game depicted in Figure 10.14 is that the strategy profile {*Subsidy, Subsidy*} is a Cournot-Nash equilibrium even though both countries would be better off without export subsidies. Both governments are caught in a prisoner's dilemma since an export subsidy constitutes a strictly dominant strategy that results in lower national welfare. This result is by no means universal, however, since it depends on demand conditions, the underlying production technologies, transportation costs, and so on. Moreover, care must be exercised when interpreting the payoffs in Figure 10.14 since they depend on our faith in the one-dollar, one-vote rule.

It was noted at the beginning of this chapter that there has been a systematic reduction or elimination of international trade barriers such as import tariffs, import quotas, and export subsidies since the end of World War II. These reductions were the result of multilateral negotiations under the auspices of the General Agreement of Tariffs and Trade (GATT), and later the World Trade Organization (WTO). In fact, export subsidies are only about one-tenth of what they were in the years leading up to World War II. These reductions in trade barriers were not the result of bilateral initiatives, despite the fact that it was in each country's best interest to do so. This may have been because trade restrictions were strictly dominant strategies and the players were caught in a prisoner's dilemma. Or, it may have been that the best interests of producers with deep pockets were more important than the welfare of many small consumers with no political influence. If the game depicted in Figure 10.14 is a reasonable approximation of what goes on in the real world, the importance of multilateral negotiations becomes self-evident.

FIGURE 10.14
National Welfare from Alternative Trade Strategy Profiles

		Japan	
		No subsidy	Subsidy
United States	No subsidy	($29.70, $29.70)	($29.10, $29.98)
	Subsidy	($29.98, $29.10)	**($29.38, $29.38)**

Payoffs: (United States, Japan)

CASE STUDY 10.2: U.S.–JAPANESE VOLUNTARY EXPORT RESTRAINTS

In this chapter we examined the use of consumer-financed export subsidies to shift economic profits away from a foreign company to a domestic manufacturer engaged in intraindustry trade. The Cournot-Nash equilibrium for this game was for the U.S. and Japanese governments to subsidize exports, which increased producer surplus, lowered consumer surplus, and resulted in a loss of national welfare. Export subsidies, however, are not the only trade policy instruments at a government's disposal. Other trade policy measures include import tariffs, import quotas, domestic content requirements, voluntary export restraints (a form of import quota), and so on. It turns out that our analysis of the effects of export subsidies can be extended to include these other forms of trade intervention. Consider, for example, the use of voluntary export restraints (VERs).

Foreign automobile manufacturers first gained a significant foothold in the U.S. market in the early 1970s following a sharp increase in oil prices engineered by the Organization of Petroleum Exporting Countries (OPEC). As gasoline prices soared, American gas guzzlers quickly fell out of favor as new car buyers switched to smaller, more fuel-efficient Japanese automobiles. With a growing reputation for low-quality products, the U.S. automobile industry took another hit in the late 1970s when OPEC again sharply raised oil prices. As a result, Japanese imports soared and American automobile production and employment fell dramatically.

Not surprisingly, the ailing U.S. automobile industry sought relief from Washington in the form of trade protection. Although several congressional bills were drafted to limit foreign imports, the Carter administration countered by offering compensation to displaced U.S. automobile workers. This approach was later abandoned by the Reagan administration, which entered into bilateral trade negotiations with the Japanese. The resulting agreement called for the Japanese to voluntarily limit automobile exports to the United States. These voluntary export restraints (VERs) remained in effect from 1981 until 1994. Initially, Japanese VERs limited Japanese exports to 1.62 million units, down from 1.8 million units in 1980. This limit was gradually increased to 2.2 million units in 1985, and then to 2.3 million units by 1992. The prices of Japanese automobile imports rose dramatically. By contrast, the prices of domestically produced automobiles actually fell somewhat. While automobiles produced in Japan and sold in the United States were subject to VERs, cars produced by Japanese plants located in the United States were unaffected. Japanese companies responded to the changed trade regime by accelerating the construction of assembly plants in the United States. By 1990, five Japanese automobile companies had American manufacturing operations.

Using a Bertrand model, Barry, Levinsohn, and Pakes (1995) analyzed the national welfare effect of the VERs. They estimated that domestic automobile manufacturers earned about $9.6 billion more with VERs while the profits of Japanese producers were largely unaffected since the loss of sales was largely offset by higher prices. By contrast, U.S. consumer welfare fell by more than $12 billion due to the higher prices of Japanese imports. The net impact of the VERs was a decline in U.S. national welfare. Assuming no retaliation, had the U.S. government imposed an equivalent tariff on Japanese imports, Barry et al. estimated that $13 billion in revenues would have flowed into U.S. government coffers, which would have more than offset the loss in consumer surplus due to VERs. At one level, the effect of the VERs was similar to the effect of an equivalent import tariff. But, unlike an import tariff, the proceeds of the levy were returned to Japanese automobile producers in the form of higher U.S. prices. Barry et al. interpreted this rather unusual arrangement as a bribe paid to the Japanese government to avoid trade retaliation.

INCREASING RETURNS TO SCALE

It was noted earlier that neoclassical trade theory, which is predicated on the assumptions of perfect competition and constant returns to scale, predicts that a country tends to produce and export goods and services that use its relatively abundant factors of production intensively and import products that use its relatively scarce factors intensively. Yet, most international trade is carried out by a relatively small number of highly industrialized countries with similar resource endowments, production technology, and consumer preferences. Moreover, much of this trade involves the two-way trade of goods and services produced by companies in the same industry.

In the previous section it was demonstrated that imperfect competition helps to explain this apparent contradiction and that it may be possible for a government to improve national welfare by shifting profits away from the foreign companies by subsidizing the exports of domestic producers. When both governments subsidize exports, they may be caught in a prisoner's dilemma in which the overall impact on a country's overall national well-being is less clear. Each country may have a strictly dominant strategy to subsidize exports, which could make both countries worse off than if neither government had intervened.

What about the other footing of neoclassical trade theory—constant returns to scale? Production characterized by **increasing returns to scale** will exhibit economies of scale in which per unit costs of production fall with an increase in total output. As discussed in Chapter 9, if a single firm's production technology exhibits economies of scale over a range of output that is sufficient to satisfy total market demand, it may be possible to drive potential rivals out of the market through price cuts. Such firms are referred to as natural monopolies. As it turns out, increasing returns to scale strengthens the conclusions of the previous section.

Increasing returns to scale When a proportional increase in the use of all inputs results in a more than proportional increase in total output.

Although the production function and the best-response functions are more complicated, this game has a Cournot-Nash equilibrium in which the firms have an even stronger incentive to engage in intraindustry trade. Moreover, increasing returns to scale in production increases a government's incentive to subsidize exports. Not only will export subsidies shift profits from the foreign to the domestic producer, but the resulting increase in output reduces the firm's per unit cost of production. Governments that subsidize exports may even find that they are no longer caught in a prisoner's dilemma since the result may be a net increase in national welfare.

CHAPTER REVIEW

This chapter applied the principles of game theory to an analysis of *strategic trade policy,* which refers to measures taken by governments to influence the decisions of consumers and producers engaged in international trade. Trade policy instruments, such as import tariffs, import quotas, export subsidies, domestic content requirements, trade regulations, and so on, redirect the flow of goods and services across national boundaries. Although the textbook objective is to maximize national welfare, trade policy can be, and is, used to benefit special-interest groups with political influence.

The principle of comparative advantage and the Heckscher-Ohlin theorem lie at the heart of our formal understanding of international trade. Although neoclassical trade theory predicts that a country will tend to produce and export those goods and services that use its relatively abundant factors of production intensively and import those products that use its relatively scarce factors intensively, an examination of the trade data reveals something quite different. Most international trade is carried out by a relatively small number of highly industrialized countries with similar resource endowments, production technology, and consumer preferences. Much of this trade involves the simultaneous export and import of goods and services produced by companies in the same industry. This is referred to as *intraindustry trade.*

An explanation for this apparent contradiction may be found in the underlying assumptions of neoclassical theory. Neoclassical trade theory assumes perfect competition, constant returns to scale, and identical production functions. These assumptions are often violated in reality. In fact, most international trade in terms of value is carried out by imperfectly competitive companies that exercise market power in both home and foreign markets.

In spite of the predictions of neoclassical trade theory, it may be possible for a government to improve a country's national welfare by subsidizing exports of domestic producers. By changing the market-clearing price and market shares in the foreign market, the export subsidy enhances the profits of the domestic producer at the expense of its foreign rivals. Even when countries are not made better off, governments may still be caught in a prisoner's dilemma since subsidizing exports may be a dominant strategy. This helps explain why the systematic reduction or elimination of trade barriers since World War II as been the result of multilateral negotiations sponsored by such multinational organizations as GATT or the WTO.

Increasing returns to scale strengthens a government's incentive to subsidize exports. Not only will export subsidies shift profits from the foreign to the domestic producer, but the resulting increase in output reduces the firm's per unit cost of production. What is more, these governments may no longer be trapped in a prisoner's dilemma since the outcome may be an increase in national welfare.

CHAPTER QUESTIONS

10.1   Suppose that there are two countries engaged in free international trade. Who are the players in this game? What are the strategies and payoffs?

10.2   Define strategic trade policy. Do you believe that strategic trade policy has a role in perfectly competitive international markets? Explain.

10.3   How are consumer and producer surplus used to assess the benefits of international trade versus autarky? What are the advantages? What are the drawbacks?

10.4   What is the one-dollar, one-vote rule? How is this rule used to assess the benefits of international trade? What are its deficiencies?

10.5 What is intraindustry trade? Is intraindustry trade possible according to the predictions of orthodox neoclassical trade theory?

10.6 According to the text, what assumptions about neoclassical trade theory need to be relaxed to permit intraindustry trade to occur? Can you think of any other explanations for the existence of intraindustry trade?

10.7 Are two pre-trade domestic monopolies better off before or after the opening of intraindustry trade? Explain your answer. Are consumers better off? Explain. Is the country better off? How sure are you of your answer?

10.8 Which group is most likely to benefit from government intervention to assist domestic producers engaged in international trade? Is the country as a whole better off or worse off? How sure are you of your answer?

10.9 Explain how export subsidies and voluntary export restraints are similar. How are they different?

10.10 What were the national welfare effects of the Japanese decision voluntarily to restrict automobile exports to the United States in the 1980s and 1990s? Would your answer have been different if voluntary export restraints had been replaced with an equivalent tariff on imports of Japanese automobiles?

10.11 In this chapter, we assumed that the objective of government trade policy was to maximize the domestic monopoly's profits. Do you believe that this assumption was reasonable?

10.12 In your view, what should be the objective of strategic trade policy? Do you believe that this is the objective of trade policy in reality? Why?

CHAPTER EXERCISES

10.1 Suppose that there are three identical firms considering producing a homogeneous product for sale in the global marketplace. Figure E10.1 summarizes the payoffs in billions of dollars.

   a. If larger payoffs are preferred, what is the Nash equilibrium strategy profile(s) for this game?

   b. Suppose that country A offers $5 billion to firm A as a production subsidy. If larger payoffs are preferred, what is the Nash equilibrium strategy profile for this game?

   c. Suppose that countries A and B offer firms A and B $5 billion in production subsidies. If larger payoffs are preferred, what is the Nash equilibrium strategy profile for this game?

   d. Suppose that countries A, B, and C offer firms A, B, and C $5 billion in production subsidies. If larger payoffs are preferred, what is the Nash equilibrium strategy profile for this game?

10.2 Suppose that intraindustry trade between the United States and Japan involves just two companies. The inverse demand equations for an identical product are:

$$P_{US} = 9 - \frac{1}{3}Q_{US} \qquad \text{(E10.1.1a)}$$

$$P_J = 9 - \frac{1}{3}Q_J \qquad \text{(E10.1.1b)}$$

FIGURE E10.1

Static Game in Chapter Exercise 10.1

Firm C

		Produce		Don't produce	
		Firm B		**Firm B**	
		Produce	Don't produce	Produce	Don't produce
Firm A	Produce	(−1, −1, −1)	(2, 0, 2)	(2, 2, 0)	(5, 0, 0)
	Don't produce	(0, 2, 2)	(0, 0, 5)	(0, 5, 0)	(0, 0, 0)

Payoffs: (Firm *A*, Firm *B*, Firm *C*)

where $Q_{US} = Q_{US}{}^{US} + Q_{US}{}^{J}$ and $Q_J = Q_J{}^{J} + Q_J{}^{US}$. Neither firm incurs fixed costs and the marginal cost of production is constant at $3 per unit. Overseas shipping costs are $1 per unit for each company. Sales are in millions of units.

a. Determine the Cournot-Nash equilibrium output levels in the U.S. and Japanese markets.

b. Using the one-dollar, one-vote rule, compare producer and consumer surpluses before and after the onset of intraindustry trade. Who are the winners and who are the losers? Are the two countries better off or worse off as a result of intraindustry trade? Can you be sure?

c. Suppose that Japanese trade officials decide to subsidize exports of the Japanese company. Determine the per unit subsidy if the objective is to maximize Japanese company profits net of the export subsidy.

d. Suppose that the Japanese government grants an "optimal " export subsidy, but the U.S. government does not. Who gains and who loses in both countries?

e. Who are the winners and losers if the United States decides to retaliate with its own export subsidy compared with the no-export-subsidy case?

f. Suppose that the objective of trade officials is to maximize consumer surplus. What is the Nash equilibrium for this game?

10.3 Suppose that in your answer to chapter exercise 10.2 the objective of strategic trade policy is to maximize consumer surplus. Using the one-dollar, one-vote rule, determine the free-trade Cournot-Nash equilibrium for each country.

10.4 Suppose that the two companies in chapter exercise 10.2 are Bertrand price-setting competitors.

a. What are the best-response functions for this game?

b. What is the Bertrand-Nash equilibrium for this game assuming no government intervention?

c. Compare producer surplus before and after the onset of intraindustry trade.

10.5 Suppose that intraindustry trade between countries *A* and *B* involves just two companies. The inverse demand equations for an identical product in the two countries are:

$$P_A = 25 - \frac{1}{5}Q_A \qquad\qquad \text{(E10.5.1a)}$$

$$P_B = 10 - \frac{1}{2}Q_B \qquad\qquad\qquad \text{(E10.5.1b)}$$

where $Q_A = Q_A{}^A + Q_A{}^B$ and $Q_B = Q_B{}^B + Q_B{}^A$. Neither firm has fixed cost, the marginal cost of production is \$2 per unit, and transportation costs are \$1 per unit.

a. What is the Cournot-Nash equilibrium for this game?
b. How much will each company earn from intraindustry trade? How do you explain the differences in your results?
c. If the objective of country B is to maximize company B profits, what is the optimal export subsidy? What is the optimal export subsidy for company A?
d. What is the Cournot-Nash equilibrium if the objective of both governments is to maximize national welfare using the one-dollar, one-vote rule?

ENDNOTES

1. This outcome is referred to as a *Pareto improvement* since the subsidy to Boeing has made at least one player better off without making any other player worse off.
2. Neoclassical trade theorists assumed international markets were perfectly competitive. Thus, firms engaged in international trade are "price takers." The constant-returns-to-scale assumption rules out the emergence of a natural monopoly. A natural monopoly is a firm that is able to satisfy total market demand at a per unit cost of production that is less than any other firm in the industry. A firm that exhibits increasing returns to scale may be able to drive its competitors out of the market by pricing below their per unit cost of production.
3. If necessary, the reader should review the discussion of integrals as the area under a curve in the Appendix to Chapter 1.

CHAPTER

11 DYNAMIC GAMES WITH COMPLETE AND PERFECT INFORMATION

In this chapter we will:

- *Analyze dynamic games with complete and perfect information;*
- *Introduce game trees;*
- *Discuss the distinction between subgame equilibria and a subgame perfect equilibrium;*
- *Find equilibria to dynamic games using the backward induction solution concept;*
- *Describe the first mover's advantage;*
- *Discuss the importance of credible threats in dynamic games;*
- *Analyze dynamic games with continuous strategies;*
- *Introduce the concepts of Stackelberg leader and follower.*

INTRODUCTION

In this chapter we explore the basic elements of dynamic games under conditions of complete and perfect information. Dynamic games are also referred to as sequential-move or multistage games. Dynamic games differ from static games in that the players move in sequence, that is, they take turns. When playing dynamic games, each player attempts to determine a rival's most likely responses, and uses that information to formulate an optimal strategy. As the game proceeds, each player acquires information about a rival's strategy based on earlier moves and countermoves. This information is used to update a player's strategy whenever possible.

GAME TREES

The diagrammatic representation of a sequential-move game is referred to as an **extensive-form game**, or, more popularly, a **game tree**. Game trees summarize the players, the information available to each player at each stage of the game, the order of the moves, and the payoffs from alternative strategy profiles. In a dynamic game, a strategy is the complete description of a player's moves at each stage of the game. In addition to the assumption of complete information, we also assume that each player has **perfect information**. By perfect information we mean that each player is aware of a rival's prior moves. A player with perfect information knows his or her location in the game tree.

> **Extensive-form game** A diagrammatic representation of a dynamic game that summarizes the players, the stages of the game, the information available to each player at each stage, player strategies, the order of the moves, and the payoffs from alternative strategies.

> **Perfect information** When each player is aware of a rival's prior moves. A player with perfect information knows his or her location in a game tree.

Board games such as chess, checkers, and backgammon are examples of games involving perfect information. By contrast, a player who moves without knowing how a rival has already moved is said to have imperfect information. In one-time, static games, players have complete, but imperfect, information. It is also possible for players in dynamic games to have imperfect information. When this happens, the players may not be aware of their position in the game tree.

Game trees are similar to **decision trees**, which are used to determine an optimal course of action in situations that do not involve strategic interaction. As an illustration, suppose that you are traveling from Dupont Circle in Washington, D.C., to the corner of Second Avenue and East 78th Street in Manhattan. A friend offers you a ride to any place you desire to begin your trip. You must make several decisions. To begin with, you will have to decide whether to travel by automobile, train, airplane, or bus. If you decide to go by car, after traveling north via U.S. Interstate 95, you must decide whether to use the Verrazano-Narrows Bridge, the Holland Tunnel, the Lincoln Tunnel, or the George Washington Bridge. If you travel by Amtrak, should you disembark at Newark and take a PATH train into lower Manhattan, or continue on to Penn Station? If you travel by air, should you purchase a ticket for Newark Liberty International, John F. Kennedy International (JFK), or La Guardia Airport? If you land at La Guardia or JFK, which are both located in Queens, should you travel into Manhattan by bus, subway, or taxi cab? From Newark International, you can take Amtrak and PATH into Manhattan. Once in Manhattan, should you walk, take a bus, subway, or taxi cab? Perhaps you even know someone who lives in Manhattan and actually owns a car!

These and other travel options can be illustrated with a diagram that resembles a tree. Choices are made at branching points. There is one branch for each decision. The idea is to move along the tree from its root (Washington, D.C.), to branch, to leaf (Second Ave. and East 78th St.). How you decide to climb the decision tree depends on what you are trying to accomplish. Traveling the distance as quickly as possible will require one choice of branches, while getting there as inexpensively as possible will require another.

> **Decision tree** A diagram used to determine an optimal course of action in a situation that does not involve strategic interaction.

The main difference between a decision tree and a game tree is the number of players. Game trees involve two or more players who alternate moves at each branching point, or **decision node**. In the game tree depicted in Figure 11.1, the decision nodes are depicted as rectangles. Within each rectangle is the name of the player moving at that decision node. At each decision node, the indicated player must decide how to move. Each move, which is represented as an arrow, is called a **branch**. A game tree is the complete collection of decision nodes and branches. The first decision node, which is located at the far left of the game tree in Figure 11.1, is called the **root of the game tree**. All subsequent decision nodes are referred to as **subroots**. At the far right of the figure are the "leaves" of the game trees. These small, closed circles are called **terminal nodes**. The game ends at the terminal nodes. To the right of the terminal nodes are the players' payoffs. The first entry in the parenthesis is the payoff to player A and the second entry is the payoff to player B.

> **Decision node** The location in a game tree where the designated player decides his or her next move.
>
> **Branches** Arrows from decision nodes in game trees indicating a player's possible moves.
>
> **Game tree** The complete collection of decision nodes and branches in an extensive-form game.
>
> **Root of a game tree** The first decision node in a game tree.

FIGURE 11.1
Game Tree

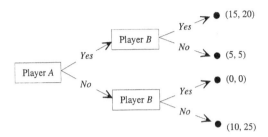

Payoffs: (Player *A*, Player *B*)

Terminal node The point on a game tree where a game ends.

In the game depicted in Figure 11.1, the order of play is from left to right. At the root of the game tree, player *A*'s move represents the first stage of the game. Player *A* must decide whether to move *yes* or move *no*. After player *A* moves, player *B* must then decide on a countermove in the second stage of the game. For example, if player *A* moves *yes,* player *B* must decide whether to respond with a *yes* move or a *no* move.

As with static games, the players' payoffs depend upon the strategies adopted. In the game depicted in Figure 11.1, player *A* moves without knowing player *B*'s intended response. Player *B*'s move, on the other hand, is contingent on how player *A* moves. While player *B* moves with the knowledge of player *A*'s move, player *A* can only anticipate how player *B* will react. The ideal order of play is for player *A* to begin with a *yes* move, followed by a *yes* move by player *B,* which results in payoffs of (15, 20). By contrast, the ideal order of play for player *B* is for player *A* to open with a *no* move, and for player *B* to follow with a *no* move. This sequence of moves will result in payoffs of (10, 25).

CASE STUDY 11.1: SUN COUNTRY VERSUS NORTHWEST AIRLINES

For sixteen years, privately owned Sun Country Airlines operated charter airline service out of the Minneapolis–St. Paul metropolitan area. On June 1, 1999, Sun Country began regularly scheduled nonstop service to Boston, New York, Washington, D.C., Los Angeles, and seven other destinations. Although Sun Country's chief competitor was Northwest Airlines, senior management believed there was enough room in the market for two airlines. Unfortunately for Sun Country, Northwest Airlines saw things quite differently.

On June 7, an executive of Northwest Airlines circulated an internal memo stating: "Effective immediately, parts support to Sun Country is terminated until further notice." A Sun Country executive claimed that Northwest's action violated an industry tradition of mutual cooperation in aircraft maintenance assistance. Without such cooperation, the cost to any individual airline of maintaining sufficient spare parts would be prohibitively expensive. Within days, several Sun Country flights experienced extensive delays because of a shortage of spare parts. Northwest's retaliation did not stop there.

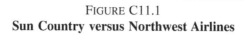

FIGURE C11.1
Sun Country versus Northwest Airlines

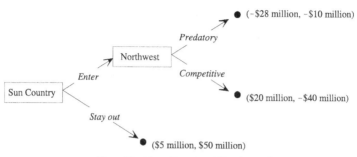

Payoffs: (Sun Country, Northwest)

In September 2000, Sun Country released a study accusing Northwest of predatory pricing by unleashing an "avalanche" of discount seats.[1] According to the study, which was underwritten by Sun Country and a grant from the Hughes foundation, Northwest systematically cut prices and added 33 percent more seats in the seven markets where Sun Country launched competitive nonstop service. The author of the study, Professor Paul Dempsey, director of the Transportation Law Program at the University of Denver, wrote that "Northwest's seat dumping dramatically reduces the potential for viable competitive service." According to Dempsey, "Northwest's actions in the Minneapolis–Milwaukee market demonstrate why Northwest has become the poster child for bad behavior in the airline industry." Not surprisingly, Northwest spokesperson Jon Austin branded the study's allegations as "baseless." Austin also questioned the study's impartiality since Dempsey had served on the board of directors of Frontier Airlines, a low-cost carrier based in Denver. In its first year and a half of operations, Sun Country reportedly lost $27.9 million.[2]

So, where did upstart Sun Country go wrong? Why wasn't the upstart airline able to anticipate Northwest's aggressive response? It would appear that executives at Sun Country failed to project forward and reason backward—a process known as backward induction. Consider the hypothetical situation depicted in Figure C11.1. Suppose that Sun Country decided to stay out of the regularly scheduled airline market. Sun Country earns $5 million in profits, while Northwest earns $50 million. On the other hand, if Sun Country enters the market, Northwest can continue its competitive pricing strategy, or engage in predatory pricing. If Northwest prices competitively, Sun Country will earn $20 million in profits, while Northwest loses $40 million. If Northwest engages in predatory pricing, Sun Country will lose $28 million while Northwest loses a lesser $10 million. Perhaps because of its "deep pockets," Northwest may be able to maintain its predatory pricing policy until Sun Country is forced out of the market, after which Northwest can recover its short-term losses by charging monopoly air fares.

The above scenario highlights the importance of predicting Northwest's most plausible response to alternative strategies and then to reason backward. Sun Country failed to do this by assuming that Northwest Airlines would continue business as usual. If it had attempted to anticipate Northwest's response, Sun Country might have modified its entry strategy accordingly, perhaps by gradually converting to a full-service airline by adding regularly scheduled, nonstop service one destination at a time.

SUBGAME PERFECT EQUILIBRIUM

As the reader may have surmised, the outcome to the game depicted in Figure C11.1 depends on whether Sun Country believes that Northwest's threat to engage in predatory pricing is credible. We can also see this in the game depicted in Figure 11.1. The outcome to that game depends on whether player *A* believes that player *B* will follow through with the threat to respond with a *no* move. Is there any reason to believe that player *B*'s threat is credible? Probably not. To see why, recall that player *A*'s best move is *yes*. It is rational for player *B* to follow up with a *yes* move as well. This sequence of moves results in payoffs to player *A* and player *B* of (15, 20). If player *B* is rational, the threat to move *no* lacks credibility since this will result in a payoff that is lower than by moving *yes*.

Note that the strategy profile in which both players move *yes* is a Nash equilibrium because neither player can do better by moving differently. In particular, player *B*'s best move is *yes* if player *A* moves *yes,* and *no* if player *A* moves *no*. Given that player *A* moves first, player *B* will be worse off by moving in any other way. The strategy profiles *yes* followed by *yes,* and *no* followed by *no,* are Nash equilibria, while the strategy profiles *yes* followed by *no,* and *no* followed by *yes,* are not.

The game depicted in Figure 11.1 has two Nash equilibria, but which combination of moves results in the most plausible outcome? Which outcome is the most reasonable? In this case, it is the Nash equilibrium corresponding to the sequence of moves *yes* by player *A,* followed by a *yes* by player *B.* This is because player *B* has no incentive to follow through with a *no* threat. This sequence of moves is a Nash equilibrium because neither player can obtain a better payoff by moving differently at any stage of the game.

The development of the idea of an equilibrium in sequential-move games owes much to the work of Reinhard Selten (1975), who, along with John Nash and John Harsanyi, was awarded the 1994 Nobel Prize in economics for his pioneering work in game theory. Selten formalized the concept of a **subgame**, which is any subset of branches and decision nodes in a game tree.

Selten's important contribution was to recognize that each subgame constitutes a game in itself. The initial decision node of a subgame is called a **subroot**. Every decision node is the subroot of a new subgame. A subgame has three basic characteristics. To begin with, all subgames have the same players as the entire game itself, although some of these players may not move in the subgame. Second, a subgame includes the subroots, branches, decision nodes, and terminal nodes of all subsequent subgames. Thus, every game is a subgame of itself, and is referred to as a **trivial subgame**. All subgames that do not include the root of the game tree are called **proper subgames**. Because each trivial subgame includes all proper subgames, the payoffs of each of the subgames are included among the payoffs of the entire game.

> **Subgame** A subset of branches and decision nodes in a game tree.
>
> **Subroot** The root of a subgame.
>
> **Trivial subgame** A subgame that begins at the root of the game tree.
>
> **Proper subgame** A subgame that begins at a decision node other than the root of the game tree.

Once a player begins a subgame, he or she must play to the end. Once a player has moved, he or she cannot turn back or exit the subgame in search of an alternative outcome. Each subgame must be treated as a complete and self-contained game with its own unique **subgame equilibrium**. As it turns out, there are at least as many subgame equilibria as there are subgames. With trivial

FIGURE 11.2

A Subgame

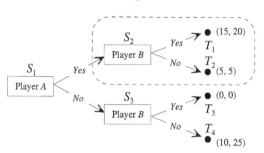

Payoffs: (Player *A*, Player *B*)

exceptions, a unique equilibrium for the entire game will be found among the complete collection of subgame equilibria. This unique outcome is called a **subgame perfect equilibrium**.

> **Subgame equilibrium** The Nash equilibrium of a proper subgame.

> **Subgame perfect equilibrium** The Nash equilibrium for a dynamic game, which is found among the subgame equilibria.

The concept of a subgame is illustrated in Figure 11.2, which consists of three subgames. The sequential-move game itself begins at the initial decision node (root of the game tree), S_1. The subgame that begins at the decision node S_1 is a trivial subgame. Two nontrivial, proper subgames begin at decision nodes (subroots) S_2 and S_3. The subgame that begins at subroot S_2, which is enclosed by the dashed, rounded rectangle, has two terminal nodes, T_1 and T_2, with payoffs of (15, 20) and (5, 5), respectively. Each of the subgames commencing at subroot S_2 and S_3 has a unique subgame equilibrium.

Example: Attack of the Clone

Nexus Corporation is planning to introduce a new online computer game into the market.[3] Nexus' management is considering two marketing strategies. The first involves a "Madison Avenue"–style advertising campaign, while the second emphasizes a word-of-mouth approach. The first marketing strategy is referred to as the "slick" approach and the second a "simple" approach. Although expensive, a *slick* advertising strategy will result in high sales volume in the first year, although sales in the second year will decline substantially as the market becomes saturated. The inexpensive *simple* approach, on the other hand, is expected to result in lower sales volume in the first year, but much higher sales volume in the second year as "word gets around." Regardless of the marketing strategy adopted by Nexus, no significant sales are anticipated after the second year. Nexus' expected profits from both campaigns are summarized in Table 11.1.

The accounting data presented in Table 11.1 suggests that Nexus should adopt the inexpensive *simple* approach because of the greater total net profits. The problem for Nexus, however, is the threat of "legal clones." In this case, a legal clone of Nexus' computer game is a competing game manufactured by Mobius LLC that is, for all outward appearances, a near-perfect substitute. The difference between the two computer games is the underlying programming code, which is sufficiently different from the original that Mobius cannot be sued for copyright infringement. In this example, Mobius is able to clone Nexus' computer game within a year at a cost of $300,000. If

TABLE 11.1
Nexus' Profits if Mobius Does Not Enter the Market

	Slick	Simple
Gross profit in year 1	$900,000	$200,000
Gross profit in year 2	$100,000	$800,000
Total gross profit	$1,000,000	$1,000,000
Advertising cost	–$570,000	–$200,000
Total net profit	$430,000	$800,000

Mobius decides to produce the clone and enter the market, the two firms will split the market in the second year. The payoffs for both companies are summarized in Tables 11.2 and 11.3. What are the optimal marketing strategies for Nexus and Mobius? The above situation may be depicted in Figure 11.3.

In Figure 11.3, Nexus moves first and must decide whether to adopt a *slick* or a *simple* marketing strategy. Mobius' strategy, on the other hand, is conditional on the strategy adopted by Nexus. Since Nexus must decide between *slick* and *simple,* Mobius has two decision nodes. If Nexus adopts a *slick* strategy, Mobius' best response is to *stay out* and earn nothing, compared with losing $250,000 by adopting an *enter* strategy. If Nexus adopts a *simple* strategy, Mobius should *enter* and earn $100,000, compared with earning nothing by adopting a *stay out* strategy. Even before Nexus moves, Mobius has determined that its optimal responses are *stay out* if *slick,* and *enter* if *simple.* Of course, which of these two strategies Mobius adopts will depend upon the advertising campaign adopted by Nexus. So, what strategy will Nexus choose, and why?

BACKWARD INDUCTION

There is no universally accepted method of determining a subgame perfect equilibrium. The most commonly used approach, which was developed by Reinhard Selten (1965), is backward induction, also known as the fold-back method, which was briefly discussed in Chapter 5. The rationale underlying this solution algorithm is simple: Determine a rival's best response to a move then use that information to formulate a strategy that will result in the best payoff. This commonsense approach to playing sequential-move games can be codified in the following principle governing strategic behavior: Project forward, reason backward.

Principle: Project forward then reason backward.

Before discussing the backward induction methodology in detail, we first consider the payoffs in Figure 11.3, which are summarized in the normal-form game depicted in Figure 11.4. This is a one-time, static game in which Nexus has a strictly dominant *simple* marketing strategy. Mobius, on the other hand, does not have a dominant strategy. This is because the strategy adopted by Mobius depends on the strategy adopted by Nexus. Nevertheless, the Nash equilibrium strategy profile for this game is {*Simple, Enter*}.

There are several drawbacks with using the normal form of a game when searching for a Nash equilibrium. For one thing, it can only accommodate at most three players and two stages. If there

TABLE 11.2
Nexus' Profits if Mobius Enters the Market

	Slick	Simple
Gross profit in year 1	$900,000	$200,000
Gross profit in year 2	$50,000	$400,000
Total gross profit	$950,000	$600,000
Advertising cost	−$570,000	−$200,000
Total net profit	$380,000	$400,000

TABLE 11.3
Mobius' Profits After Entering the Market

	Slick	Simple
Gross profit in year 1	$0	$0
Gross profit in year 2	$50,000	$400,000
Total gross profit	$50,000	$400,000
Cloning cost	−$300,000	−$300,000
Total net profit	−$250,000	$100,000

are more than three players and two stages then this setup won't work. To make matters worse, if the players have two or more possible countermoves, eliminating dominated strategies may result in multiple Nash equilibria. While this is not unusual for static games, dynamic games with complete and perfect information typically have unique subgame perfect equilibria.

More important, the normal form is not sensitive to the sequential nature of the game. Recall that Nexus moves first and can choose between two strategy options: *Slick* or *simple*. Mobius, on the other hand, cannot move until after Nexus has made its move. This does not mean, however, that Mobius cannot plan ahead. In this game, Mobius has four possible responses: [*Enter, Enter*], [*Enter, Stay out*], [*Stay out, Enter*], or [*Stay out, Stay out*]. The first strategy in the square brackets corresponds to Mobius' move when Nexus adopts a *slick* strategy, while the second entry is its response to a *simple* strategy. While this collection of possible response strategies is exhaustive, they are not all equally likely. For example, the strategy response profile [*Enter, Stay out*] says that if Nexus adopts a *slick* strategy, the countermove by Mobius is to *enter*. This will result in a payoff to Mobius of −$250,000, instead of $0 by staying out of the market. This strategy response profile also says that if Nexus adopts a *simple* strategy, the countermove by Mobius is to *stay out*. This will result in a payoff to Mobius of $0, instead of $100,000 by entering. If Mobius is rational, these countermoves are highly unlikely. Thus, the strategy response profile [*Enter, Stay out*] would be ruled out as implausible.

FIGURE 11.3
Attack-of-the-Clone Game

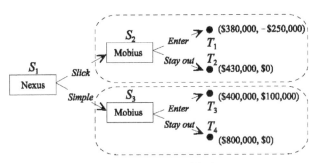

Payoffs: (Nexus, Mobius)

FIGURE 11.4
Two-Player, Static Game

Mobius

		Enter	Stay out
Nexus	Slick	($380, –$250)	($430, **$0**)
	Simple	(**$400, $100**)	(**$800**, $0)

Payoffs: (Nexus, Mobius)

FIGURE 11.5
Strategic Form of the Attack-of-the-Clone Game

Mobius

		[Enter, Enter]	[Enter, Stay out]	[Stay out, Enter]	[Stay out, Stay out]
Nexus	Slick	($380, –$250)	($380, –$250)	(**$430, $0**)	(**$430, $0**)
	Simple	(**$400, $100**)	(**$800**, $0)	($400, **$100**)	(**$800**, $0)

Payoffs: (Nexus, Mobius)

Is it possible to modify Figure 11.4 to preserve the sequential nature of the game depicted in Figure 11.3? Consider Figure 11.5, which combines Nexus' advertising strategies with the complete set of Mobius' strategy response profiles. Figure 11.5 is referred to as the **strategic form of the sequential-move game**.

> **Strategic form of a sequential-move game** Summarizes the payoffs to each player from every possible strategy response profile.

The strategic form game in Figure 11.5 resembles a normal-form game with payoffs in thousands of dollars from all possible strategy profiles. Recall that the first entry in the brackets corresponds to Mobius' strategic response to a *slick* strategy, while the second entry corresponds to Mobius' response to a *simple* strategy. Although Mobius must wait for Nexus to move, this does not mean that Mobius cannot plan ahead. The complete set of Mobius' response strategies is listed in Figure 11.5.

How do we interpret the entries in Figure 11.5? If Nexus adopts a *slick* strategy, and Mobius responds with *enter,* the payoffs in thousands of dollars are ($380, –$250). If Nexus adopts a *slick* strategy, a *stay out* response by Mobius will result in payoffs of ($430, $0). Clearly, the optimal response by Mobius to a *slick* strategy is to *stay out* since this will result in a payoff of $0, instead of –$250,000 by entering the market. The relevant strategy profiles are {*Slick,* [***Stay out,*** *Enter*]} or {*Slick,* [***Stay out,*** *Stay out*]} (Remember: the second entry in the square brackets is Mobius' response to a *simple* strategy, which will not be considered once Nexus has committed itself to a *slick* strategy.) For both of these strategy profiles the payoffs are ($430, $0).

Likewise, if Nexus adopts a *simple* strategy, the appropriate response is the second entry in the square brackets. If Mobius responds with an *enter* move, the payoffs are ($400, $100). If Mobius responds by staying out of the market, the payoffs are ($800, $0). The optimal response by Mobius to a *simple* strategy is to *enter* since this will result in a payoff of $100,000, instead of $0 by staying out. The appropriate strategy profiles are {*Simple,* [*Enter,* ***Enter***]} or {*Simple,* [*Stay out,* ***Enter***]}.

We will now proceed to search for the Nash equilibrium for the game in Figure 11.5. We will begin by looking at the problem from Nexus' perspective. Using the method first introduced in Chapter 1, the reader should verify that this game appears to have two Nash equilibria: {***Slick,*** [***Stay out,*** *Enter*]} and {***Simple,*** [*Enter,* ***Enter***]}. Although the notation appears complicated, the interpretation is straightforward. The strategy profile {*Slick,* [*Stay out, Enter*]} says that if Nexus adopts a *slick* strategy, Mobius will counter by staying out of the market. The strategy profile {*Simple,* [*Enter, Enter*]} says that if Nexus adopts a *simple* strategy, Mobius will counter by entering the market. In fact, these two strategy profiles represent the subgame equilibria for the subgames commencing at subroots S_2 and S_3 in Figure 11.3. The unique subgame perfect equilibrium for this game must be one of these two strategy profiles. Which one depends on Nexus. Since Nexus moves first, it will clearly choose *slick* because of the larger payoff. Thus, the subgame perfect equilibrium for the game in Figure 11.3 is {*Slick,* [*Stay out, Enter*]}.

The above approach to finding a subgame perfect equilibrium has several shortcomings. To begin with, although it utilizes the familiar payoff matrix of the normal-form game, the notation is complicated. Moreover, if the game has more than two players or stages, the notation becomes exponentially more complicated. Its main strength is that it explicitly recognizes the forward planning of the players, which is captured by the strategy response profiles. Fortunately, there are far simpler and more efficient methods for finding subgame perfect equilibria for games of the type depicted in Figure 11.3. The most widely used of these is backward induction, which involves five steps (Bierman and Fernandez 1998, pp. 129–130):

Step 1: Start at the terminal nodes. Trace each node to its immediate predecessor decision node. These decision nodes may be described as "basic," "trivial," or "complex." A basic decision node has branches that lead only to terminal nodes. A decision node is trivial if it has only one branch. A decision node is complex if at least one branch leads to a terminal node. If a trivial decision node is reached, continue moving through the game tree until you get to a complex or a nontrivial decision node.

Step 2: From the decision node identified in step 1, compare each of the payoffs for that player. The branch that leads to the best payoff represents that player's optimal move. The branches representing all other moves are nonoptimal and should be eliminated from further consideration.

Step 3: After pruning away all nonoptimal branches identified in step 2, move to the decision node immediately preceding the decision node identified in step 1. From among the remaining branches, choose the move that leads to the best payoff for the player at that decision node. Prune away all other nonoptimal branches.

FIGURE 11.6

Finding a Subgame-Perfect Equilibrium Using Backward Induction

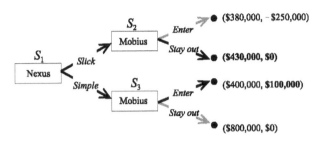

Payoffs: (Nexus, Mobius)

Step 4: Keep repeating steps 1–3 until the root of the tree has been reached.

Step 5: From the root of the game tree, identify the sequence of optimal moves that leads to the best payoff for that player. The ordered collection of moves from root to leaf represents the subgame perfect equilibrium.

The backward induction solution algorithm will now be applied to the dynamic game depicted in Figure 11.3. From each terminal node, move to the two Mobius decision nodes. Each of these decision nodes is basic since the branches lead to exactly one terminal node. If Nexus adopts a *slick* marketing strategy, the optimal response by Mobius is to *stay out* since the payoff is $0, compared with a payoff of –$250,000 by entering the market. The *enter* branch following *slick* should now be disregarded from any further consideration. If Nexus adopts a *simple* marketing strategy, the optimal response by Mobius is an *enter* strategy since the payoff is $100,000, compared with a payoff of $0 with a *stay out* move. Thus, the *stay out* branch following *simple* should be eliminated. The optimal strategy response profile for Mobius is [*Stay out, Enter*]. The resulting extensive-form game is illustrated in Figure 11.6. The root of the game tree has, in effect, been transformed into a basic decision node. The remaining heavily shaded branches lead directly to terminal nodes. We are now ready to identify Nexus' optimal advertising strategy.

An examination of the extensive-form game in Figure 11.6 clearly suggests that Nexus should adopt a *slick* strategy since this will lead to a payoff of $430,000. Adopting a *simple* strategy will lead to a lower payoff of $400,000. Thus, the subgame perfect equilibrium for this game is {*Slick, [Stay out, Enter]*}. Nexus will adopt a *slick* marketing strategy and Mobius will respond with a *stay out* strategy. The resulting payoffs from this strategy profile are $430,000 for Nexus and $0 for Mobius.

The backward induction methodology has several advantages over simple payoff matrices. For one thing, backward induction is straightforward and comparatively easy to follow. More importantly, game trees and backward induction allow us to consider games involving multiple players, strategies, and stages.

What is remarkable about backward induction in games with complete and perfect information is that it will almost always result in a unique Nash equilibrium. The only exception is when a player is indifferent between two or more moves at a basic decision node. Even when this is the case, randomly selected moves may be optimal. Since the subgame perfect equilibrium strategy profile depends on the moves made at each decision node, this indifference may affect how players move at earlier stages of the game. This could, in turn, affect the moves made by other players at subsequent stages. As a result, backward induction involving indifferent moves may not result in a unique subgame perfect equilibrium. Happily, this situation is rarely encountered in practice.

—————————————— **Demonstration Problem 11.1** ——————————————

Consider the extensive-form game depicted in Figure D11.1, where larger profits are preferred. Using the backward-induction solution concept, determine the optimal strategy profile for this game. Illustrate your answer.

FIGURE D11.1
Subgame-Perfect Equilibrium for Demonstration Problem 11.1

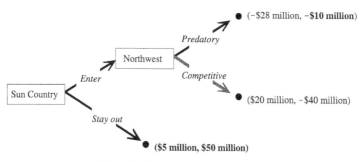

Payoffs: (Sun Country, Northwest)

Solution

Begin at the terminal nodes to the far right. The Northwest decision node is basic since the branches lead to exactly one terminal node. If Northwest adopts a *predatory* pricing strategy, it will earn a profit of –$10 million. If Northwest adopts a *competitive* pricing strategy, the payoff is –$40 million. *Predatory* pricing is Northwest's optimal strategy because it results in the smallest loss. Thus, Northwest's *competitive* pricing branch will be eliminated from further consideration. At the root of the game tree, Sun Country must decide whether to adopt an *enter* or a *stay out* strategy. If Sun Country decides to *enter,* this will lead to a payoff of –$28 million. If it decides to *stay out,* its payoff will be $5 million. Since a *stay out* results in the largest payoff, Sun Country will not enter the market. Northwest has no need to respond, and the game ends. The subgame perfect equilibrium for this game is for Sun Country to *stay out* of the market, which is depicted in Figure D11.1.

—————————————— **Demonstration Problem 11.2** ——————————————

Consider, again, the attack-of-the-clone game from Table 11.3. Suppose that the cost of cloning Nexus' computer game is $10,000 instead of $300,000.

    a. Diagram this new sequential-move game.
    b. Using the backward induction methodology, determine the subgame perfect equilibrium for this game.

Solution

    a. Mobius' profits after entering the market when Nexus adopts a *slick* or a *simple* marketing strategy are summarized in Table D11.2. The extensive form of this game is illustrated in Figure D11.2.

TABLE D11.2
Mobius' Profits After Entering the Market

	Slick	Simple
Gross profit in year 1	$0	$0
Gross profit in year 2	$50,000	$400,000
Total gross profit	$50,000	$400,000
Cloning cost	–$10,000	–$10,000
Total net profit	$40,000	$390,000

FIGURE D11.2
Subgame-Perfect Equilibrium for Demonstration Problem 11.2

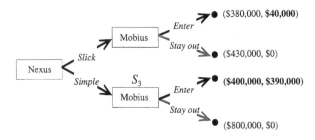

Payoffs: (Nexus, Mobius)

b. Using the backward induction methodology, begin by moving from each terminal node to each of Mobius' two basic decision nodes. If Nexus chooses a *slick* campaign, the optimal move for Mobius is to *enter* since the payoff is $40,000 compared with a payoff of $0 by staying out of the market. Thus, the *stay out* branch should be disregarded. If Nexus chooses a *simple* marketing strategy, the best move for Mobius is to *enter* since the payoff is $390,000 compared with a payoff of $0 by adopting a *stay out* strategy. Thus, the *stay out* branch should also be disregarded. The resulting extensive-form game shows that the optimal strategy response profile for Mobius is [*Enter, Enter*]. This game has two subgame equilibria: {*Slick*, [*Enter, Enter*]} and {*Simple*, [*Enter, Enter*]}. Which of these is the subgame perfect equilibrium? To determine this, follow the remaining branches to Nexus' best payoff, which is $400,000 from a simple advertising strategy. Thus, the subgame perfect equilibrium for this game is {*Simple*, [*Enter, Enter*]}. The payoffs for both companies are highlighted in boldface.

───────────────── **Demonstration Problem 11.3** ─────────────────

Suppose that the cost of cloning Nexus' computer game in Table 11.3 is $500,000 instead of $300,000.

a. Diagram this sequential-move game.
b. Using backward induction, determine the subgame perfect equilibrium for this game.

TABLE D11.3
Mobius' Profits After Entering the Market

	Slick	Simple
Gross profit in year 1	$0	$0
Gross profit in year 2	$50,000	$400,000
Total gross profit	$50,000	$400,000
Cloning cost	$500,000	$500,000
Total net profit	−$450,000	−$100,000

FIGURE D11.3
Subgame-Perfect Equilibrium for Demonstration Problem 11.3

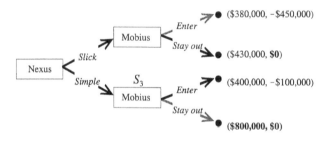

Payoffs: (Player A, Player B)

Solution

a. Mobius' profits after entering the market are summarized in Table D11.3. The extensive form of this game is illustrated in Figure D11.3.
b. Using backward induction, begin at the terminal nodes and Mobius' basic decision nodes. If Nexus chooses *slick*, the optimal response by Mobius is to *stay out* since the payoff of $0 is greater than −$450,000 by entering. The *enter* branch should be disregarded. If Nexus chooses *simple*, again the optimal move for Mobius is to stay out since the payoff of $0 is greater than a payoff of −$100,000. Thus, the *enter* branch should be disregarded. The resulting extensive-form game shows that the subgame equilibria are {*Slick*, [*Stay out, Stay out*]} and {*Simple*, [*Stay out, Stay out*]}. To determine which of these is the subgame perfect equilibrium, follow the remaining branches from the root of the game tree in Figure D11.3. The best payoff for Nexus is to adopt a *simple* marketing strategy. Thus, the subgame perfect equilibrium is {*Simple*, [*Stay out, Stay out*]}. The payoffs for both companies are highlighted in boldface.

─────────────────────────── **Demonstration Problem 11.4** ───────────────────────────

Consider the sequential-move game in Figure D11.4. Using backward induction, determine the subgame perfect equilibrium. Illustrate your answer.

FIGURE D11.4
Subgame-Perfect Equilibrium for Demonstration Problem 11.4

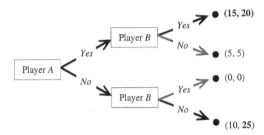

Payoffs: (Player *A*, Player *B*)

Solution

From the terminal nodes, move to player *B*'s basic decision nodes. If player *A* moves *yes*, the optimal move for player *B* is to move *yes* because of the larger payoff. The *no* branch should now be ignored. If player *A* moves *no*, player *B*'s best response is *no*, again because of the larger payoff. The *yes* branch should also be ignored. Player *B*'s optimal strategy response profile is [*Yes, No*]. Thus, the subgame equilibria for this game are {*Yes*, [*Yes, No*]} and {*No*, [*Yes, No*]}. To determine which of these is the subgame perfect equilibrium, follow the remaining branches in Figure 11.4. from player *A*'s decision node to the two remaining terminal nodes. It is in player *A*'s best interest to move *yes* because this will lead to the larger payoff. Thus, the subgame perfect equilibrium for this game is {*Yes*, [*Yes, No*]}. The payoffs are highlighted in boldface.

───

SIMPLIFYING THE NOTATION

The notation used in this chapter for identifying subgame perfect equilibria becomes exponentially more complicated as we increase the number of stages. Suppose, for example, that the game depicted in Figure 11.6 included a third stage in which Nexus responds to Mobius' decision to enter the market by adopting, say, a *passive, competitive,* or *predatory* pricing strategy. In terms of our notation, this means that for each strategy response by Mobius there is an associated strategy response by Nexus. One possible strategy profile might be {*Slick,* [*Stay out* [*Passive, Passive, Predatory*], *Enter* [*Competitive, Predatory, Competitive*]}. If there were a fourth stage, for each strategy response by Nexus in the third stage there would be an associated strategy response by Mobius, and so on.

The notation in the two-stage examples was valuable because it not only preserved the sequential nature of the decision-making process, but it also identified the subgame equilibria from which the subgame perfect equilibrium was chosen. Unfortunately, this notation is complicated and can be very confusing. Recall that the subgame perfect equilibrium for the game in Figure 11.6 was {*Slick,* [*Stay out, Enter*]}, which resulted in payoffs of ($430,000, $0). When there is no risk of confusion, we will simplify the notation by identifying strategy profiles in terms of the branches traveled to reach a particular terminal node. Thus, we will write the strategy profile for the sub-game perfect equilibrium for the game depicted in Figure 11.6 as {*Slick → Stay out*}. Suppose that this game had a third stage in which Nexus chooses *predatory* pricing. In this case, we would use the notation {*Slick → Stay out → Predatory*}. Although this notation is unconventional, it is reasonably straightforward and far less confusing than the alternative of including reaction functions at each stage of the game.

FIRST-MOVER ADVANTAGE

The reason why backward induction results in a unique Nash equilibrium is precisely because the moves are sequential. Once the first mover commits to a move, choosing a best response is no longer a problem for the second mover. This explains why most board games usually result in a single winner. It also helps to explain why simultaneous-move games often have multiple Nash equilibria, and sometimes none.

Anyone who is familiar with chess knows that the advantage of playing with the white pieces is that you get to move first. When players are evenly matched, white has the advantage in the opening stages of the game because black is immediately placed on the defensive. In principle, if neither player makes a mistake, white should be able to exploit this early advantage and eventually checkmate black.

Many, although not all, games involving strategic behavior exhibit what is known as the **first-mover advantage**. In the parlance of game theory, a player is a first-mover (sometimes referred to as a *Stackelberg leader*) if he or she can commit to a strategy before the game begins. By irrevocably committing to a game plan, a player has a first-mover advantage if the payoff is no worse than if the players move simultaneously. In a dynamic game, if a player can commit to a strategy, he or she should do so. On the other hand, if more than one player is in a position to commit to a strategy, it may not be in his or her best interest to do so. The reason for this is that in games with complete and perfect information, the ability to commit to a strategy guarantees that the payoff in a sequential-move game is no worse than in a simultaneous-move game, but the payoff may be better by not moving first.

> **First-mover advantage** When a player who can commit to a strategy first enjoys a payoff that is no worse than if all players move simultaneously.

CASE STUDY 11.2: LEADER OF THE PACK

It is sometimes argued that one key to success in business is to be the first to market with a new product or innovation. Amazon.com, Bayer, Coca-Cola, IBM, Intel, Kodak, Microsoft, Standard Oil, and Xerox are well-known examples of companies that were first to introduce a new product or innovation to the market. But being first does not guarantee success. Many people believe that Proctor & Gamble, the IBM of diapers, was the first company to introduce the resealable, disposable diaper (Pampers) in 1961. In fact, this honor belongs to the now defunct Chux, which was first introduced by Kimberly Clark. Proctor & Gamble's success resulted from its ability to produce a cheaper product with a wider market appeal. The same story can be told of video cassette recorders, which were originally introduced in 1956 by the American company, Ampex. Twenty years later this company fell victim to cheaper models by such Japanese manufacturers as Matsushita and Sony.

This pattern of first-mover being outdone by imitators has been repeated time and time again. A study by Tellis and Golder (1996) found that in more than fifty markets, first-movers were industry leaders only 10 percent of the time. Moreover, current industry leaders on average entered the market thirteen years after the first-mover. One explanation for this is strategic behavior. Industry leaders have little incentive to risk their market share by bringing an unproven product to market. In general, their dominant strategy is to "follow the pack." IBM, for example, is not known for product innovation, but rather for its ability to standardize technology for mass-market consumption. In fact, most new ideas came from then start-up companies as Sun Microsystems and Apple because risky innovations were their best hope of capturing market share.

Does this mean that the first-mover advantage is a myth? Not at all, and there are several possible explanations why first movers failed to maintain their industry lead. For one thing, having a first-mover advantage only says that a player who can commit to a strategy will enjoy a payoff that is no worse than if all players moved simultaneously. It does not say that the player will always win the game. Moreover, the first-mover advantage is only guaranteed in games with complete and perfect information. Finally, technologically innovative managers are not always good game players. Depending on the company's objective, a failure to project forward and reason backward often leads to business decisions that result in lower profits, disappointed stakeholders, and corporate oblivion.

To illustrate the first-mover advantage, consider an industry that consists of two firms producing an identical product. The only decision variable for each firm is the total quantity produced. We will designate the output strategy of each firm as *high, medium, low,* or *none.* The market price is inversely related to the total quantity produced by both firms. Output decisions of all firms are made simultaneously. Finally, we will assume that each firm will not alter its level of output in response to a change in a rival's output. The basic set up of this game may sound familiar. It is the Cournot output-setting game that was introduced in Chapter 8. Figure 11.7 is the normal form of this noncooperative, simultaneous-move, one-time game. The payoffs are the firms' profits, which are in thousands of dollars.

In this game, neither firm has a dominant strategy. After successively eliminating all dominated strategies, this game simplifies to Figure 11.8. The reader should verify that the iterated strictly dominant strategy for both players is *medium.* Thus, the Nash equilibrium strategy profile for this game is {*Medium, Medium*}, with each firm earning $50,000, which are highlighted in boldface.

FIGURE 11.7
Cournot Output-Setting Game

Firm 2

		High	Medium	Low	None
	High	(0, 0)	(37.5, 25)	(56.25, **28.125**)	(**112.5**, 0)
Firm 1	*Medium*	(25, 37.5)	(**50, 50**)	(**62.5**, 46.875)	(100, 0)
	Low	(**28.125**, 56.25)	(46.875, **62.5**)	(56.25, 56.25)	(84.375, 0)
	None	(0, **112.5**)	(0, 100)	(0, 84.375)	(0, 0)

Payoffs: (Firm 1, Firm 2)

FIGURE 11.8
Iterated Strictly Dominant Strategy Equilibrium for the Cournot Duopoly Game

		Firm 2	
		Medium	*Low*
Firm 1	*Medium*	**(50, 50)**	**(62.5**, 46.875)
	Low	(46.875, **62.5**)	(56.25, 56.25)

Payoffs: (Firm 1, Firm 2)

FIGURE 11.9
Extensive Form of the Cournot Output-Setting Game in Figure 11.7

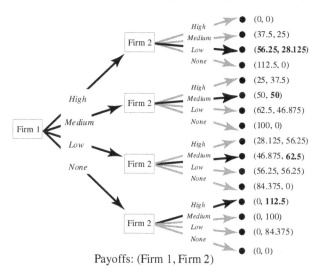

Payoffs: (Firm 1, Firm 2)

Will the outcome of the game in Figure 11.8 change if either firm is able to commit to a particular strategy? To answer this question, let us assume that firm 1 moves first. The sequential version of this game is depicted in Figure 11.9. It is left as an exercise for the reader to verify that the subgame perfect equilibrium for this game is for firm 1 to adopt a *high* output strategy and for firm 2 to adopt a *low* output strategy, or more compactly {*High → Low*}. In this case, firm 1 earns $56,250 and firm 2 earns $28,125, compared with profits of $50,000 for each firm when moves are made simultaneously. It would appear that firm 1 has a clear first-mover advantage.

———————————— **Demonstration Problem 11.5** ————————————

Homer and Marge are playing a noncooperative, simultaneous-move, one-time game in which larger payoffs are preferred. This game is depicted in Figure D11.5.1.

a. Does either player have a strictly dominant strategy?
b. Does this game have a Nash equilibrium?
c. Illustrate the game depicted in Figure D11.5.1 as a sequential-move game in which Homer moves first.
d. Identify the subgames.

FIGURE D11.5.1
Static Game for Demonstration Problem 11.5

Marge

		Bart	Lisa	Maggie
	Bart	(150, 150)	(**200**, 250)	(250, **350**)
Homer	Lisa	(250, 125)	(175, 150)	(**275, 245**)
	Maggie	(**350**, 250)	(150, 275)	(200, **300**)

Payoffs: (Homer, Marge)

FIGURE D11.5.2
Subgame-Perfect Equilibrium for Demonstration Problem 11.5 in Which Homer Moves First

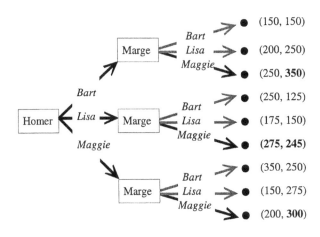

Payoffs: (Homer, Marge)

e. Use backward induction to determine the subgame perfect equilibrium. How does this compare with your answer to part b?

f. Suppose that Marge moves first. What is the subgame perfect equilibrium? What can you conclude from your answers to parts e and f?

Solution

a. The reader should verify that Marge has a strictly dominant *Maggie* strategy. By contrast, Homer does not have a dominant strategy. Since Marge will play *Maggie,* Homer will play *Lisa.*

b. The Nash equilibrium strategy profile for this game is {*Lisa, Maggie*} with payoffs of (275, 245).

c. The extensive form of this game if Homer moves first is illustrated in Figure D11.5.2.

FIGURE D11.5.3
**Subgame-Perfect Equilibrium for Demonstration Problem 11.5 in Which
Marge Moves First**

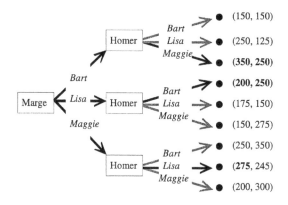

Payoffs: (Homer, Marge)

d. The dynamic game in Figure D11.5.2 has four subgames. The trivial subgame begins at the root of the game tree, S_1. The three proper subgames begin at subroots S_2, S_3, and S_4. These proper subgames are enclosed within the dashed, rounded rectangles.

e. Using the method of backward induction, the subgame perfect equilibrium for this game is {*Lisa*, [*Maggie, Maggie, Maggie*]}, or more compactly {*Lisa → Maggie*}, which results in payoffs of (275, 245). These payoffs are the same as the noncooperative, simultaneous-move, one-time game in part b. Because Homer's payoff is no worse off, he has a first-mover advantage.

f. The extensive-form of this game in which Marge moves first is illustrated in Figure D11.5.3. For ease of comparison, the order of payoffs is the same as in Figure D11.5.2. Homer's strategy response profile is [*Maggie, Bart, Lisa*]. Thus, this game has three subgame equilibria: {*Bart → Maggie*}, {*Lisa → Bart*}, and {*Maggie → Lisa*}. Surprisingly, this game has *two* subgame perfect equilibria: {*Bart → Maggie*} and {*Lisa → Bart*}, with payoffs of (350, 250) and (200, 250). This unusual outcome results from Marge's indifference between *Maggie* and *Bart* due to identical payoffs. Regardless of the strategy adopted, however, Marge has a first-mover advantage because in both cases her payoff is higher than in the simultaneous-move game depicted in Figure D11.5.2.

—————————————————— **Demonstration Problem 11.6** ——————————————————

Suppose Homer and Marge are playing the one-time, static game depicted in the Figure D11.6.1.

a. Does either player have a dominant strategy?
b. Does this game have a Nash equilibrium?
c. Suppose that this static game is modeled as a dynamic game with Homer moving first. Does Homer have a first-mover advantage?
d. Does Marge have a first-mover advantage?

FIGURE D11.6.1

Static Game for Demonstration Problem 11.6

Marge

		Bart	Lisa	Maggie
Homer	Bart	(150, 150)	**(275, 245)**	(240, 240)
	Lisa	(225, 150)	(200, 250)	(200, **350**)
	Maggie	(**350**, 250)	(150, 200)	(250, 275)

Payoffs: (Homer, Marge)

FIGURE D11.6.2

Subgame-Perfect Equilibrium for Demonstration Problem 11.6 in Which Homer Moves First

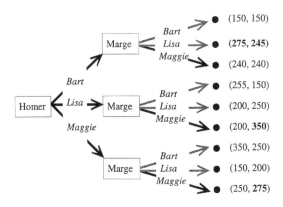

Payoffs: (Homer, Marge)

Solution

a. In this game, neither player has a dominant strategy, both Homer and Marge each have strictly dominated strategies. Regardless of the strategy adopted by Marge, Homer will never adopt a *Lisa* strategy. Regardless of the strategy adopted by Homer, Marge will never adopt a *Bart* strategy.

b. The reader should verify that once the dominated strategies are eliminated, this game has two Nash equilibrium strategy profiles: {*Bart, Lisa*} and {*Maggie, Maggie*}. The payoffs for these strategy profiles are highlighted in boldface.

c. The extensive form of this game in which Homer moves first is depicted in Figure D11.6.2. Using backward induction, the subgame perfect equilibrium for this game is the strategy profile {*Bart → Lisa*}, with payoffs to Homer and Marge of (275, 245). Unlike the game in Figure D11.6.1, this sequential-move game has a unique Nash equilibrium. Homer has a first-mover advantage because he is guaranteed to obtain the higher of the two Nash equilibrium payoffs in Figure D11.6.1.

d. The extensive form of this game in which Marge moves first is illustrated in Figure D11.6.3. The order of payoffs in Figure D11.6.1 have been preserved. The subgame perfect equilibrium for this game is the strategy profile {*Maggie → Maggie*}, with payoffs to Homer and Marge of (250, 275). This is the other Nash equilibrium in Figure

FIGURE D11.6.3
Subgame-Perfect Equilibrium for Demonstration Problem 11.6 in Which Marge Moves First

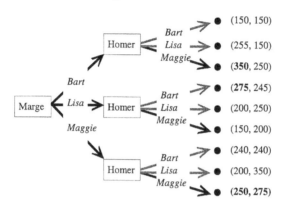

Payoffs: (Homer, Marge)

D11.6.1. Maggie also enjoys a first-mover advantage since she receives the higher of the two payoffs.

———————————— **Demonstration Problem 11.7** ————————————

Homer and Marge are playing the one-time, static game in Figure D11.7.1.

a. Does either player have a dominant strategy?
b. Does this game have a Nash equilibrium?
c. Does Homer have a first-mover advantage if this game is modeled as a sequential-move game?
d. Does Marge have a first-mover advantage?

Solution

a. Neither player has a dominant strategy, but both have strictly dominated strategies. Marge will never play *Bart* and Homer will never play *Lisa*.
b. The reader should verify that even after all dominated strategies have been eliminated, the static game depicted in Figure D11.7.1 does not have a Nash equilibrium.
c. The extensive form of this game in which Homer moves first is depicted in Figure D11.7.2. The subgame perfect equilibrium is the strategy profile {*Bart* → *Maggie*}, with payoffs to Homer and Marge of (240, 245). Without additional information about how the players might move in a simultaneous-move game, it is not possible to say for certain whether Homer enjoys a first-mover advantage.
d. The extensive form of this game in which Marge moves first is depicted in Figure D11.7.3. The order of the payoffs are preserved. The subgame perfect equilibrium for this game is the strategy profile {*Bart* → *Maggie*}, with payoffs to Homer and Marge of (350, 250). Without additional information about how the players might move in a simultaneous-move game, it is not possible to say for certain whether Marge enjoys a first-mover advantage.

FIGURE D11.7.1

Static Game for Demonstration Problem 11.7

Marge

		Bart	Lisa	Maggie
	Bart	(150, 150)	(**275**, 240)	(240, **245**)
Homer	Lisa	(225, 150)	(200, 250)	(200, **350**)
	Maggie	(**350**, 250)	(150, **275**)	(**250**, 200)

Payoffs: (Homer, Marge)

FIGURE D11.7.2

Subgame-Perfect Equilibrium for Demonstration Problem 11.7 in Which Homer Moves First

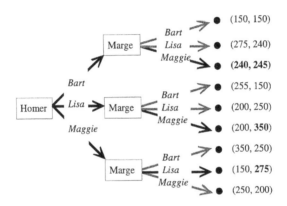

Payoffs: (Homer, Marge)

FIGURE D11.7.3

Subgame-Perfect Equilibrium for Demonstration Problem 11.7 in Which Marge Moves First

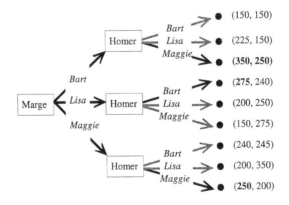

Payoffs: (Homer, Marge)

CREDIBLE THREATS

The reader will recall that a threat is an example of a strategic move that is designed to alter a rival's behavior. A threat is credible if it is in a player's best interest to follow through with his or her threat, otherwise it is an empty threat. So far, we have only considered **credible threats** in static games. We are now in a position to use backward induction to identify credible and empty threats in sequential-move games with complete and perfect information.

| **Credible threat** When it is in a player's best interest to follow through with his or her threat.

Consider again the game depicted in Figure 11.1. The reader will recall that the subgame perfect equilibrium for that game was {*Yes* → *Yes*} with payoffs of (15, 20). But, the optimal payoff for player *B* comes from the strategy profile {*No* → *No*}. How can player *B* get player *A* to move *no* at the beginning of the game?

Suppose that before the game begins, player *B* makes the following threat: "No matter how player *A* moves, I will move *no*." Player *B*'s threat is summarized in the strategy response profile [*No, No*]. Thus, if player *A* opens with a *yes* move and player *B* follows through with his or her threat, the strategy profile for this game will be {*Yes* → *No*} with payoffs of (5, 5). On the other hand, if player *A* opens with a *no* move, the strategy profile will be {*No* → *No*} with payoffs of (10, 25). If player *A* believes that player *B*'s threat is credible, it will be in player *A*'s best interest to move *no*. The problem is that player *B*'s strategy response profile [*No, No*] is not credible. Why? If player *A* moves *yes*, it is in player *B*'s best interest to move *yes* because of the larger payoff. If player *B* is rational, his or her threat to move *no* is simply not credible.

The importance of credibility can be seen in the attack-of-the-clone game depicted in Figure 11.3. The reader will recall that the subgame perfect equilibrium for this game was {*Slick* → *Stay out*}. Clearly, Mobius would prefer that Nexus adopt a *simple* strategy. How can Mobius accomplish this? Mobius could announce that it plans to enter the market regardless of what Nexus does. If the threat by Mobius is credible, it will be in the best interest of Nexus to adopt a *simple* strategy. Mobius' strategy response profile [*Enter, Enter*], however, is an empty threat. To see why, consider the subgame that begins at S_2. The equilibrium for this subgame is {*Slick* → *Stay out*}. The equilibrium for the subgame that begins at S_3 is {*Simple* → *Stay out*}. The subgame perfect equilibrium for this game must come from one of these two strategy profiles, which {*Simple* → *Enter*} does not. Thus, Mobius' threat to *enter* is just not credible.

In general, empty threats should be eliminated whenever we require that equilibrium strategies apply to any subgame. The reasoning is straightforward. When we examine subgames that begin at the decision nodes of the player who moves last, the only subgame equilibrium is the move that results in the largest payoff. But, this is precisely the move that is selected using backward induction. Moving toward the root of the game tree using backward induction eliminates all empty threats. For this reason, equilibria for sequential-move games with complete and perfect information are said to be a *subgame perfect*.

Example: Entry Deterrence Game

To illustrate backward induction and games involving credible and empty threats, consider the dynamic game depicted in Figure 11.10. This game involves a start-up computer software company, which we will call David, and the established industry leader, which we will refer to as Goliath. David, which has developed a faster and more efficient operating system, must decide whether to *enter* or *stay out* of the market. If David enters the market, Goliath must decide whether to adopt a *predatory* pricing strategy or share the market by adopting a *competitive* pricing strategy.

FIGURE 11.10
David Versus Goliath

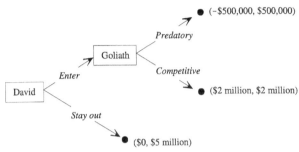

Payoffs: (David, Goliath)

If Goliath adopts a *predatory* pricing policy, David will lose $500,000 per month while Goliath will earn $500,000 per month. On the other hand, if Goliath adopts a *competitive* pricing policy, both firms will earn profits of $2 million per month. If David stays out of the market, Goliath will continue to earn profits of $5 million per month.

Using backward induction, the reader should verify that the subgame perfect equilibrium for the game depicted in Figure 11.10 is {*Enter → Competitive*}. Both companies will earn profits of $2 million per month. Of course, Goliath would prefer that David *stay out* so that it can continue to earn $5 million in monthly profits. Goliath could try to persuade David not to enter the market by threatening to pursue a *predatory* pricing strategy, but this will not work since this threat is not credible. Is there a way for Goliath to make a credible threat? If there is, Goliath will successfully deter David's entry into the market. Analytically, the problem for Goliath is that David is a first mover. The challenge confronting Goliath is to reverse these roles.

In order to preempt David's entry into the market, the CEO of Goliath circulated an internal memo informing senior managers that he has entered into a legal, binding, and irrevocable contract with several leading research universities to ensure that the company remains at the cutting edge of the computer software industry. As part of this agreement, Goliath will commit 100 percent of its operating profits to the development of advanced computer hardware and software should its dominant position in the market be challenged. This memo, which was intentionally leaked to David's senior managers, can be viewed as a serious and credible threat. The effect of this change in Goliath's corporate policy is depicted in Figure 11.11.

The reader should note that the only difference between the upper and lower halves of Figure 11.11 is Goliath's profits with and without the contract. If Goliath adopts a *no contract* strategy, company profits will be $500,000 and $2,000,000 from a *predatory* and *competitive* pricing policy, respectively. On the other hand, if Goliath adopts a *contract* strategy, it earns zero profit regardless of the pricing strategy.

The reader should verify that the equilibrium for the subgame beginning at S_2 is {*Enter → Competitive*}. The reader should also verify that the equilibrium to the subgame beginning at S_3 is {*Stay out*}. The reason for this is that Goliath has a weakly dominant *predatory* pricing strategy. Thus, if David enters the market, it will suffer a loss of $500,000. If it stays out, it will earn zero profit. Using the method of backward induction, the payoff to Goliath by not signing a contract is $2 million. By signing a contract and adopting a *predatory* pricing strategy, Goliath will successfully deter David's entry into the market and earn monthly profits of $5,000,000. Thus, the subgame perfect equilibrium for this game is {*Contract → Stay out*}.

FIGURE 11.11
Entry Deterrence Game with a Credible Threat by Goliath

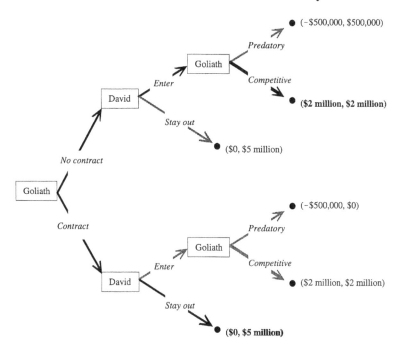

Payoffs: (Sun Country, Northwest)

DYNAMIC GAMES WITH CONTINUOUS STRATEGIES

In each of the sequential-move games considered thus far, we considered a finite number of moves at each decision node. In business, however, many managerial decisions involve choices from a continuous range of options. In the attack-of-the-clone game, for example, rather than a choice between two advertising strategies, it is far more likely that total sales are a continuous function of advertising expenditures. In dynamic games involving continuous strategies, finding a subgame perfect equilibrium using a game tree with an infinite number of branches and decision nodes is implausible. An alternative approach was suggested in the early twentieth century before the genesis of modern game theory.

THE STACKELBERG MODEL

A player who has the ability to commit to a strategy has a first-mover advantage. This player is sometimes referred to as a **Stackelberg leader**. This sobriquet can be traced to the groundbreaking work of Heinrich von Stackelberg (1934) who attempted to describe the dynamic behavior of firms in imperfectly competitive industries. Stackelberg's approach is a variation of the Cournot output-setting game discussed in Chapter 8.

Stackelberg leader A player who has the ability to commit to a strategy in a dynamic, output-setting game.

The model proposed by von Stackelberg is similar to the Cournot output-setting model with one notable exception: Firms make their output decisions sequentially, rather than simultaneously. To make the discussion more concrete, assume a two-stage game in an industry consisting of just two firms in which firm 2 is the Stackelberg leader and decides first how much to produce. This is followed by the output decision of firm 1, which is called the *Stackelberg follower*. Since output decisions are made simultaneously in the Cournot model, each firm engages in a process of circular reasoning (move and countermove) before deciding how much to produce. Here, firm 1 must wait for firm 2 to decide how much to produce, even though each firm knows its rival's reaction function. By projecting forward and reasoning backward, firm 2 can exploit firm 1's best response to maximize its profits.

> **Stackelberg model** A theory of strategic interaction in which one firm, the Stackelberg leader, believes that its rival, the Stackelberg follower, will not alter its level of output. The production decisions of the Stackelberg leader will exploit the anticipated behavior of the Stackelberg follower.

To illustrate the implications of the Stackelberg assumption of sequential decision making, assume that the firms' profit equations are given by Equations (8.13). The firms' reaction functions in the Cournot output-setting game were given by Equations (8.15). The primary difference between the Cournot and Stackelberg models is that firm 2, the Stackelberg leader, does not have a reaction function. Firm 2 moves first, followed by firm 1.

The challenge confronting firm 2, as the Stackelberg leader, is to select a level of output that will maximize its profits given firm 1's best response. Since firm 2 moves first, firm 1's reaction function, which is given by Equation (8.15a), constitutes an optimal strategy. Whereas the solution profile to the Cournot output-setting game was $\{Q_1*(Q_2*), Q_2*(Q_1*)\}$, the subgame perfect equilibrium for the sequential move Stackelberg game is $\{Q_1*(Q_2*), Q_2*\}$. A Nash equilibrium for the Stackelberg game requires that firm 2 first select the output level Q_2* that maximizes its profit, after which firm 1 uses its reaction function to select the output level $Q_1*(Q_2*)$ that maximizes its profit. To do this, firm 2 will incorporate firm 1's reaction function directly into its profit equation. Firm 2's profit equation is:

$$\pi_2 = 15Q_2 - 0.5Q_2^2 - 0.5(15 - 0.5Q_2)Q_2 = 7.5Q_2 - 0.25Q_2^2 \tag{11.1}$$

The first-order (necessary) condition for profit maximization is:

$$\frac{d\pi_2}{dQ_2} = 7.5 - 0.5Q_2 = 0 \tag{11.2}$$

It is left as an exercise for the reader to verify that the second-order (sufficient) condition for profit maximization is satisfied.

Solving Equation (11.2), the profit-maximizing level of output for firm 2 is $Q_2* = 15$, or 15,000 units per month. Substituting this result into firm 1's best-response function, the profit-maximizing output level for firm 1 is $Q_1* = 7.5$, or 7,500 units per month. Using these results, firm 2 earns a profit of $56,250 and firm 1 earns a profit of $28,125. Enjoying a first-mover advantage, firm 2 earns twice as much as firm 1.

In our example, firm 2's leadership position may have been the result of superior management or a research and development program that enables it to introduce new or improved products ahead of its competition. Perhaps its leadership position stems from a nexus of long-term contractual arrangements with suppliers, which allows it to bring its product to market more quickly.

The existence of long-term supplier contracts, however, could work to firm 2's disadvantage by making it difficult to respond quickly to changes in market conditions. Frequent shifts in consumer tastes or rapidly emerging product technology could make the leader vulnerable to challenges from rivals seeking to take over the number one spot. This implies that the kind of Stackelberg competition described above is much more likely to occur in mature industries in which product demand is stable, or when changes in product technology are infrequent and predictable. Stackelberg competition is far less likely in situations in which predicting changes in consumer tastes is akin to hitting a moving target, or where product development is very rapid and unpredictable, such as in the computer software or personal communications industries.

─────────────── **Demonstration Problem 11.7** ───────────────

Consider the Cournot model in Demonstration Problem 8.2 where the inverse demand equation for two profit-maximizing firms is:

$$P = 200 - 2(Q_1 + Q_2) \qquad \text{(D11.7.1)}$$

The firms' total cost functions are:

$$TC_1 = 4Q_1 \qquad \text{(D11.7.2a)}$$

$$TC_2 = 4Q_2 \qquad \text{(D11.7.2b)}$$

where Q_1 and Q_2 represent the output levels of firm 1 and firm 2, respectively. Assume that firm 2 is a Stackelberg leader and firm 1 is a Stackelberg follower. What are the equilibrium price, profit-maximizing output levels, and profits for each firm?

Solution

From the above solution to the Cournot duopoly problem, the reaction function of firm 1 is:

$$Q_1 = 49 - 0.5Q_2 \qquad \text{(D11.7.3)}$$

Substituting firm 1's reaction function into firm 2's profit equation we get:

$$\pi_2 = 196Q_2 - 2Q_2^2 - 2Q_2(49 - 0.5Q_2) = 98Q_2 - Q_2^2 \qquad \text{(D11.7.4)}$$

The necessary condition for profit maximization is:

$$\frac{\partial \pi_2}{\partial Q_2} = 98 - 2Q_2 = 0 \qquad \text{(D11.7.5)}$$

Solving Equation (D11.7.5), firm 2's profit-maximizing level of output is 49,000 units of output. Substituting this result into firm 1's reaction function, the profit-maximizing output level is 24,500 units of output. Thus, total industry output is 73,500 units of output. As such, the market-clearing price is:

$$P^* = 200 - 2(49 + 24.5) = \$53 \qquad \text{(D11.7.6)}$$

The maximum profits of firms 1 and 2 are:

$$\pi_1^* = 196(24.5) - 2(24.5)^2 - 2(49)(24.5) = \$1{,}200.50 \qquad \text{(D11.7.8)}$$

$$\pi_2^* = 196(49) - 2(49)^2 - 2(24.5)(49) = \$2{,}401.00 \qquad \text{(D11.7.9)}$$

Compare these results with those obtained in the Cournot output-setting model in demonstration problem 8.2. In the Stackelberg model, total industry output is higher (73,500 > 65,340) and market price is lower ($53 < $69.3). Moreover, profits are greater for the Stackelberg leader and less for the Stackelberg follower.

CHAPTER REVIEW

Dynamic (sequential-move or multistage) games differ from *static (simultaneous-move)* games in that the players take turns. Sequential-move games may be depicted as a *game tree*, which summarizes the players, the information available to each player at each stage, the order of the moves, and the payoffs from alternative strategies.

In a game with complete and perfect information, a *subgame* is any subset of branches and decision nodes of the entire game. The initial decision node of a subgame is called a *subroot*. Once a player begins to play a subgame, that player will continue to play the subgame until the end of the game.

The Nash equilibrium of a sequential-move game, which is called a *subgame perfect equilibrium*, is found among the complete set of subgame equilibria. There are as many subgame equilibria as there are subgames, and there are as many subgames as there are decision nodes. *Backward induction,* also known as the *fold-back method,* is the most commonly used method for finding a subgame perfect equilibrium.

A player with a *first-mover advantage* can commit to a strategy first and will enjoy a payoff that is no worse than if all players move simultaneously. If more than one player can commit to a strategy, it may not be in that player's best interest to do so since a better payoff might be possible by not moving first.

A *threat* is an example of a *strategic move* that is designed to alter a rival's behavior. A threat is credible if it is in a player's best interest to follow through with the threat; otherwise it is an empty threat. In general, empty threats should be eliminated whenever we require that equilibrium strategies apply to any subgame. The backward induction solution algorithm eliminates empty threats.

The *Stackelberg model* is a dynamic variation of the Cournot output-setting model that involves continuous strategies. The *Stackelberg leader* has a first-mover advantage over a *Stackelberg follower*. By incorporating the reaction function of the Stackelberg follower into its own production decisions, the Stackelberg leader will capture a larger share of the market and earn greater profits.

CHAPTER QUESTIONS

11.1   A subgame perfect equilibrium is impossible in games with multiple Nash equilibria. Do you agree or disagree? Explain.

11.2   It is not possible to have multiple Nash equilibria in dynamic games with complete and perfect information. Do you agree with this statement? If not, why not?

11.3   Each subgame of a sequential-move game has a unique Nash equilibrium. Do you agree with this statement? If not, why not?

11.4   Suppose that a noncooperative, simultaneous-move, one-time game has a weakly dominant-strategy equilibrium. If this game is modeled as a sequential-move game, the payoffs associated with a subgame perfect equilibrium are the same, regardless of which player moves first. Do you agree? Explain.

11.5   Describe what is meant by the strategic form of a sequential-move game?

11.6   Explain how a strategy involving advertising expenditures can be used to deter the entry of potential rivals.

11.7   Coca-Cola and Pepsi-Cola have managed to maintain their dominant market positions for nearly a century, while General Motors and Ford have not. In what way can game theory be used to explain this?

11.8   Which of the following firms is likely to have a first-mover advantage, and why?

   a.   Pfizer patents Celebrex as the first effective remedy for the relief of pain, swelling, and stiffness from osteoarthritis and adult rheumatoid arthritis, and dysmennhorea.
   b.   Home Depot opens a super store in Eagle, Colorado.
   c.   Panasonic is the first company to announce that it plans to manufacture DVD players.
   d.   Maxwell House is the first company to sell freeze-dried coffee.

11.9   Why is the outcome of a sequential-move game with complete and perfect information using backward induction subgame perfect?

11.10   Explain how the method of backward induction can be used to eliminate all incredible threats.

11.11   Having a first-mover advantage is never detrimental. Do you agree, and why?

11.12   How does the Stackelberg duopoly model modify the Cournot duopoly model?

11.13   What is the first-mover advantage in the Stackelberg game? What are the possible sources of the first-mover advantage? Apart from the ones cited in the text, can you think of any other possible sources of the first-mover advantage?

11.14   What types of industries are more conducive to Stackelberg competition? What types of industries are least conducive?

CHAPTER EXERCISES

11.1   Suppose that an industry consists of two firms, Magna Company and Summa Corporation, that produce an identical product. Magna and Summa are trying to decide whether to *expand* or *not expand* its production capacity. If the firm expands, it must also decide whether the expansion should be *moderate* or *extensive*. Each firm is currently operating at full capacity. The trade-off confronting each firm is that expansion will result in a larger market share, although increased output will put downward pressure on price. Consider the noncooperative, one-time, static game depicted in Figure E11.1. The payoffs represent company profits in millions of dollars. What is the Nash equilibrium for this game?

11.2   Suppose that the static game in chapter exercise 11.1 is modeled as a sequential-move game in which Magna moves first.

   a.   Illustrate the extensive form of this game.
   b.   What are the subgames for this game?
   c.   What is the Nash equilibrium for each subgame?
   d.   Use backward induction to find the subgame perfect equilibrium.

11.3   Figure E11.3 is a modified version of the static game in chapter exercise 11.1.

FIGURE E11.1
Static Game for Chapter Exercise 11.1

		Summa		
		None	*Moderate*	*Extensive*
Magna	*None*	(25, 25)	(15, 30)	(10, 25)
	Moderate	(30, 15)	(20, 20)	(8, 13)
	Extensive	(25, 10)	(12, 8)	(0, 0)

Payoffs: (Magna, Summa)

FIGURE E11.3
Static Game for Chapter Exercise 11.3

		Summa		
		None	*Moderate*	*Extensive*
Magna	*None*	(25, 20)	(15, 16)	(0, 22)
	Moderate	(30, 14)	(10, 21)	(8, 14)
	Extensive	(25, 9)	(12, 5)	(10, 3)

Payoffs: (Magna, Summa)

FIGURE E11.4
Static Game for Chapter Exercise 11.4

		Blue Dragon		
		Barnacle Bottom	*Old Toby*	*Southern Star*
Red Pony	*Barnacle Bottom*	(150, 150)	(200, 250)	(250, 350)
	Old Toby	(250, 125)	(175, 200)	(270, 245)
	Southern Star	(350, 250)	(150, 275)	(200, 300)

Payoffs: (Red Pony, Blue Dragon)

a. Does this game have a Nash equilibrium?
b. If this is a sequential-move game, what is the subgame perfect equilibrium for this game if Magna moves first?

11.4 Suppose that the static game depicted in Figure E11.4 is modeled as a sequential-move game with Red Pony moving first.

a. Illustrate the extensive form of this game.
b. Identify the trivial and proper subgames.
c. What is the Nash equilibrium for each proper subgame?
d. Use backward induction to find the subgame perfect equilibrium.

11.5 Consider the centipede game depicted in Figure E11.5, which involves two players who alternately have the opportunity to take a larger portion of an ever-increasing amount of money.[4] If both players prefer larger payoffs, use backward induction to identify the subgame perfect equilibrium for this game.

FIGURE E11.5
Centipede Game for Chapter Exercise 11.5

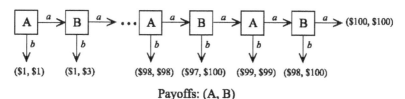

Payoffs: (A, B)

FIGURE E11.6
Centipede Game for Chapter Exercise 11.6

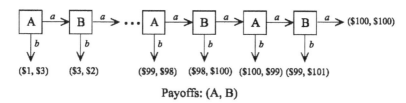

Payoffs: (A, B)

11.6 Consider the version of the centipede game depicted in Figure E11.6.

a. If both players prefer larger payoffs, use backward induction to identify the subgame perfect equilibrium for this game.

b. How does your answer to part a differ from your answer to chapter exercise 11.5?

11.7 Alex, Andrew, and Adam are playing the sequential-move game depicted in Figure E11.7.

a. What are the proper subgames for this game?

b. What is the Nash equilibrium for each subgame?

c. Use backward induction to find the subgame perfect equilibrium.

11.8 Suppose that the sequential-move game involving Alex, Andrew, and Adam is depicted in Figure E11.8.

a. What are the proper subgames for this game?

b. What is the Nash equilibrium for each subgame?

c. Use backward induction to find the subgame perfect equilibrium.

11.9 Suppose that the market for a specialized type of microprocessor is dominated by just two firms. The market-clearing price of the microprocessor depends on total industry output. All microprocessors produced are sold at the market-clearing price. The only decision variable for each firm is how many microprocessors to produce. Each firm must decide whether to produce a high, medium, or small volume of microprocessors, or to produce no microprocessors at all. Once a decision is made, each firm believes that its rival will not alter its level of output in response. High, medium, and low volume are 10, 8, and 3 million microprocessors, respectively. The profit per microprocessor is $20 - Q_T$ dollars, where Q_T is the total number of microprocessors produced.

a. What is the Nash equilibrium for this static game?

b. Suppose that one of the two firms is a Stackelberg leader. What is the Nash equilibrium for this game? How does this differ from your answer to part a? Explain.

FIGURE E11.7
Game Tree for Chapter Exercise 11.7

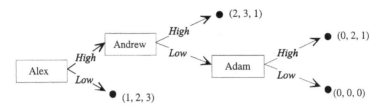

Payoffs: (Alex, Andrew, Adam)

FIGURE E11.8
Game Tree for Chapter Exercise 11.8

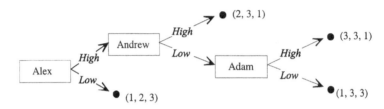

Payoffs: (Alex, Andrew, Adam)

11.10   Suppose that an industry consisting of two firms produces a differentiated product. The demand equation for the output of the industry is:

$$P = 145 - 5(Q_1 + Q_2) \qquad \text{(E11.10.1)}$$

where Q_1 and Q_2 represent the outputs of firm 1 and firm 2, respectively. The total cost equations of the two firms are:

$$TC_1 = 3Q_1 \qquad \text{(E11.10.2a)}$$

$$TC_2 = 5Q_2 \qquad \text{(E11.10.2b)}$$

Suppose that firm 2 believes that firm 1 will take the output of firm 2 as constant. By contrast, firm 2 will attempt to exploit the behavior of firm 1 by incorporating firm 1's reaction of the follower into its own production decisions. Calculate the equilibrium price, output levels, and profits of each firm.

11.11   Consider the static game depicted in Figure E11.11.

a.   Does either player have a dominant strategy?
b.   Does this game have a unique Nash equilibrium strategy profile? If so, what is it?
c.   Suppose that this is a multistage game and player A moves first. Illustrate the extensive form of this game.
d.   What are the subgames for this game?
e.   What are the subgame equilibria?

FIGURE E11.11
Static Game for Chapter Exercise 11.11

Player B

		B1	B2	B3
	A1	(3, 3)	(4, 1)	(5, 4)
Player A	A2	(2, 2)	(3, 1)	(3, 3)
	A3	(2, 2)	(3, 2)	(2, 4)

Payoffs: (Player A, Player B)

f. Use backward induction to find the subgame perfect equilibrium. Illustrate your solution.

ENDNOTES

1. Predatory pricing is the practice of setting a low price to deter entry into the market by potential rivals. A price that deters entry is sometime referred to as a *limit price*. A limit price prevents potential entrants from entering the market while allowing the incumbent firm to earn some profit. In general, predatory pricing only works if the incumbent firm sets a price below average variable cost to force competitors out of the industry, after which it raises price to recover lost revenues. Can a predatory-pricing strategy be successful in the long run? Wouldn't the prospect of potential new entrants into the industry keep the incumbent firm from raising prices? Part of the answer depends on whether investors believe that the incumbent's threat always to increase price to deter potential challengers is credible. Knowledge that the incumbent has "deep pockets" and can afford to follow through with this threat may discourage investors from getting embroiled in a price war that they could not possibly hope to win.

2. The aggressive response by Northwest Airlines to the Sun Country challenge is not an isolated case. In 1995, Spirit Airlines attempted to enter the Detroit–Philadelphia market, which was also dominated by Northwest. As in the case of Sun Country, Spirit began by offering round-trip fares for prices ranging between $49 and $139. Northwest retaliated by matching Spirit's fares. Although Northwest incurred sizable losses, by late 1996 Spirit raised the white flag and exited the market. Within a few months, Northwest's predatory pricing had driven from the market almost every other low-cost air carrier operating along that route. According to a study by Oster and Strong (2001), airlines frequently use their near-monopoly status at hub airports to drive out potential competition. The study cited twelve instances in which a new entrant attempted to enter a market dominated by a major carrier by offering low fares. In every case, the incumbent adopted a predatory-pricing strategy. Within two years, half the challengers had been forced out of the market, while most of the incumbents were able to recoup their losses by raising prices.

3. This example was adapted from Bierman and Fernandez (1998, Chapter 6).

4. This game was first proposed by Rosenthal (1982). It is known as the centipede game because the original version consisted of a sequence of a hundred moves with linearly increasing payoffs.

CHAPTER

12 BARGAINING

In this chapter we will:

- *Introduce the concept of bargaining;*
- *Analyze the bargaining process as a static game;*
- *Analyze the bargaining without impatience as a dynamic game with a finite number of counteroffers;*
- *Discuss the last-mover's advantage;*
- *Analyze the bargaining process as a dynamic game with symmetric and asymmetric impatience.*

INTRODUCTION

Perfectly competitive markets are characterized by a large number of buyers and sellers entering into verbal, tacit, and written contracts for the purpose of exchanging goods and services for something of value, usually money. The monetary value of a single unit is its price. Firms in perfectly competitive industries are price takers because they are unable to influence the market-clearing price through their individual production decisions. Consumers may similarly be described as price takers because their individual purchasing power is too small to extract better terms from sellers. Theoretically, since neither buyer nor seller has "market power," the ability to bargain, or "haggle," over the terms of the contract is nonexistent. By contrast, a profit-maximizing monopolist sets the price of its product; buyers, having nowhere else to go, can either accept or reject the asking price. They lack the ability to negotiate the terms of the transaction. Even in those instances where neither the buyer nor the seller may be thought of as a "price taker," such as a monopsonist buying from an oligopolist, economists until recently had little to say about the possibility of negotiating, or **bargaining**, over the terms of the contract. A contract is a written or oral agreement that obligates the parties to perform, or refrain from performing, a specified act in exchange for some valuable benefit or consideration.

> **Bargaining** A process whereby individuals or groups of individuals negotiate over the terms of a contract.

Whether in business, politics, law, international relations, or everyday life, bargaining is a fact of life. Whether negotiating with your boss for an increase in salary and benefits, or haggling over the price of a new car, such interactions are commonplace. In business, contract negotiations between producer and supplier, contractor and subcontractor, wholesaler and distributor, retailer and wholesaler, and so on, is the norm, rather than the exception. As an exercise, the reader is asked to consider why market power, and the ability to bargain with suppliers, allows large retail outlets such as Home Depot, Sports Authority, or Costco to offer prices that are generally lower than those charged by local hardware, sporting goods, or department stores. Even in markets characterized by many buyers and sellers, it is often possible to find "pockets" of local monopoly or monopsony power that permits limited bargaining over contract terms. Game theory is a useful tool for analyzing and understanding the dynamics of the bargaining process.

It is useful to think of bargaining as a process whereby rivals negotiate over the division of a pie. Common sense and experience suggest that two individuals with equal bargaining power should each receive exactly one-half of the pie. In general, n individuals with equal bargaining power will each receive a one-nth share. In fact, the method of backward induction discussed in the previous chapter will lead to precisely this outcome in most negotiations.

In general, there are two questions that must be answered when analyzing bargaining scenarios. First, what are the bargaining rules? Second, what happens if the players fail to reach an agreement? In most retail establishments, for example, the seller posts a fixed asking price. The customer must decide whether to accept or reject this price. This is an example of a take-it-or-leave-it bargaining rule. In the case of collective bargaining, union representatives may propose a wage and benefit package. Management may accept the offer, reject the offer and wait for the union to modify its proposal, or reject the proposal and make a counteroffer. In some cases, the order of play is determined by custom or law. In other cases, strategic considerations may determine the sequencing of the bargaining process. In the next section, we will examine the bargaining process as a static game in which the players negotiate the distribution of a divisible object of value. In subsequent sections, we will analyze the bargaining process as a sequential move game in which the length of time it takes to reach an agreement is an important determinant in the outcome of the game.

NASH BARGAINING

We will begin our discussion of the bargaining process by going back to something a bit more basic. In a **Nash bargaining** game, the players "haggle" over the distribution of a divisible object of value, such as a cash amount. In this game, the players agree to submit their bids simultaneously. If the sum of the players' bids is less than the available amount, each player receives the bid amount and the game ends. If the sum of the players' bids is greater than the amount available, the players receive nothing.

> **Nash bargaining** A noncooperative static game in which the players bargain over the distribution of a divisible object of value.

Consider the following hypothetical example of a Nash bargaining game. In their most recent collective bargaining agreement, the senior management of Matrix Corporation, a semiconductor manufacturer, agreed with local union representatives to negotiate the distribution of any profits in excess of a specified amount as an incentive to increase worker productivity. At the conclusion of the last fiscal year, Matrix realized excess profits amounting to $200 per worker. As per the collective bargaining agreement, each side submits a one-time, sealed bid that represents their respective proposals for the distribution of excess profits. To ensure fairness, management and union agreed to bring in an outside, independent arbitrator to monitor the proceedings.

To keep things simple, we will assume that bids must be submitted in $100 increments. Once submitted, the arbitrator will open the bids and declare the distribution of profits. According to the rules of this Nash bargaining game, if the sum of the players' bids does not exceed $200, each worker will be awarded their bid, and the remainder will be distributed to Matrix shareholders. If the sum of the bids exceeds $200, bargaining ends and the game is declared a draw, in which case the players will submit a new bid. To avoid the possibility of this becoming an infinitely repeated

FIGURE 12.1

Nash Bargaining Game

		Union		
		$0	*$100*	*$200*
	$0	($0, $0)	($0, $100)	**(0, 200)**
Management	*$100*	($100, $0)	**($100, $100)**	(−$5, −$5)
	$200	**($200, $0)**	(−$5, −$5)	(−$5, −$5)

Payoffs: (Management, Union)

game, we will also assume that to defray the cost of retaining the arbitrator's services, each player will be penalized $5 per worker each time the sides fail to come to an agreement. The normal form of this Nash bargaining game is depicted in Figure 12.1.

Consider the Nash bargaining game from management's perspective. Suppose management submits a bid for $200. The only way that an agreement can be reached is if union representatives submit a bid of $0. In this case, the strategy profile is {*$200, $0*}, which is a Nash equilibrium. Is the union likely to be so altruistic? Hardly. By the same token, the solution profile {*$0, $200*} is also a Nash equilibrium, but, by the same reasoning, Matrix management is equally unlikely to be so magnanimous. So, what is each player's most likely bid?

It is clearly in both players best interest to submit a bid of $100. The strategy profile {*$100, $100*} is a focal-point equilibrium. This outcome is also in keeping with our earlier discussion of the minimax theorem in zero-sum games. This is because the two players are bargaining over the distribution of a $200-per-worker pie where one player's gain is the other player's loss. Recall that if both players simultaneously attempt to maximize their gains and minimize their losses, the game will have a Nash equilibrium in which the minimum of the maximum (minimax) and the maximum of the minimum (maximin) payoffs are equal.

The coordination game in Figure 12.1 also illustrates the difficulty of finding a solution to noncooperative, static games with multiple Nash equilibria since they often lead to inefficient outcomes. Six of the nine strategy profiles result in total payoffs that are less than the amount of excess profits available for distribution. In fact, three of these six strategy profiles result in negative payoffs in the event of a draw. This, in part, explains why collective bargaining disputes sometimes fail to immediately produce a mutually beneficial outcome: Both sides want a slice of the pie that, when combined, is larger than the pie itself. Fortunately, experimental evidence and real-world experience confirms the wisdom of the minimax theorem. There is a tendency for players in bargaining games to accept an equitable split as "fair," even in the presence of multiple Nash equilibria.

DYNAMIC BARGAINING AND THE LAST-MOVER ADVANTAGE

We will continue our discussion of the bargaining process by considering something perhaps more familiar than the Nash bargaining game discussed in the previous section. Suppose that Andrew wants to buy Adam's car. Adam knows that Andrew is willing to pay up to $1,000; Andrew knows that Adam will not accept anything less than $500. The maximum price that Andrew is willing to pay is called the **buyer's reservation price** and the minimum price that Adam is willing to accept is called the **seller's reservation price**. If Andrew and Adam come to an agreement, the gain to both will add up to the difference between the buyer's and the seller's reservation prices, which is $500.

FIGURE 12.2
Used-Car Bargaining Game Without Impatience

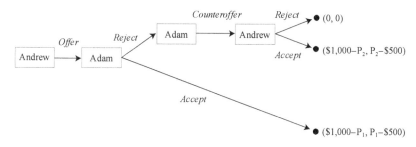

Payoffs: (Andrew, Adam)

> **Buyer's reservation price** The maximum price that a buyer in a bargaining scenario is willing to pay.
>
> **Seller's reservation price** The minimum price that a seller in a bargaining scenario is willing to accept.

Negotiations between Andrew and Adam may be modeled as the extensive form game depicted in Figure 12.2. We will assume for simplicity's sake that negotiations involve only two rounds—an offer and a counteroffer—and that Andrew makes the first offer, which is designated P_1. Adam, who moves second (and last), can either accept, or reject, Andrew's offer. If the offer is accepted, negotiations are concluded and the payoffs to Andrew and Adam are ($1,000 – P_1, P_1 – $500). If the offer is rejected, Adam can make a counteroffer of P_2, where $P_2 > P_1$. If the counteroffer is accepted, the payoffs to Andrew and Adam are ($1,000 – P_2, P_2 – $500), respectively. If Andrew rejects Adam's counteroffer, the game comes to an end, and the payoffs are (0, 0).

As an illustration, suppose that Andrew makes an offer of $800. If Adam accepts Andrew's offer, the game ends. The gains to Andrew and Adam are $1,000 – $800 = $200 and $800 – $500 = $300, respectively. The sum of the payoffs to each player is equal to the difference between the players' reservation prices. Suppose that Adam does not accept Andrew's offer and makes a counteroffer of $900. If Andrew accepts Adam's counteroffer, the game ends. The gain to each player is $1,000 – $900 = $100 and $900 – $500 = $400. Once again, the sum of the payoffs equals $500. The used-car bargaining game is illustrated in Figure 12.3.

Using the method of backward induction, it is easy to see that the subgame perfect equilibrium for the game in Figure 12.3 is for Adam to reject Andrew's offer of $800 and for Andrew to accept Adam's counteroffer of $900. The subgame perfect equilibrium for this game is {*Offer →* *Reject → Counteroffer → Accept*}. Andrew will earn a surplus of $100 and Adam will earn a surplus of $400. On the other hand, a counteroffer of $900 is not an optimal because Adam can do better by making an even larger counteroffer. To see why, suppose that Adam's counteroffer is $950. In this case, if Adam rejects Andrew's offer, the payoffs are ($50, $450). If Adam accepts, the payoffs are ($200, $300). It is left as an exercise for the reader to illustrate this new bargaining game and use backward induction to show that Adam will reject Andrew's offer. In fact, Adam's best counteroffer is Andrew's reservation price of $1,000. The reason for this is that the payoffs if Adam rejects Andrew's offer are ($0, $500). If Adam accepts, the payoffs are ($200, $300).

The subgame perfect equilibrium in this bargaining game is for Adam to reject any offer that is less than his reservation price, and for Andrew to accept any counteroffer up to and including Adam's reservation price. In this game, Adam will extract the entire surplus of $500. No matter what Andrew's initial offer may be, he will end up paying Adam $1,000. As long as Adam has the

FIGURE 12.3

Used-Car Bargaining Game in Which Andrew Makes an Offer of $800 and Adam Makes a Counteroffer of $900

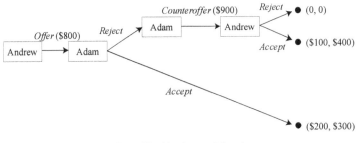

Payoffs: (Andrew, Adam)

ability to make a counteroffer, Adam will never accept any offer below Andrew's reservation price as final! Thus, in the two rounds of negotiation depicted in Figure 12.3, Adam "holds all the cards." The ability of Adam to dictate the final terms of the negotiations is referred to as a **last-mover advantage**. By projecting forward and reasoning backward, Andrew might just as well save his breath and offer Adam $1,000 at the outset of the negotiations.

> **Last-mover advantage** When a player is able to dictate the final terms of a negotiated agreement.

The above example suggests that the final outcome of this class of bargaining situations depends crucially on who makes the first offer, and on the number of negotiating rounds. If Andrew makes the first offer, and there are an odd number of rounds of negotiations, then Andrew has the last-mover's advantage and will be able to extract the entire surplus of $500. In this case, Adam should accept Andrew's initial offer of $500 and save both players the time, effort, and aggravation of the bargaining process. Similarly, if Andrew has the first move and there are an even number of rounds of negotiations, it will be in both players' best interest for Andrew to offer Adam $1,000. In this case, Adam will extract the entire surplus of $500.

DYNAMIC BARGAINING WITH SYMMETRIC IMPATIENCE

If negotiations of the type described in the previous section were that simple, they would never take place. Something must be missing. In this section we will make the underlying conditions of the bargaining process somewhat more realistic by assuming that there are multiple rounds of offers and counteroffers, and that there are opportunity costs associated with not immediately reaching an agreement. When negotiations drag on, foregone investment opportunities shrink the size of the pie. In this section, we will demonstrate how the time value of money plays an important role in determining the outcome of protracted negotiations.

In the preceding section we assumed only two bargaining rounds. In fact, the bargaining process is likely to involve multiple rounds of offers and counteroffers lasting days, weeks, months, or even years. Failure to quickly reach an agreement may impose considerable costs on the bargainers. Consider, for example, the rather large opportunity costs incurred by an individual who discovers that his or her car has been stolen. The theft has introduced a higher than usual level of anxiety into the situation. Failure to quickly reach an agreement on the purchase price of the car may not only result in significant psychological costs, but could also result in lost income.

In the above scenario, the buyer can take one of two possible approaches when negotiating with the used-car dealer. On the one hand, the buyer can withhold knowledge of his or her ill fortune from the seller and negotiate with a "cool head." Alternatively, the buyer may admit to the theft in an attempt to garner sympathy to obtain a lower price. As we will soon see, looking for sympathy from a rival in the bargaining process is not without cost. When one person's gain is another's loss, a buyer will have better luck finding a sympathetic ear from a priest, pastor, rabbi, psychologist, or mom, than from a used-car dealer. To see this, consider a situation in which the buyer and the seller enter into negotiations without any knowledge of the opportunity costs that may be imposed on the other by failing to immediately reach an agreement. This is equivalent to the situation where the buyer negotiates with the used-car dealer with a "cool head."

Let us return to the used-car example. Recall that Andrew is willing to pay up to $1,000 and Adam will not accept any offer below $500. Instead of only one offer and counteroffer, that is, just two negotiating rounds, suppose that there are thirty negotiating rounds. We assume that any delay in reaching an agreement reduces the gains to both players from a settlement by 5 percent per period. We will refer to this as the **discount factor**. This assumption is equivalent to assuming that both players have **symmetric impatience**. The higher the discount factor, the more impatient a player will be to come to an agreement. For simplicity, each negotiating round takes one period. We will also assume that the players have complete and perfect information. Each player is aware of the opportunity cost imposed on his or her rival by failing quickly to come to an agreement. With 30 negotiating rounds, it is impractical to illustrate the bargaining process with a game tree. Nevertheless, it is still possible to use backward induction to determine each player's negotiating strategy. Consider the information summarized in Table 12.1.

> **Discount factor** The rate used for finding the present value of the gains from a bargaining agreement. The greater the players' impatience (the higher the discount rate) the less advantageous will be the gains from bargaining.
>
> **Symmetric impatience** The equal reduction in the gains to all players by failing to quickly reach a bargaining agreement. Each player has the same discount factor.

Since Andrew makes the first offer and there are an even number of negotiating rounds, Adam has the last-mover's advantage. Thus, if negotiations drag on until the thirtieth round, Adam will sell his car for $1,000.00 and extract the entire surplus of $500. Andrew, of course, knows this. He also knows that Adam is indifferent between receiving a surplus of $500 in the thirtieth round and receiving a surplus of $500/1.05 = $476.19 in the twenty-ninth round because this delay reduces the present value of Adam's surplus by 5 percent. In capital budgeting terminology, the time value of $452.38 in the twenty-ninth round is the discounted value of $500.00 in the thirtieth round. Thus, it makes sense for Andrew to offer Adam $976.19 in the twenty-ninth round. If this offer is accepted, Andrew will realize a surplus of $500 – $476.19 = $23.81. Of course, this is not the end of the bargaining story.

Adam also knows that any delay in reaching an agreement will reduce Andrew's surplus by 5 percent per round. Thus, Andrew is indifferent between receiving a surplus of $23.81 in the twenty-ninth round $23.81/1.05 = $22.68 in the twenty-eighth round. Thus, Adam will make a counteroffer of $977.32. This will increase Adam's surplus to $477.32. The first four rounds of the used-car bargaining game are illustrated in Figure 12.4. Proceeding in the same manner, the reader can verify using backward induction that Andrew's best offer in the first round is $803.16, which Adam will accept. Adam and Andrew will enjoy a surplus of $303.16 and $196.84, respectively. It should be intuitively clear that as the number of negotiating rounds increases, the division of the surplus becomes more equitable. It depends on the discount factor, the number of negotiating rounds, and who makes the last counteroffer.

TABLE 12.1

Subgame-Perfect Equilibrium for the Used-Car Bargaining Game with Symmetric Impatience with a Discount Factor for Both Players of 5 Percent

Round	Offer maker	Offer price	Adam's surplus	Andrew's surplus
30	Seller	$1,000.00	$500.00	$0.00
29	Buyer	$976.19	$476.19	$23.81
28	Seller	$977.32	$477.32	$22.68
27	Buyer	$954.59	$454.59	$45.41
26	Seller	$956.75	$456.75	$43.25
⋮	⋮	⋮	⋮	⋮
5	Buyer	$819.93	$319.93	$184.07
4	Seller	$824.70	$324.70	$175.30
3	Buyer	$809.24	$309.24	$190.76
2	Seller	$818.32	$318.32	$181.68
1	Buyer	$803.16	$303.16	$196.84

FIGURE 12.4

First Four Rounds in the Used-Car Bargaining Game

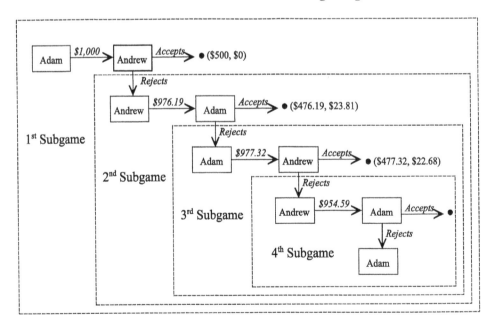

DYNAMIC BARGAINING WITH ASYMMETRIC IMPATIENCE

Suppose that instead of maintaining an "even keel," the buyer reveals to the used-car dealer the importance of quickly replacing the stolen car. The used-car dealer will immediately recognize the higher opportunity cost to the buyer. To demonstrate the impact that this knowledge has on the bargaining process, consider again the negotiations between Andrew and Adam. We will continue

TABLE 12.2

Subgame-Perfect Equilibrium for the Used-Car Bargaining Game in Which Adam's Discount Factor Is 5 Percent and Andrew's Discount Factor Is 10 Percent

Round	Offer maker	Offer price	Adam's surplus	Andrew's surplus
30	Seller	$1,000.00	$500.00	$0.00
29	Buyer	$976.19	$476.19	$23.81
28	Seller	$978.35	$478.35	$21.65
27	Buyer	$955.57	$455.57	$44.43
26	Seller	$959.61	$459.61	$40.39
⋮	⋮	⋮	⋮	⋮
5	Buyer	$849.85	$349.85	$150.15
4	Seller	$863.50	$363.50	$136.50
3	Buyer	$846.19	$346.19	$153.81
2	Seller	$860.17	$360.17	$139.83
1	Buyer	$843.02	$343.02	$156.98

to assume that there are thirty negotiating rounds, but that the opportunity cost to Andrew from delaying an agreement increases his discount factor to 10 percent, while Adam's opportunity cost remains unchanged at 5 percent. Proceeding as before, the information in Table 12.2 summarizes the gains to Andrew and Adam from **asymmetric impatience**, that is, when different discount factors are applied to each player. The more impatient the player to reach an agreement, the greater the discount factor for finding the present value of the gains from a bargaining agreement.

> **Asymmetric impatience** The unequal reduction in the gains to all players by failing to quickly reach a bargaining agreement. The players have different discount factors.

Utilizing backward induction, the reader should readily verify from Table 12.2 that Andrew's best first round offer is $843.02. This will give Adam a surplus of $343.02, which is more than double the gain enjoyed by Andrew. The results presented in Table 12.2 demonstrate that the negotiating party with the lowest opportunity cost has the advantage in the negotiating process. Clearly patience and secrecy are virtues when bargaining. "Crying the blues" has put the buyer at a bargaining disadvantage. Other things being equal, looking for sympathy from a rival during negotiations is a very poor bargaining tactic.

The discount rate for finding the present value of gains from trade is used as a measure of a player's bargaining impatience. The greater the player's impatience (the higher the discount rate) the less advantageous will be the gains from bargaining. Ariel Rubenstein (1982) demonstrated that in a two-player bargaining game there exists a unique subgame perfect equilibrium.[1] Suppose that player A and player B are bargaining over the division of a surplus, that player B makes the first offer, that there is no limit to the number of negotiating rounds, and that both players accept offers when indifferent between accepting and rejecting the offer. If we denote player A's discount factor as $0 \leq \delta_A \leq 1$ and player B's discount factor as $0 \leq \delta_B \leq 1$, this bargaining game has a unique subgame perfect equilibrium if the share of the surplus retained by player B is:

$$\omega_B = \frac{\theta_B(1-\theta_A)}{1-\theta_A\theta_B} \tag{12.1}$$

The fraction of the surplus retained by player A is:

$$\omega_A = \frac{1-\theta_B}{1-\theta_A\theta_B} \tag{12.2}$$

In Equations (12.1) and (12.2), $\theta_A = 1 - \delta_A$ and $\theta_B = 1 - \delta_B.^2$

─────────────────────── **Demonstration Problem 12.1** ───────────────────────

Players A and B are bargaining over the price of a service contract. The highest price that player B is willing to pay is $250. The lowest price that player A is willing to accept is $100. Suppose that each player's discount factor is $\delta_A = \delta_B = 0.06$.

    a. For a subgame perfect equilibrium to exist, what price should player B offer for the contract in the first round? What portion of the surplus will player A receive? What portion of the surplus will player B keep?

    b. Suppose that player B's discount factor remains $\delta_B = 0.06$, but player A's discount factor increases to $\delta_A = 0.08$. What price should player B offer for the contract in the first round? What portion of the surplus will player A receive? What portion of the surplus will player B keep?

Solution

    a. $\theta_A = 1 - \delta_A = 0.94$; $\theta_B = 1 - \delta_B = 0.94$. Substituting these values into Equation (12.1), the fraction of the surplus going to player B is:

$$\omega_B = \frac{\theta_B(1-\theta_A)}{1-\theta_A\theta_B} = \frac{0.94(1-0.94)}{1-(0.94)(0.94)} = \frac{0.0564}{0.1164} = 0.4845 \tag{D12.1.1}$$

The amount of the surplus retained by player B is 0.4845($150) = $72.68. In the first round, player B should offer $177.32 for the service contract. The fraction of the surplus received by player A is:

$$\omega_A = \frac{1-\theta_B}{1-\theta_A\theta_B} = \frac{(1-0.94)}{1-(0.94)(0.94)} = \frac{0.06}{0.1164} = 0.5155 \tag{D12.1.2}$$

    The amount of the surplus going to player A is 0.5155($150) = $77.32. The sum of the shared surpluses is $72.68 + $77.32 = $150. Player A receives a greater share of the surplus than player B.

    b. $\theta_A = 1 - \delta_A = 0.92$; $\theta_B = 1 - \delta_B = 0.94$. Substituting these values into Equation (12.1) we obtain

$$\omega_B = \frac{\theta_B(1-\theta_A)}{1-\theta_A\theta_B} = \frac{0.94(1-0.92)}{1-(0.92)(0.94)} = \frac{0.0752}{0.1352} = 0.5562 \tag{D12.1.3}$$

The amount of the surplus retained by player B is 0.5562($150) = $83.43. In the first round, player B should offer $166.57 for the service contract. The fraction of the surplus going to player A is:

$$\omega_A = \frac{1-\theta_B}{1-\theta_A\theta_B} = \frac{(1-0.94)}{1-(0.94)(0.92)} = \frac{0.06}{0.1352} = 0.4438 \tag{D12.1.4}$$

The share of the surplus received by player A is $0.4438(\$150) = \$66.57$. The sum of the shared surpluses is $\$83.43 + \$66.57 = \$150$.

CHAPTER REVIEW

Bargaining is the process whereby individuals or groups of individuals negotiate over the terms of a *contract*. A contract is a formal agreement that obligates the parties to perform, or refrain from performing, a specified act in exchange for something of value.

In a *Nash bargaining* game, the players "haggle" over the distribution of a divisible object of value. In this game, the players agree to submit their bids simultaneously. If the sum of the players' bids is less than the available amount, each player receives the bid amount and the game ends. If the sum of the players' bids is greater than the amount available, the players receive nothing.

The most likely Nash equilibrium strategy profile solution may be found by applying the *minimax theorem*. According to this theorem, if both players simultaneously attempt to maximize their gains and minimize their losses, the minimum of the maximum (minimax) and the maximum of the minimum (maximin) payoffs are equal.

Most bargaining games are modeled as dynamic games. *Bargaining without impatience* assumes that the players do not incur opportunity costs by failing to reach an agreement quickly. The final outcome of this class of bargaining processes depends on who makes the first offer, and on the number of negotiating rounds. The player who makes the final offer has a *last-mover advantage* and is able to dictate the final terms of a negotiated agreement.

Bargaining with impatience asserts that players incur opportunity costs by failing to reach an agreement quickly. Impatience may be symmetric or asymmetric. *Symmetric impatience* assumes that the players' discount factors are the same. The *discount factor* is the rate used for finding the present value of the gains from a bargaining agreement. The greater the players' impatience (the higher the discount rate) the less advantageous will be the gains from bargaining. With *asymmetric impatience,* the discount factor is different for each player. Players with greater patience (lower discount factor) have the advantage in the negotiating process. In these games, the player who makes the last counteroffer will receive the larger fractional share of the bargaining surplus. The amount of this gain will depend on the discount factor of each player and the number of negotiating rounds. In general, the more patient the player, the larger will be that player's share of the surplus.

CHAPTER QUESTIONS

12.1 In bargaining scenarios without player impatience, explain how the number of bargaining rounds and who makes the first offer affect the final outcome.

12.2 Explain what is meant by the last-mover's advantage.

12.3 What is the difference between bargaining with symmetric and asymmetric impatience? How is bargaining impatience measured?

12.4 What is the discount factor in bargaining games without impatience?

CHAPTER EXERCISES

12.1 Suppose that two players are bargaining over the distribution of $1,000 in which payoffs must be in increments of $1. Each player submits a one-time bid. If the sum of the bids is less than or equal to $1,000, each player gets the amount of his or her bid and the game ends. If the sum of the bids is greater than $1,000, the game ends and the players go home empty-handed.

   a. Does this game have a Nash equilibrium?

   b. What is the most likely Nash equilibrium strategy profile for this game?

12.2 Two individuals are bargaining over the distribution of $100 in which payoffs must be in increments of $5. Each player must submit a one-time bid. If the sum of the bids is less than or equal to $100, each player gets the amount of the bid and the game ends. If the sum of the bids is greater than $100, the game ends and the players get nothing.

   a. Does this game have a Nash equilibrium?

   b. What is the most likely equilibrium strategy profile for this game?

12.3 Suppose that Alex wants to purchase a boat from Rosette. Alex is willing to pay up to $18,000, while Rosette is not willing to accept any offer below $15,000. Assume that there are a finite number of negotiating rounds.

   a. If the discount factors for Rosette and Alex are $\delta_R = 0.05$ and $\delta_A = 0.05$, respectively, how much should Alex offer for the boat?

   b. Suppose that Rosette's discount factor is $\delta_R = 0.20$, and Alex's discount factor is $\delta_A = 0.15$. How much should Alex offer for the boat? How does this offer differ from your answer to part a, and why?

12.4 Suppose that Andrew wants to purchase Adam's car. Adam knows that Andrew is prepared to pay a maximum of $2,000. Adam is unwilling to accept anything lower than $1,000. If Andrew and Adam can come to an agreement, the gain to both will add up to $1,000, which is the difference between the buyer's and the seller's reservation prices. Andrew's and Adam's discount rates per round are $\delta_A = 0.10$ and $\delta_B = 0.10$, respectively.

   a. For a subgame perfect equilibrium to exist, what portion of the surplus should Adam offer Andrew in the first round? What portion of the surplus should Adam keep for himself?

   b. Suppose that Adam's discount rate remains $\delta_B = 0.10$, but Andrew's discount rate increases to $\delta_A = 0.20$. What portion of the surplus should Adam offer Andrew in the first round and what portion should he keep for himself?

12.5 Suppose in demonstration problem 12.2 that player A is perfectly patient and player B is perfectly impatient.

   a. How much of the surplus will be received by each player?

   b. How much should player B offer for the service contract?

ENDNOTES

1. The proof of this result is quite difficult. A reader-friendly explanation can be found in Osborne and Rubenstein (1990) and Binmore (1992).

2. In dynamic bargaining games without impatience, the discount factor for both players is zero. The reader should verify using Equations (12.1) and (12.2) that Andrew will pay $1,000 for Adam's car and receive a zero surplus, while Adam will receive a surplus of $500.

CHAPTER

13

PURE STRATEGIES WITH UNCERTAIN PAYOFFS

In this chapter we will:

- *Introduce uncertainty and risk into payoffs of pure-strategy static and dynamic games;*
- *Discuss static games with incomplete information—also known as static Bayesian games;*
- *Briefly examine Harsanyi transformations and Bayesian Nash equilibria;*
- *Examine players' attitudes toward risk and the effect this may have on strategy choices and the search for Nash equilibria;*
- *Review consumer and firm behavior under conditions of uncertainty;*
- *Introduce the concept of an information set, and briefly consider how this may be used to analyze dynamic games involving uncertain payoffs.*

INTRODUCTION

We have thus far assumed that the payoffs from alternative pure-strategy profiles were known with certainty. In fact, the choice of strategies often involves situations in which outcomes are uncertain and where decisions must be made with less than complete information. Many business decisions are made under a cloud of uncertainty. In most cases, managers do not know how the public will react to the introduction of a new product line, or how fluctuations in macroeconomic activity, shifting consumer tastes, the behavior of rival companies, changes in resource availability, changes in input prices, labor unrest, political instability, and so forth will affect sales, revenues, and profits. On the other hand, experience and market or economic analyses may make it possible to reduce uncertainty by assigning probabilities to each possible outcome.

Companies submitting competitive bids for the right to extract mineral deposits on public lands or develop a new type of military aircraft are not privy to a rival's feasibility studies, accounting data, or profit estimates. A college graduate entering the labor market has to convince prospective employers that the job advertised is a perfect fit despite a low starting salary. The employer must not only attempt to uncover the applicant's true motives, but how he or she will perform on the job. In these and many other cases, the payoffs and strategies of players are not common knowledge. In this chapter, we will expand our discussion to include decision making under conditions of uncertainty by examining games with **incomplete information**.

> **Incomplete information** Games in which players' strategies and payoffs are not common knowledge.

STATIC GAMES WITH UNCERTAIN PAYOFFS

We will begin our discussion by introducing uncertainty into the oil-drilling game depicted in Figure 2.6, which is recreated below. Recall that this game involved two oil companies, PETROX

FIGURE 2.6
Static Game with an Iterated Strictly Dominant Strategy Equilibrium

GLOMAR

		Don't drill	Narrow	Wide
	Don't drill	(0, 0)	(0, **132**)	(0, 104)
PETROX	Narrow	(**132**, 0)	(52, 52)	(12, **70**)
	Wide	(104, 0)	(**70**, 12)	(**24, 24**)

Payoffs: (PETROX, GLOMAR)

FIGURE 13.1
Oil-Drilling Game in Which Oil Is Not Present

GLOMAR

		Don't drill	Narrow	Wide
	Don't drill	(**0, 0**)	(**0**, −68)	(**0**, −98)
PETROX	Narrow	(−68, **0**)	(−68, −68)	(−98, −98)
	Wide	(−98, **0**)	(−98, −98)	(−98, −98)

Payoffs: (PETROX, GLOMAR)

and GLOMAR, deciding whether to drill for oil that lies below adjacent tracts of land leased from the government. In that game, the amount of oil under the ground and the payoffs from alternative pure drilling strategies were known with certainty. The Nash equilibrium pure-strategy profile {Wide, Wide} was identified by iteratively eliminating all dominated strategies. Figure 13.1, on the other hand, is the normal form of the oil-drilling game in which it is known with certainty that no oil is present.

Unfortunately, certainty of outcomes and payoffs is a happy state of affairs that is rarely encountered in reality. This is especially true in the oil-drilling business despite technologically sophisticated seismic testing, exploration, and recovery methods. We will introduce uncertainty into the oil-drilling game by assuming that there is a 60 percent chance of striking oil and a 40 percent chance of drilling a dry well. How can we incorporate this uncertainty into the strategic decision-making process?

John Harsanyi (1968) devised a method for analyzing games with incomplete information. His solution is known as a **Harsanyi transformation**. Harsanyi recognized that players form beliefs about payoffs, players, and rival strategies. To help resolve this uncertainty problem, Harsanyi introduced a new player into the game, which he called Nature. In the case of the oil-drilling game, this player has no vested interest in the outcome of the game, but moves first by randomly selecting the *state of nature* and/or *player type* according to fixed probabilities that are common knowledge. As a result, a game with incomplete information is transformed into a game with complete, but imperfect, information. In a game with **imperfect information**, a player must move without knowing a rival's intentions, which is characteristic of all static games. A static game with incomplete information is referred to as a **static Bayesian game**.

> **Harsanyi transformation** Transforms games with incomplete information into games with complete but imperfect information by having Nature move first to determine states of nature and/or player types.

FIGURE 13.2

Oil-Drilling Game with Expected Payoffs and a 60 Percent Chance of Striking Oil

		GLOMAR		
		Don't drill	*Narrow*	*Wide*
	Don't drill	(0, 0)	(0, **52**)	(**0**, 23.2)
PETROX	*Narrow*	(**52**, 0)	(**4, 4**)	(–20, 2.8)
	Wide	(23.2, **0**)	(2.8, –20)	(–24.8, –24.8)

Payoffs: (PETROX, GLOMAR)

Imperfect information When a player moves without knowledge of a rival's intentions. A static game with complete information is an example of a game with imperfect information.

Static Bayesian game A static game with incomplete information.

We will now apply a Harsanyi transformation to the oil-drilling games depicted in Figures 2.6 and 13.1. Figure 13.2 summarizes the expected payoffs in millions of dollars from alternative strategy profiles. The objective of each player is to choose a strategy that maximizes expected payoffs. Suppose, for example, that both companies adopt a *wide* strategy. The expected payoff for each company is $0.6(24) + 0.4(-98) = -\$24.8$ million. The unique Nash equilibrium strategy profile for this game is {*Narrow, Narrow*}. This strategy profile is also referred to as a **Bayesian Nash equilibrium**. Although expected payoffs have been used to find the pure-strategy Nash equilibrium for this game, the reader must not confuse expected payoffs with actual payoffs, which are only known after the results of the companies' drilling efforts are known. If oil is found, the payoff for both companies is \$52 million. If no oil is found, each company will lose \$68 million.

Bayesian Nash equilibrium The Nash equilibrium for a static Bayesian game.

Example: The Slumlord's Dilemma

Consider the following variation of the prisoner's dilemma, which was discussed by Davis and Whinston (1962). The owners of adjacent slum tenements, Slumlord Larry and Slumlady Sally, are considering renovating their buildings. This game involves positive externalities since investment by either landlord will cause the property values, and market-determined rents, of both buildings to rise, although the increase will be greater if both renovate. In this game, if only one tenement owner renovates, he or she will suffer a decline in net income, presumably because the increase in rental income is not sufficient to offset the increase in operating costs. On the other hand, if both slumlords renovate, rental income will be significantly higher and each will experience an increase in monthly profits. The normal form of this static game in pure strategies and certain payoffs is depicted in Figure 13.3. The payoffs are in thousands of dollars. Despite the increased profits if both renovate, the players are caught in a prisoner's dilemma. Each property owner has a strictly dominant *don't renovate* strategy. Although both slumlords are better off by renovating, the Nash equilibrium strategy profile {*Don't renovate, Don't renovate*} results in no change in profits. These payoffs are highlighted in boldface.

It is clearly in both players' best interest to cooperate and renovate their properties. Unfortunately, there is an obvious incentive for Slumlord Larry and Slumlady Sally to violate any such agreement and free ride on the other's investment. As difficult as it may be for Slumlord Larry and Slumlady Sally to trust each other, this problem is magnified as the number of tenement owners

FIGURE 13.3
Slumlord's Dilemma

Slumlady Sally

Slumlord Larry		Renovate	Don't renovate
	Renovate	(5, 5)	(−3, **7.5**)
	Don't renovate	(**7.5**, −3)	**(0, 0)**

Payoffs: (Slumlord Larry, Slumlady Sally)

in a given neighborhood increases. The inherent difficulty in voluntarily forming alliances and cooperative agreements helps to explain the perpetuation of socially detrimental slum conditions in many urban areas. One public policy solution to this dilemma is for municipal authorities to exercise their powers of eminent domain by condemning, then acquiring, these properties. The government could then redevelop these properties at public expense, or sell the properties to a single developer who agrees to do so. This is one approach to urban renewal.

Now, let us consider the slumlord's dilemma when uncertainty is introduced into the calculations. We will do this by assuming that these tenements are located in an urban area known for its high arson rate. Each slumlord is assumed to carry fire insurance for the market value of the property, which varies depending on whether one, both, or neither property has been renovated. Figure 13.4 summarizes the slumlord's certain change in monthly profits from alternative pure-strategy profiles in the event of arson. The strictly dominant Nash equilibrium strategy profile for this game is {*Renovate, Renovate*}. The payoffs are highlighted in boldface.

How is the outcome of the slumlord's dilemma altered if strategy choices are based on expected payoffs? Let us assume that actuarial studies indicate that there is a 20 percent probability that a tenement will be the target of arson. We will use Nature to randomly select which buildings are "torched." Figure 13.5 summarizes the expected payoffs from alternative strategy profiles. The expected payoffs from the strategy profile {*Don't renovate, Renovate*}, for example, are 0.8(7.5) + 0.2(−7.5) = 4.5 and 0.8(−3) + 0.2(12) = 0, respectively. The Nash equilibrium strategy profile for this game is {*Renovate, Renovate*} if both slumlords carry fire insurance, whereas without fire insurance it is rational not to renovate. The introduction of uncertainty, and the purchase of fire insurance, has enabled the players in this game to escape the slumlord's dilemma.

So far we have assumed that Nature randomly chooses which buildings are torched according to fixed probabilities. But, there is something else going on here. While expected payoffs allow us to identify the Nash equilibrium strategy profile, the actual payoffs depend on the actual state of nature. In the no arson scenario, if both slumlords renovate, monthly payoffs increase by $5,000. In the arson scenario, payoffs increase by $5,500. Although the no arson scenario is the most likely outcome, the actual payoffs provide the players with an incentive to torch their own buildings! Of course, the decision to commit arson must be tempered by the risk associated with getting caught, paying hefty fines, and doing serious jail time. Whether or not the players in this game are willing to take that risk will depend on a number of factors, including the probability of getting caught and the players' attitudes toward risk. It is to this topic that we turn next.

RISK AND UNCERTAINTY

Up to this point we have assumed that players' strategy choices were made under conditions of complete certainty and sure payoffs, or expected payoffs where the probability distribution of random events was common knowledge. Although the introduction of alternative states of nature has enhanced our understanding of strategic behavior, we have examined only part of the problem.

FIGURE 13.4
Slumlord's Dilemma in the Event of Arson

Slumlady Sally

Slumlord Larry		Renovate	Don't renovate
	Renovate	**(5.5, 5.5)**	(12, –7.5)
	Don't renovate	(–7.5, 12)	(–1, –1)

Payoffs: (Slumlord Larry, Slumlady Sally)

FIGURE 13.5
Slumlord's Dilemma with Expected Payoffs

Slumlady Sally

Slumlord Larry		Renovate	Don't renovate
	Renovate	**(5.1, 5.1)**	(0, 4.5)
	Don't renovate	(4.5, 0)	(–0.2, –0.2)

Payoffs: (Slumlord Larry, Slumlady Sally)

While a formal treatment of decision-making under conditions of uncertainty is beyond the scope of this book, in this section we will briefly explore the concepts of **risk** and **uncertainty**.

Many people confuse risk with uncertainty. Both risk and uncertainty involve choices from among multiple possible payoffs. The fundamental difference is that the probability of a risky outcome is either known or can be estimated. By contrast, the probability of an uncertain outcome is either unknown, cannot be estimated, or is meaningless. As a practical matter, decision makers do not discriminate between the two terms. In fact, the phrase *decision making under conditions of uncertainty* is often used to describe the probability distribution of multiple payoffs, regardless of whether or not these probabilities are known, meaningful, or can be estimated.

> **Uncertainty** Multiple payoffs from alternative states of nature with unknown or meaningless probabilities.

> **Risk** Multiple payoffs from alternative states of nature with known probabilities.

It is important, however, not to focus on expected values alone when comparing two or more probabilistic payoffs. Formally, an expected value is defined as the payoff times the probability of its occurrence. If the probability is known, the payoff is risky, otherwise it is uncertain. Does this mean that a payoff with a lower expected value is riskier than one with a higher expected payoff? Not at all. In fact, riskier payoffs are typically associated with higher expected returns, but this need not be the case. Expected values alone tell us nothing about the relative riskiness of alternative payoffs. Different strategies under alternative states of nature may have the same expected payoff, but may not be equally risky.

As an illustration, consider two financial investments. Each promises an expected payoff of $100. The actual payoff depends on two states of nature: economic prosperity or recession. Suppose that both states of nature have an equal probability of occurring. In a prosperous economy, the actual payoff from an investment in project *A* is $101, while the actual payoff in a recession is $99. The expected payoff from the investment is 0.5($101) + 0.5($99) = $100. By contrast, the actual payoff from an investment in project B in a prosperous economy is $200, but $0 during a recession. Once again, the expected payoff from this investment is 0.5($200) + 0.5($0) = $100. Even though project B has the same expected payoff as project A, it is a riskier investment. The reason is the volatility of the actual payoffs. In a worse-case scenario, a person could lose $1 by investing in project A, but could lose everything by investing in project B.

The most commonly used statistic to measure the volatility of payoffs is the standard deviation (see the appendix to this chapter). When comparing alternative payoffs, a player must balance its expected value with the volatility of payoffs across different states of nature. As a general rule, increased volatility must be compensated with a greater expected payoff. How much additional compensation is required depends on an individual's attitude toward risk.

CASE STUDY 13.1: GLOBAL WARMING: UNCERTAINTIES, CERTAINTIES, AND URGENCIES

Recall that the probability of a risky outcome is known, or can be estimated. By contrast, the probability of an uncertain outcome is unknown, cannot be estimated, or is meaningless. Decision making under conditions of risk is difficult and may be costly, but the underlying facts are indisputable. Decisions made under conditions of uncertainty, on the other hand, are controversial at best, and possibly fatal at worst. Nobel laureate Thomas Schelling (2007) discussed the role of uncertainty in the debate over climate change, and the potential long-run consequences of doing nothing.

First, consider the uncertainties. With the onset of the industrial revolution, the level of greenhouse gases in the atmosphere began to rise exponentially. Many scientists believe that this increase is a natural phenomenon, but many others believe that this is the result of the burning of fossil fuels, particularly coal and petroleum by-products. The buildup of greenhouse gases may be causing the planet to heat up at an unacceptable rate (see Case Study 7.1). How much has been the subject of much debate for a quarter century. According to Schelling, "the range of uncertainty has been about a factor of three" (Schelling 2007, p. 1). How will rising temperatures affect rainfall, sunlight, cloud cover, humidity, the polar ice caps, and ocean levels? What effect will global warming have on productivity, especially in agriculture, fishing, forestry, population migration, public health, famine, social and political unrest, and military conflict?

Next, consider the certainties. Scientists have known for more than a century that trapped greenhouse gases lead to temperature increases. The greenhouse effect on Venus is so severe, with surface temperatures hundreds of degrees higher than on the Earth, that surface water does not exist. Greenhouse gases are so deficient on Mars that it is too cold for surface water to exist in liquid form. We know for certain that the Earth is warming, but we do not know how fast, how far, the precise causes, and what should be done.

Finally, there are the urgencies. According to Schelling, at the top of the list is to keep studying the problem, especially through government-sponsored energy research and development. "We need, urgently, to better understand what alternatives to fossil fuels there will be, how much energy can be conserved, how to extract carbon dioxide from the atmosphere, and if necessary how to increase the earth's . . . reflectance of incoming sunlight" (Schelling 2007, p. 2). Essential research and development, says Schelling, will not be undertaken by the private sector since the benefits cannot be "captured" by investors. Another area of research goes under the name of "geoengineering," which deals with the amount of sunlight that is absorbed by oceans, forests, the plains, and urban areas, and the amount that is reflected back into space. This is not a task for the private sector. Multinational sponsorship may be necessary to escape this apparent prisoner's dilemma.

A peculiar aspect of the public discourse on global warming is that the uncertainties of the problem are being used to postpone action until more is known. According to Schelling, ". . . it is interesting that this idea that costly actions are unwarranted if the dangers are uncertain is almost unique to climate. In other areas of policy, such as terrorism, nuclear proliferation, inflation, or vaccination, some 'insurance' principle seems to prevail: if there is a sufficient likelihood of sufficient damage we take some measured anticipatory action" (Schelling 2007, p. 4). Left unsaid is the influence of special interest groups on the public decision-making process. According to Schelling, when it comes to possible consequences of global warming, there are few actions that uncertainty makes infeasible and not worth trying.

ATTITUDES TOWARD RISK

In the first part of this chapter we considered the search for a Nash equilibrium in the oil-drilling game in which certain payoffs were replaced with expected values. At no time, however, did we consider the possibility that the payoffs from, say, a *narrow* drilling strategy might be riskier than the payoffs from a *wide* drilling strategy. By focusing only on the expected values of each strategy, we essentially asserted that PETROX and GLOMAR were indifferent to drilling risk. Unfortunately, considering risk makes the search for a Nash equilibrium more complicated. The reason for this is that while each company may know how much risk it is willing to assume, it may not be possible to know how much risk a rival is willing to incur. Because of this, it is far easier to assume that all players are indifferent to risk and that decisions are based on expected payoffs alone. Be that as it may, a player's attitude toward risk is an important element in the decision-making process.

The analysis of the oil-drilling game in Figure 13.2 is acceptable as far as it goes. Unfortunately, our analysis was incomplete because it was based on the assumption that both companies were indifferent to risk and that strategy choices were based only on expected payoffs without regards to the volatility of these payoffs. How might a firm's attitude toward risk be incorporated into a player's choice of strategies?

We will consider this question by considering another version of the oil-drilling game in which the probability of striking oil is now 85 percent, with a 15 percent chance of drilling a dry well. This game is depicted in Figure 13.6. The reader should verify that the pure-strategy Nash equilibrium for this game is {*Wide, Wide*}. The firms are apparently caught in a prisoner's dilemma since it seems to be in both firms' best interest to adopt a *narrow* strategy. Of course, the actual payoffs depend on the state of nature. If the companies drill wide wells and strike oil, the payoffs to both companies are $24 million. If the well is dry, both companies lose $98 million.

Now suppose that both companies cooperate and agree to drill narrow wells. Although this strategy profile is not a Nash equilibrium, it may nonetheless constitute a focal-point equilibrium. The reason may be gleaned from Figures 2.6 and 13.1. If the firms cooperate and the well is dry, both companies will lose $68 million. If they strike oil, the payoff to both companies is $52 million, or a difference of $52 − (−$68) = $120 million. On the other hand, if they do not cooperate and drill *wide* wells, the difference in payoffs is $24 − (−$98) = $122 million. Drilling a *narrow* well is not only less risky, but it offers a higher expected payoff. The only way that the strategy profile {*Wide, Wide*} makes sense is if both players are oblivious to risk and base their strategy choices exclusively on expected payoffs.

FIGURE 13.6

Oil-Drilling Game with Expected Payoffs and an 85 Percent Chance of Striking Oil

GLOMAR

		Don't drill	Narrow	Wide
	Don't drill	(0, 0)	(0, **102**)	(0, 73.7)
PETROX	Narrow	(102, 0)	(34, 34)	(−4.5, **44.8**)
	Wide	(73.7, 0)	(**44.8**, −4.5)	(**5.7, 5.7**)

Payoffs: (PETROX, GLOMAR)

FIGURE 13.7

Increasing, Constant, and Decreasing Marginal Utility of Money

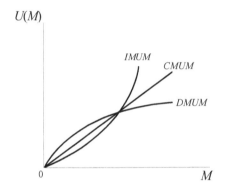

In the game depicted in Figure 13.6, a *narrow* strategy not only has a greater payoff, but is also less risky than a *wide* strategy. Thus, there is little question which drilling strategy should be adopted. In many cases, however, the strategy with the greatest expected payoff is also the riskiest. When this happens, which strategy should a player adopt? The answer depends on the players' attitudes toward risk. One way to illustrate differences in these attitudes is to consider a concept that is familiar to most economics students.

The **marginal utility of money** is the additional satisfaction that an individual receives from an additional dollar of money income or wealth. Figure 13.7 illustrates three total utility of money functions, where money is measured along the horizontal axis, and a cardinal index of utility (satisfaction) is measured along the vertical axis. These total utility of money curves illustrate the concepts of constant, increasing, and diminishing marginal utility of money. Increasing marginal utility of money (*IMUM*) occurs when additional dollars received yield positive and increasing incremental units of satisfaction. Constant marginal utility of money (*CMUM*) occurs when additional dollars received yield positive, but unchanging, incremental units of satisfactions. Finally, decreasing marginal utility of money (*DMUM*) describes the situation in which additional dollars received yield positive and decreasing incremental units of satisfaction.

> **Marginal utility of money** The additional satisfaction received from an additional dollar of money income or wealth

More formally, the total utility of money function is:

$$U = U(M) \tag{13.1}$$

Total utility is assumed to be an increasing function of money, that is, $dU/dM > 0$. If Equation (13.1) exhibits constant marginal utility of money then $d^2U/dM^2 = 0$. If this function exhibits increasing marginal utility of money, $d^2U/dM^2 > 0$. Finally, if this function exhibits diminishing marginal utility of money then $d^2U/dM^2 < 0$.

Suppose that a person is offered the following wager. In exchange for a bet of $1,000, a person can win $2,000 by flipping a coin that comes up "heads," but wins nothing if it comes up "tails." If the coin is "fair," there is an equal probability of flipping a heads or tails. The expected value of this wager (M) is:

$$E(M) = p_W W + p_L L \tag{13.2}$$

In Equation (13.2), p_W is the probability of winning some amount W, and p_L is the probability of losing some amount L. The probability of winning $2,000 on the toss of a fair coin is $p_W = 0.5$, while the probability of losing $1,000 is $p_L = 1$ since this is the certain cost of playing this game. Substituting these values into Equation (13.2) we obtain $E(M) = 0.5(\$2,000) + 1(-\$1,000) = 0$. This is an example of a **fair gamble** since the expected value of the wager is $E(M) = 0$. Alternatively, this is a fair gamble because the cost of the wager is equal to the expected value of winning the bet. An **unfair gamble** is when the expected value of a wager is negative ($E(M) < 0$).

> **Fair gamble** When the expected value of winning a wager is equal to the expected value of losing.

> **Unfair gamble** When the expected value of winning a wager less than the expected value of losing.

Sometimes, the price of the wager is incorporated into the gamble itself. Suppose, for example, that the above individual is offered the following wager. Flip a fair coin. If it comes up heads, the individual wins $2,000. If it comes up tails, the individual loses $2,000. In this case, the person is not charged a price to play the game, but can lose an amount by flipping tails. Since the expected value of this gamble is $E(M) - 0.5(\$2,000) + 0.5(\ \$2,000) = 0$, this is a fair gamble. That is, the expected value of winning ($p_W W$) equals the expected value of losing ($p_L L$).

─────────── **Demonstration Problem 13.1** ───────────

Lugg Hammerhands has been offered the following wager. Blindfolded, Lugg can draw a single marble from an urn containing ten marbles. Nine of the marbles are green and one is red. If Lugg draws the red marble, he wins $450. If he draws a green marble, he loses $50. Is this a fair gamble?

Solution

Substituting these values into Equation (13.2), the expected value of the gamble is:

$$E(M) = 0.1(\$450) - 0.9(\$50) = 0 \tag{D13.1.1}$$

Since the expected value of winning equals the expected value of losing, this is a fair gamble.

Demonstration Problem 13.2

In the United States, many state governments sponsor lotteries,. In New York State, for example, $1 will buy two chances to win the game of "Lotto." Each game involves selecting six of 59 numbers. The New York State Lottery Commission randomly draws six numbers, and whoever selects the correct combination wins, or shares, the top prize. According to the New York State Lottery Commission, on a bet of $1, the odds of winning the top prize is 1 in 22,528,737. Suppose that the top prize is $20 million. Is this a fair gamble?

Solution

The probability of losing $1 is $p_L = 1$ because this is the price of a lottery ticket. Assuming that the number selection process is fair, the expected value of this wager is:

$$E(M) = \left(\frac{1}{22,528,737}\right)\$20,000,000 + 1(-\$1) = -\$0.11 \qquad (D13.2.1)$$

Since the expected value is negative, Lotto is an unfair gamble.

Economists have found that most people, groups, and organizations are risk averse. A risk-averse person will generally decline a fair, but risky, gamble. Risk aversion was identified by the eighteenth-century Dutch-born mathematician Daniel Bernoulli (1738) who theorized that people are generally more concerned with the utility of a wager than with its expected value. He speculated that when the payoff of a fair gamble does not mirror an individual's utility, that person will refuse to participate. The reason for this is that most people exhibit diminishing marginal utility of money. To illustrate this, suppose that an individual's total utility of money function is $U = 100M^{0.5}$. We will assume that utility is measurable in hypothetical units called *utils*. This total utility of money function exhibits diminishing marginal utility of money since $dU/dM = 50M^{-0.5} > 0$ and $d^2U/dM^2 = -25M^{-1.5} < 0$. A risk-averse individual will generally not accept a fair gamble. The reason for this is that the psychic pain of losing, say, $1,000 on the flip of a fair coin is greater than the psychic gain of winning $1,000 because dollars received earlier yield greater additional satisfaction than dollars received later.

To see why a risk-averse individual might not accept this bet, suppose that $M = \$50,000$. The total utility received by this person is $U = 100(50,000)^{0.5} = 22,361$ utils. Now, suppose that the individual accepts the wager and wins $1,000. This person now has $51,000. The total utility of money is $U = 100(51,000)^{0.5} = 22,583$ utils. In this case, the individual has gained 222 utils of extra satisfaction. Suppose, on the other hand, the individual loses $1,000. The person now has $49,000. The total utility of money is $U = 100(49,000)^{0.5} = 22,136$ utils. In this case, the individual has lost 225 utils of satisfaction. The expected change in utility by accepting this fair gamble is:

$$E(\Delta U) = \sum_{i=1}^{2}(\Delta U_i)p_i = (\Delta U_1)p_1 + (\Delta U_2)p_2$$

$$= (222)0.5 + (-225)0.5 = -1.5 \text{ utils} \qquad (13.3)$$

Since the expected change in total utility is negative, a rational person would not accept this fair gamble.

Alternatively, the individual's expected utility from this fair gamble is $E(U) = 0.5(22,583) + 0.5(22,136) = 22,359.50$ utils. The difference between the utility from this fair gamble is less than

FIGURE 13.8
Risk Aversion

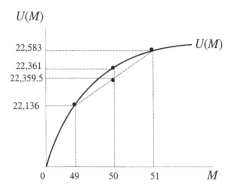

the utility from not accepting (22,359.50 < 22,361). Since the change in the individual's expected utility is negative, a risk-averse individual would probably not accept this wager. This situation is depicted in Figure 13.8.

In general, diminishing marginal utility of money is typical of **risk aversion**. The risk-averse individual in Figure 13.8 would not accept this fair gamble because the utility gained from $1,000 is greater than the utility gained from an expected payoff of the same amount. Alternatively, a risk-averse person would prefer the certainty of keeping the wager to winning its expected value. In general, a risk-averse person would not accept a fair gamble since $M \succ E(M)$.[1]

> **Risk aversion** When an individual prefers the certain value of a wager to its expected value ($M \succ E(M)$). In general, a risk-averse individual would not accept a fair gamble.

While it is generally assumed that people are risk averse, we cannot rule out the possibility that some individuals behave differently when confronted with a risky situation. An individual is said to be **risk loving** if the expected value of a payoff is preferred to its certainty equivalent, that is $F(M) \succ M$. A risk-loving individual has an increasing marginal utility of money. An individual is said to be **risk neutral** when he or she is indifferent between the value of a certain payoff and its expected value, that is, $E(M) \sim M$. Such an individual has a constant marginal utility of money. These last two situations are depicted in Figures 13.9 and 13.10.

> **Risk loving** When an individual prefers the expected value of a wager to its certainty equivalent, that is, ($E(M) \succ M$).

> **Risk neutral** When an individual is indifferent between a certain payoff and an expected value of the same amount, that is, ($M \sim E(M)$).

──────────────── **Demonstration Problem 13.3** ────────────────

An individual is offered the following fair gamble. If the flip of a fair coin comes up heads, the person wins $1,000. If the coin comes up tails, the person loses $1,000. The individual's total utility of money function is:

$$U = M^{1.1}$$
(D13.3.1)

a. For a positive level of money income, what is this person's attitude toward risk?

FIGURE 13.9
Risk Loving

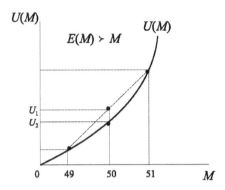

FIGURE 13.10
Risk Neutrality

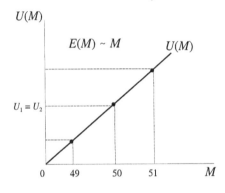

    b. Will this person accept this fair gamble? Assume that this person has an income of $50,000.

Solution

    a. The first derivative of the total utility of money function is:

$$\frac{dU}{dM} = (1.1)M^{0.1} > 0 \qquad\qquad (D13.3.2)$$

This individual's total utility is an increasing function of money. The second derivative of the total utility function is:

$$\frac{d^2U}{dM^2} = 0.011M^{-0.9} > 0 \qquad\qquad (D13.3.3)$$

Since the second derivative is positive, this individual is a risk lover.

    b. At $M = \$50,000$, the individual's total utility is $U = (\$50,000)^{1.1} = 147,525.47$ utils. If the individual wins $1,000, total utility is $U = (\$51,000)^{1.1} = 150,774.26$ utils.

The corresponding change in total utility is $\Delta U = 3{,}248.79$ utils. If the individual loses \$1,000, total utility is $U = (\$49{,}000)^{1.1} = 144{,}283.17$ utils. The corresponding change in total utility is $\Delta U = -3{,}242.30$ utils. The expected change in total utility from taking the bet is:

$$E(\Delta U) = \sum_{i=1}^{2}(\Delta U_i)p_i = (\Delta U_1)p_1 + (\Delta U_2)p_2$$

$$= (3{,}248.79)0.5 + (-3{,}241.70)0.5 = 3.55 \qquad (D13.3.4)$$

This risk-loving individual will accept this fair gamble since the expected change in total utility is positive.

RISK-AVERSE CONSUMER BEHAVIOR

Knowledge of risk-averse behavior by consumers has a wide range of applications in game theory. Suppose, for example, that a firm is considering the introduction of a new brand of toothpaste. The market is presently dominated by an established brand. Will knowledge of risk-averse consumer behavior influence the firm's marketing strategy? The challenge is to persuade potential customers to give the new product a try. If the price of both brands is the same, a risk-averse consumer will tend to stay with the established brand rather than switch to a potentially inferior product. This suggests at least two possible marketing strategies. Either the firm can offer the product at a lower price to compensate the consumer for the risk of sampling the new brand, or the firm can adopt an advertising campaign designed to convince consumers that the new brand is superior. Either marketing strategy will raise the expected value to the consumer of sampling the new product.

Another example of the consequences of risk-averse behavior relates to the benefits enjoyed by chain stores and franchise operations over independently owned and operated retail outlets. A risk-averse American tourist visiting a foreign country for the first time, for example, is more likely to have his or her first meal at McDonald's rather than sample the native cuisine at a neighborhood bistro. The reason for this is that a risk-averse tourist may initially prefer a familiar meal of predictable quality to more exotic fare of unpredictable quality. This will, of course, very likely change over time as the tourist becomes familiar with indigenous delicacies, customs, and the reputations of local dining establishments. This also explains why large retail chain stores or franchise operations are typically found in urban areas where there are a relatively large number of out-of-town visitors.

A familiar example of risk-averse consumer behavior is the purchase of insurance. One reason why people purchase insurance is to protect themselves against the possibility of catastrophic financial loss. Many people purchase health insurance, for example, in the event that they incur unexpected medical bills. Insurance premiums are said to be actuarially fair if they are equal to the expected financial loss from an illness or other health-care problem. A risk-averse individual purchases health insurance because the utility from a certain income, which is equal to his or her total income less premium payments, is greater than the utility from facing the world uninsured.

RISK-AVERSE FIRM BEHAVIOR

In the previous section we saw how risk-averse consumer behavior can influence a firm's marketing strategy. In fact, consideration of risk and uncertainty enter into almost every aspect of a firm's operations, including pricing, output, and capital investment decisions. In this section, we will briefly explore the interplay of risk and expected return when a firm contemplates alternative capital investment projects.

A common approach to evaluating two or more capital investment projects is to compare the difference between the present value of projected cash inflows and outflows. If the net present value of a project is negative, the project should be rejected. If the net present value is positive, the project should be considered for adoption. In general, projects with higher net present values are preferred. Equation (13.4) summarizes the net present value of a project as the difference between projected revenues (R_t) and expenditures (O_t).

$$NPV = \sum_{t=1}^{n} \frac{R_t}{(1+k)^t} - \sum_{t=1}^{n} \frac{O_t}{(1+k)^t} \tag{13.4}$$

In Equation (13.4), k is the **discount rate**, which is the rate used to discount future cash flows to the present. The calculations of net present value using Equation (13.4) are made under conditions of certainty. No account is made for the relative riskiness of the investments under consideration. One way to handle this shortcoming is to use risk-adjusted discount rates, which introduce risk directly into the net present value calculations.[2]

Discount rate The rate used to discount future cash flows to the present.

Three risk-return trade-off functions are illustrated in Figure 13.11. The riskiness of a capital investment project is measured as the standard deviation (σ) of returns, which is measured along the horizontal axis. The expected rate of return on an investment (k) is measured along the vertical axis.

The risk-return trade-offs in Figure 13.11 are called **investor indifference curves**. They summarize the expected rates of return that an investor must receive in excess of the expected rate of return from a risk-free investment (k_{rf}) to compensate for the risk associated with a particular investment project. Investor risk-return indifference curves also reflect the investor's attitude toward risk. To see this, consider a risk-free investment ($\sigma = 0$). For a risky investment ($\sigma > 0$), the investor must be compensated with a **risk premium**.

Investor indifference curve Summarizes the combinations of risk and expected rate of return for which the investor is indifferent between a risky and a risk-free investment.

Risk premium The difference between the expected rate of return on a risky investment and the expected rate of return on a risk-free investment.

The risk premium on an investment is the difference between the expected rate of return on a risky investment and the expected rate of return on a risk-free investment. The size of the risk premium depends on the investor's attitude toward risk. Consider, for example, the investor indifference curve I in Figure 13.11. In this case, a risk premium of $k - k_{rf}$ is required to make this investor indifferent between an investment with risk of $\sigma_1 > 0$ and a risk-free investment. On the other hand, the indifference curve I' illustrates the risk-return trade-offs of a more risk-averse investor. In this case, the investor will require a large risk premium $(k' - k_{rf}) > (k - k_{rf})$ as compensation for the same level of risk. Similarly, the indifference curve I summarizes the risk-return trade-offs for a less risk-averse investor. Here, a risk premium of $(k - k_{rf}) < (k - k_{rf})$ is required to make this investor indifferent to a risk-free investment.

The risk-return indifference curve may be used to evaluate **mutually exclusive** and **independent investments**. Investment projects are mutually exclusive if acceptance of one means rejection of all others. Investment projects are independent if the cash flows from alternative projects are unrelated to each other. Figure 13.12 illustrates an investor's risk-return indifference curve and

FIGURE 13.11
Investor Risk-Return Indifference Curves

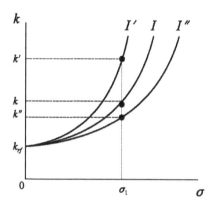

FIGURE 13.12
Investor Risk-Return Indifference Curve and Alternative Investment Projects

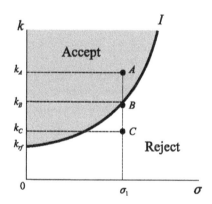

three mutually exclusive investments. As measured by the standard deviation, projects *A, B,* and *C* are equally risky.

> **Mutually exclusive investments** When acceptance of one investment project means rejection of all others.

> **Independent investments** When cash flows from alternative projects are unrelated to each other.

When confronted with alternative, mutually exclusive investment projects of equivalent risk, it is rational to choose the project with the highest expected rate of return, but only if the risk is acceptable. Only investments with risk-return combinations in the shaded region are acceptable. The expected rate of return from project A (k_A) is greater than the expected rate of return (k_B) required to make the investor indifferent between accepting or rejecting the project of equivalent risk. By contrast, the investor is indifferent between a risk-free rate of return (k_{rf}) and the rate of return on risky project B (k_B). If the projects are mutually exclusive, the investor will prefer project A to project B. Project C is unacceptable and will be rejected outright because the rate of return

FIGURE 13.13
Assessing Alternative Investment Projects

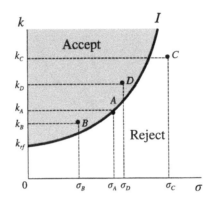

(k_C) is not sufficient to compensate the investor for the risk incurred. Any risk-return combination in the unshaded region will be rejected by this risk-averse investor. By contrast, if the projects under consideration are independent, the investor will choose project A, is indifferent to project B, but will reject project C.

Knowledge of the expected rate of return and the riskiness of a project is not sufficient to identify an optimal investment strategy. We must also know the investor's attitude toward risk, which is summarized in the risk-return indifference curve. To amplify this point, consider the situation in Figure 13.13. The expected rate of return from project C is greater than the rate of return from project D, which is greater than the expected rate of return from project A. Project B has the lowest expected rate of return, but is also the least risky. Project C has the highest rate of return, but is also the riskiest. If the projects are independent, this investor will select projects B and D, will reject project C, and will be indifferent to project A. If the projects are mutually exclusive, should the investor choose project B or D?

An investor's risk-return indifference curve summarizes the combinations of risk and return for which an investor is indifferent between a risky and a risk-free investment. Each investor has a "map" of such curves, such as that depicted in Figure 13.14. An investor's risk-return indifference map is similar to the indifference map in consumer theory.[3] The higher the risk-return indifference curve, the greater the level of investor satisfaction. In Figure 13.14, for example, the risk-return combinations summarized by indifference curve I_3 are preferred to those of I_2 because for any given level of risk, the investor receives a higher expected rate of return. Each investor has an infinite number of such risk-return indifference curves, and each investor has a unique indifference map. Returning to Figure 13.13, if the projects are mutually exclusive and project B lies on a higher risk-return indifference curve, project D will be rejected, and vice versa.

DYNAMIC GAMES WITH UNCERTAIN PAYOFFS

We have used game trees to analyze multistage games involving certain outcomes. In this section, we will use game trees to search for subgame perfect equilibria in games involving risky outcomes. To analyze the decision-making process under conditions of uncertainty, consider the multistage game depicted in Figure 13.15, which involves the pricing strategy of two firms. In this game, firm A must decide whether to charge a *high price* or a *low price*. Using backward induction, the certain payoffs for the subgame perfect equilibrium {*Low price → Low price*} are ($250, $250,000), which are highlighted in boldface.

FIGURE 13.14
Investor Indifference Map

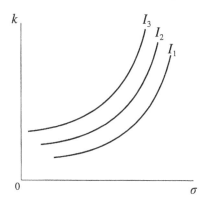

FIGURE 13.15
Game Tree for the Pricing Game

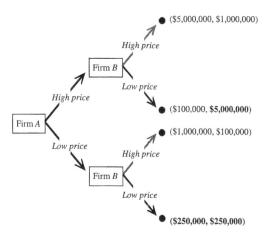

We will now modify the game in Figure 13.15 by introducing uncertainty into the decision-making process. We will assume that both firms are risk neutral, so that the only thing that matters are the expected payoffs. Suppose, for example, that firm A believes that by adopting a *high price* strategy, there is a 40 percent chance that firm B will charge a *high price* and a 60 percent chance that it will charge a *low price*. Similarly, if firm A adopts a *low price* strategy, there is an 80 percent chance that firm B will charge a *high price* and a 20 percent chance that it will charge a *low price*. The resulting extensive form of this multistage game is depicted in Figure 13.16.

The first entry in the parentheses at the terminal nodes is the *expected* payoff to firm A, while the second entry indicates a *certain* payoff to firm B. The reason for this is that while firm A is uncertain as to whether firm B will charge a *high price* or a *low price*, firm B knows for certain how it will react to firm A's move. In the game depicted in Figure 13.16, firm B's response profile remains [*Low price, Low price*]. That is, regardless of the strategy adopted by firm A, both subgame equilibria involve firm B charging a *low price*. By contrast, firm A's optimal strategy is based on expected payoffs. If firm A adopts a *high price* strategy, its expected payoff is $E(\pi_{HP}) =$

FIGURE 13.16
Subgame-Perfect Equilibrium with Uncertain Payoffs

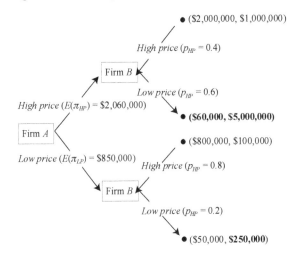

$0.4(\$5,000,000) + 0.6(\$100,000) = \$2,060,000$. If firm A adopts a *low price* strategy, the expected payoff is $E(\pi_{LP}) = 0.8(\$1,000,000) + 0.2(\$250,000) = \$650,000$. If both firms are risk neutral, the subgame perfect equilibrium is {*High price* → *Low price*}. Expected payoffs are highlighted in boldface. Actual payoffs are ($100,000, $5,000,000).

If firm A is risk averse, however, its choice of pricing strategy depends on relative volatility of the payoffs. The standard deviation of firm A's expected payoff from a *high price* strategy is:

$$\sigma_{HP} = \sqrt{\left[\pi_{HP} - E\left(\pi_{HP}\right)\right]^2 p_{HP} + \left[\pi_{LP} - E\left(\pi_{LP}\right)\right]^2 p_{LP}}$$
$$= \sqrt{0.4\left(\$2,940,000\right)^2 + 0.6\left(-\$1,960,000\right)^2} = \$2,400,000 \tag{13.5}$$

The standard deviation of firm A's expected payoff from a *low price* strategy is:

$$\sigma_{LP} = \sqrt{\left[\pi_{HP} - E\left(\pi\right)\right]^2 p_{HP} + \left[\pi_{LP} - E\left(\pi\right)\right]^2 p_{LP}}$$
$$= \sqrt{0.8\left(\$150,000\right)^2 + 0.2\left(-\$600,000\right)^2} = \$300,000 \tag{13.6}$$

Equations (13.5) and (13.6) indicate that the higher expected return from a *high price* strategy is also the riskiest. If the greater expected payoff is insufficient to compensate the firm for the increased volatility, risk-averse firm A will reject the *high price* strategy in favor of a lower, but more predictable, payoff from a *low price* strategy.

———————————————— **Demonstration Problem 13.4** ————————————————

Consider the extensive form game in Figure 13.15. If firm A adopts a *high price* strategy, there is a 5 percent chance that firm B will charge a *high price* and a 95 percent chance that it will charge a *low price*. Similarly, if firm A adopts a *low price* strategy there is a 10 percent chance that firm B will charge a *high price* and a 90 percent chance that it will charge a *low price*.

FIGURE D13.4
Game Tree for Demonstration Problem 13.4

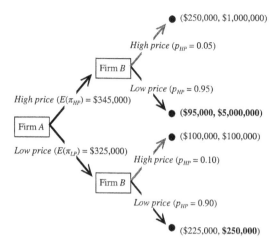

a. If firm A is risk neutral, what is the subgame perfect equilibrium for this game?
b. If firm A is risk averse, what, if anything, can you say about its pricing strategy?

Solution

a. The revised extensive form of this game is depicted in Figure D13.4. The optimal strategy for firm B is to charge a *low price* regardless of the strategy adopted by firm A. Firm A's optimal strategy is based on the expected payoff from charging a *high price* versus a *low price*. If firm A adopts a *high price* strategy, the expected payoff is $E(\pi_{HP}) = 0.05(\$5,000,000) + 0.95(\$100,000) = \$345,000$. If firm adopts a *low price* strategy, the expected payoff is $E(\pi_{LP}) = 0.10(\$1,000,000) + 0.90(\$250,000) = \$325,000$. If firm A is risk neutral, the subgame perfect equilibrium for this game is {*High price* → *Low price*} with actual payoffs of ($100,000, $5,000,000).
b. Substituting the above information into Equation (13.5), the standard deviation of firm A's payoffs from a *high price* strategy is:

$$\sigma_{HP} = \sqrt{0.05(\$4,655,000)^2 + 0.95(-\$245,000)^2} = \$1,067,935 \qquad \text{(D13.4.1)}$$

The standard deviation of firm A's payoffs from a *low price* strategy is:

$$\sigma_{LP} = \sqrt{0.10(\$675,000)^2 + 0.90(-\$75,000)^2} = \$259,000 \qquad \text{(D13.4.2)}$$

The above calculations suggest that for firm A *high price* is a riskier strategy than a *low price* strategy. A risk-neutral firm A would adopt a high price strategy because of the larger expected payoff. A risk-averse firm A probably would not. The reason for this is that the expected payoff from a *high price* strategy is only marginally better, but is considerably riskier. While the subgame perfect equilibrium for a risk-neutral firm A is {*High price* → *Low price*} with actual payoffs of ($95,000, $5,000,000), the subgame perfect equilibrium for a risk-averse firm A is probably {Low *price* → Low *price*} with actual payoffs of ($250,000, $250,000).

FIGURE 13.17

Extensive Form of the Static Oil-Drilling Game Depicted in Figures 2.6 and 13.1 with Certain Payoffs

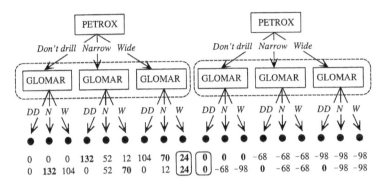

0	0	0	132	52	12	104	70	24	0	0	0	-68	-68	-68	-98	-98	-98
0	132	104	0	52	70	0	12	24	0	-68	-98	0	-68	-68	0	-98	-98

Payoffs: PETROX
              GLOMAR

Example: Oil-Drilling Game Under Uncertainty

Up to this point we have analyzed static games by means of the normal format using payoff matrices. In fact, static games may also be illustrated using the extensive format employing a convenient analytic device known as an **information set**, which is a collection of decision nodes. A player within an information set is unable to distinguish one decision node from another.[4] Consider the oil-drilling games in Figures 2.6 and 13.1. These reformatted games are depicted in Figure 13.17. In these games, there are two information sets, which are enclosed by rounded dashed rectangles. Each information set contains three GLOMAR decision nodes.

> **Information set** A collection of decision nodes. A player within an information set is unable to distinguish one decision node from another.

Although illustrating the oil-drilling game in extensive form suggests that moves made by GLOMAR are contingent on moves made by PETROX, the use of information sets allows us to preserve the static nature of the original game. The reason for this is that GLOMAR cannot distinguish between decision nodes within an information set and must therefore move without knowledge of the move made by PETROX. As with the use of a referee in the rock, paper, scissors game discussed in Chapter 1, it does not matter which player moves first. The only thing that matters is that both players are ignorant of a rival's strategy until all moves have been made.

Despite appearances to the contrary, the game depicted in Figure 13.17 is not solved through backward induction. Rather, the Nash equilibria are found by directly examining the payoffs at each decision node. To see how, consider the game tree on the left-hand side of Figure 13.17. Suppose that PETROX adopts a *don't drill* strategy. GLOMAR's payoffs are the same as those in the first row of the payoff matrix in Figure 2.6. If PETROX adopts a *narrow* strategy, GLOMAR's payoffs are represented by the second row of the payoff matrix, and so on. On the other hand, suppose that GLOMAR adopts a *don't drill* strategy. PETROX's payoffs in Figure 13.17 are the same as those corresponding to the first column of Figure 2.6. That is, the payoffs for PETROX from the strategy profiles {*Don't drill, Don't drill*}, {*Narrow, Don't drill*} and {*Wide, Don't drill*} are 0, 132 and 104, respectively. Likewise, if GLOMAR adopts a *narrow* strategy, the payoffs to

FIGURE 13.18
Extensive Form of the Static Oil-Drilling Game with Expected Payoffs

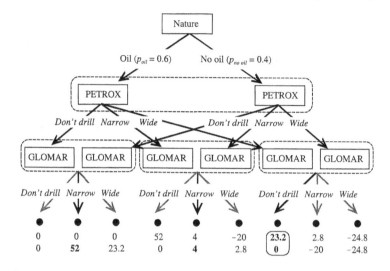

Payoffs: PETROX
          GLOMAR

PETROX are 0, 52, and 70 from the strategy profiles {*Don't drill, Narrow*}, {*Narrow, Narrow*} and {*Wide, Narrow*}. These payoffs are indicated in the second column of Figure 2.6, and so on. The payoffs for the Nash equilibria in Figures 2.6 and 13.1 are enclosed in the rounded rectangles in Figure 13.17.

We will now modify Figure 13.17 to illustrate the static oil-drilling game with uncertain payoffs. Once again, we will do this by introducing Nature, who moves first by selecting a state of nature according to fixed probabilities. For simplicity, we will assume that PETROX and GLOMAR are risk neutral so that moves are based only on expected payoffs. In this new game, which is depicted in Figure 13.18, the two decision nodes of PETROX are nested in the same information set since the company does not know for certain whether there is oil in the ground. Thus, PETROX's choice of moves is based on expected payoffs. GLOMAR is confronted with the same problem. For this reason, GLOMAR's six decision nodes are nested in three information sets. Each information set corresponds to one of PETROX's strategy choices. The data in Figures 2.6 and 13.1 were used to calculate the expected payoffs in Figure 13.18. It is left as an exercise for the reader to verify that the Nash equilibria for the static game in Figure 13.18 is {*Narrow, Narrow*} (not shown).

We will now assume that the oil-drilling game is dynamic, with PETROX moving after Nature has randomly determined the presence or absence of oil. It is left as an exercise for the reader to verify that the subgame perfect equilibria for the games depicted in Figure 13.17 are {*Wide* → *Wide*} when oil is present and {*Don't drill* → *Don't drill*} when oil is not present. In the game depicted in Figure 13.18, the presence of oil is probabilistic. To find the subgame perfect equilibrium in the oil-drilling game when Nature moves first and randomly determines the presence of oil, we will assume that both players are risk neutral.

When strategies are based on expected values and PETROX moves first, we find that there are three subgame equilibria: {*Don't drill* → *Narrow*}, {*Narrow* → *Narrow*}, and {*Wide* → *Don't drill*}. The subgame perfect equilibrium strategy profile for this game is {*Wide* → *Don't drill*} because it offers the highest expected payoff for PETROX. Of course, the actual payoffs depend on which state of nature prevails. If oil is present, PETROX is the big winner by earning profits of $104 million, while GLOMAR earns nothing since it decided not to drill. On the other hand, if oil is not present, PETROX loses $98 million. In this case, GLOMAR is the winner since it decided not to drill. What is interesting about these results is that a risk-neutral PETROX has a first-mover advantage in terms of expected payoffs. But, would a risk-averse PETROX exploit this advantage? The answer, of course, depends on the company's attitude toward risk.

CHAPTER REVIEW

Many business decisions are made under a cloud of uncertainty. In this chapter we considered static and dynamic games in which payoffs are not known with certainty, but where the choice of strategies is based on the probability distribution of each possible outcome.

Decisions are made under conditions of *risk* and *uncertainty*, which involve choices from among multiple payoffs. The fundamental difference is that the probability of each risky outcome is either known or can be estimated. By contrast, the probability of an uncertain outcome is either unknown, cannot be estimated, or is meaningless. The phrase *decision making under conditions of uncertainty* is often used to describe the probability distribution of multiple payoffs, regardless of whether or not these probabilities are known, meaningful, or can be estimated.

John Harsanyi devised a method for analyzing games under conditions of uncertainty by introducing another player, which he called Nature. Nature does not have a vested interest in the outcome of the game, but moves first by randomly selecting the *state of nature* and/or *player type* according to fixed probabilities that are common knowledge. This analytical approach is referred to as a *Harsanyi transformation*.

These probabilities are used by each player to calculate *expected payoffs* from each pure-strategy profile. As a result, a game with incomplete information is transformed into a game with complete, but imperfect, information. In games with *imperfect information*, players move without knowing a rival's intentions. This is a characteristic of all static games. A static game with incomplete information is referred to as a *static Bayesian game*.

The objective of each player is to choose a risk-adjusted strategy that maximizes expected payoffs. The choice of strategies depends on a player's attitude toward risk. A *risk-averse* individual prefers a certain payoff to a risky outcome with the same expected value. A *risk-loving individual* prefers the expected value of a risky payoff to its certainty equivalent. A *risk-neutral* individual is indifferent between the expected value of a risky outcome and its certainty equivalent. Risk-averse individuals exhibit *diminishing marginal utility of money*. Risk-loving and risk-neutral individuals exhibit *increasing marginal utility of money* and *constant marginal utility of money,* respectively.

CHAPTER QUESTIONS

13.1   What is the difference between risk and uncertainty?

13.2   Explain the difference between static games involving incomplete information and games involving imperfect information.

13.3   Explain how John Harsanyi was able to transform games with incomplete information into games with complete, but imperfect, information.

13.4   Greater volatility of payoffs must be compensated by a higher expected payoff. Do you agree with this statement? For what type of person is this statement likely to be true?

13.5   What is the standard statistical measure of risk? What does it measure? If two payoffs have the same expected value, why is the payoff with the large standard deviation considered riskier than the payoff with a small standard deviation?

13.6   Can uncertainty be measured? Explain.

13.7   Explain how the marginal utility of money can be used to explain an individual's attitude toward risk.

13.8   Why do risk-averse individuals tend to reject fair gambles?

13.9   Risk-averse individuals always reject fair gambles. Do you agree with this statement? Can you think of an example in which individuals regularly accept unfair gambles?

13.10   Consider demonstration problem 13.2. Why do so many risk-averse people play government-sponsored lotteries if they are unfair gambles? Would a local government ever sponsor a fair lottery? Why?

13.11   The New York State Lottery Commission sold the idea of a lottery to taxpayers by claiming that the proceeds would be used to support public education. In fact, lottery revenues are not earmarked for public education but flow into general government coffers. Why to you believe that it was necessary for elected officials to mislead taxpayers?

CHAPTER EXERCISES

13.1   Rosie Hemlock offers Robin Nightshade the following wager. For a payment of $10, Rosie will pay Robin the dollar value of any card drawn from a standard 52-card deck. For example, for an ace of any suite, Rosie will pay Robin $1. For an 8 of any suit, Rosie will pay Robin $8. A ten or picture card of any suit is worth $10.

   a.   What is the expected value of Rosie's offer?
   b.   Will risk-neutral Rosie accept this offer?

13.2   Suppose that capital investment project X has an expected value of $\mu_X = \$1,000$ and a standard deviation of $\sigma_X = \$500$. Suppose that capital investment project Y has an expected valuc of $\mu_Y = \$1,500$ and a standard deviation of $\sigma_Y = \$750$. Which is the relatively riskier project?

13.3   Senior management of Rubicon & Styx is trying to decide whether to advertise its world-famous hot sauce "Sergeant Garcia's Revenge" on television (campaign A) or in magazines (campaign B). The marketing department of Rubicon & Styx has estimated the probabilities of alternative sales (net of advertising costs) using each of the two media outlets, which are summarized in Table E13.3.

   a.   Calculate the expected revenues from each advertising campaign.
   b.   Calculate the standard deviation of sales revenues from each advertising campaign.
   c.   Calculate the coefficient of variation for each advertising campaign. On this basis, which advertising campaign should Rubicon & Styx adopt? (Hint: See the appendix to this chapter.)

13.4   Using investor indifference curves, illustrate three investment projects in which the expected rates of return are $k_C > k_A > k_B$, and the risks associated with each project are $\sigma_C > \sigma_A > \sigma_B$. (Hint: Draw an indifference curve for each investment project.)

TABLE E13.3
Probabilities of Alternative Revenues for Chapter Exercise 13.3

Campaign A	(Television)	Campaign B	(Magazines)
Sales (S_A)	Probability	Sales (S_B)	Probability
$5,000	0.20	$6,000	0.15
$8,000	0.30	$8,000	0.35
$11,000	0.30	$10,000	0.35
$14,000	0.20	$12,000	0.15

13.5   Using investor indifference curves, illustrate three investment projects in which the expected rates of return are $k_A > k_B > k_C$, and the risks associated with each project are $\sigma_C > \sigma_A > \sigma_B$. (Hint: Draw an indifference curve for each investment project.)

13.6   Suppose that Ted Sillywalk offers Walt Wobble the fair gamble of receiving $500 of the flip of a fair coin showing heads, and losing $500 if the coin comes up tails. Suppose that Will's total utility of money function is:

$$U = M^{1.2} \qquad\qquad (E13.3.1)$$

a.   For some positive level of money income, what is Will's attitude toward risk?
b.   If $M = \$5,000$, will Walt accept Ted's offer? Explain.

13.7   Consider the static games depicted in Figures 2.6 and 13.1. What is the Nash equilibrium for this game if PETROX and GLOMAR are risk neutral and there is a 50-50 chance of finding oil?

13.8   Consider the game in demonstration problem 13.4. Suppose that firm A believes that by charging a *high price* there is a 2 percent chance that firm B will charge a *high price* and a 98 percent chance that it will charge a *low price*. Similarly, firm A believes that by adopting a *low price* strategy there is a 25 percent chance that firm B will charge a *high price* and a 75 percent chance that it will charge a *low price*.

a.   What is the subgame perfect equilibrium for this game if firm A is risk neutral?
b.   Would your answer to part *a* have been different if firm A were risk averse?

13.9   Calculate the standard deviation of the payoffs to the subgame perfect equilibrium depicted in Figure 13.18.

ENDNOTES

1.   The symbol "≻" is read "is preferred to." The symbol "~" is read "is indifferent to."
2.   See, for example, Eugene F. Brigham and Joel F. Houston, *Fundamentals of Financial Management,* 2nd edition. New York: Dryden Press, 1999, chapter 9.
3.   For a detailed discussion of consumer indifference curves see Walter Nicholson, *Microeconomic Theory: Basic Principles and Extensions,* 6th edition. New York: Dryden Press, 1995, chapter 3.
4.   The concept of an information set will be discussed at greater length in Chapter 16 when we analyze sequential-move games with incomplete information.

APPENDIX: MEASURING RISK

MEAN (EXPECTED VALUE)

The most commonly used summary measures of risky, random payoffs are the **mean (expected value)** and the **variance**. These random payoffs may refer to profits, capital gains, prices, unit sales, etc. In risky situations, the expected value of these random payoffs is called the mean. The mean is the weighted average of all possible random outcomes. The weights are the probabilities of each outcome. For discrete random variables, the expected value may be calculated as:

$$E(x) = \mu = \sum_{i=1}^{n} x_i p_i \qquad (13A.1)$$

In Equation (13A.1), x_i is the value of the outcome, and p_i is the probability of its occurrence where:

$$\sum_{i=1}^{n} p_i = 1 \qquad (13A.2)$$

When the probability of each outcome is the same, the expected value is the sum of the outcomes divided by the number of observations. In this case, the expected value of a set of uncertain outcomes may be calculated using Equation (13A.3)

$$E(x) = \mu = \left(\frac{1}{n}\right) \sum_{i=1}^{n} x_i \qquad (13A.3)$$

> **Mean (expected value)** The weighted average of the set of random outcomes, where the weights are the probabilities of each outcome.

——————————— **Demonstration Problem 13A.1** ———————————

Suppose that the chief economist of Bombay Emerald, Ltd. believes that there is a 40 percent ($p_1 = 0.4$) probability of a recession in the next operating period and a 60 percent ($p_2 = 0.6$) probability that a recession will not occur. The COO of Bombay Emerald believes that the firm's profit will be $\pi_1 = \$100,000$ in the event of a recession and $\pi_2 = \$1,000,000$ otherwise. What is Bombay Emerald's expected profit?

Solution

Bombay Emerald's expected profit is:

$$E(x) = \sum_{i=1}^{2} \pi_i p_i = 0.4(\$100,000) + 0.6(\$1,000,000) = \$640,000 \quad (D13A.1.1)$$

——————————— **Demonstration Problem 13A.2** ———————————

Suppose that Bob offers his brother Nob the following offer. For a payment of $3.50, Bob will pay Nob the dollar value of any roll (v_i) of a fair die. For example, for a roll of $v_1 = 1$,

Bob will pay Nob $1. For a roll of $v_6 = 6$, Bob will pay Nob $6. How much can Nob expect to earn if he accepts Bob's offer?

Solution

Since the probability of any number between 1 and 6 is 1/6, Bob's expected payout is:

$$E(v) = \left(\frac{1}{n}\right)\sum_{i=1}^{6} v_i = \left(\frac{1}{6}\right)(1+2+3+4+5+6) = \$3.50 \qquad \text{(D13A.2.1)}$$

Since it will cost $3.50 to play this game, Nob's can expect to earn $E(v) - 3.50 = \$0$. Whether or not Nob should accept Bob's offer will depend on Nob's attitude toward risk. If Nob is risk averse, he will not accept this fair gamble.

VARIANCE

The strength of the mean is its simplicity. In a single number, the mean (expected value) summarizes important information about the most likely outcome of a set of random payoffs. Unfortunately, this strength hides other important information that is valuable to the decision maker. For example, suppose that an individual is offered the following fair gamble. If an individual flips a fair coin that comes up "heads," he or she wins $10. On the other hand, if the coin comes up "tails," the individual loses $10. The expected value of the wager is $0. Suppose, on the other hand the payoffs were $1,000 and –$1,000 for a head and tail, respectively. Once again, the expected value of the wager is $0.

While the expected values of the two fair gambles are the same, clearly the wagers themselves are different. The potential payoff of the second wager is much greater, but so too is the potential loss. An individual may be prepared to accept the first bet, but may not be willing to accept the second because the possibility of such a large loss may be unacceptable. For this individual, the second fair gamble is too risky.

The second wager is riskier because the spread, or dispersion, of the possible payoffs is greater. Each has the same expected value, but the swing between a gain and a loss is considerably greater. It is this dispersion of possible payoffs that is the distinguishing characteristic of risk. One of the most commonly used measures of the dispersion of a set of random outcomes is the variance. The variance is the weighed average of the squared deviations of all possible random outcomes from its mean. The variance of a set of random payoffs may be calculated using Equation (13A.4)

$$\sigma^2 = E\left[(x - E(v))^2\right] = \sum_{i=1}^{n}(x_i - E(v))^2 \, p_i \qquad \text{(13A.4)}$$

> **Variance** A measure of the dispersion of a set of random outcomes. It is the sum of the products of the squared deviations of each outcome from its mean and the probability of each outcome.

When the probability of each outcome is the same, the variance is simply the sum of the squared deviations divided by the number of outcomes.

$$\sigma^2 = E\left[(x - E(v))^2\right] = \left(\frac{1}{n}\right)\sum_{i=1}^{n}(x_i - E(v))^2 \qquad \text{(13A.5)}$$

Denoting a win and a loss as x_1 and x_2, respectively, the variances of the two wagers (σ_1^2 and σ_2^2) are:

$$\sigma_1^2 = \sum_{i=1}^{n} \left(x_i - E(v) \right)^2 p_i$$

$$= 0.5 \left(\$10 - 0 \right)^2 + 0.5 \left(-10 - 0 \right)^2 = \$100 \qquad \text{(13A.6)}$$

$$\sigma_2^2 = \sum_{i=1}^{n} \left(x_i - E(v) \right)^2 p_i$$

$$= 0.5 \left(\$1,000 - 0 \right)^2 + 0.5 \left(-\$1,000 - 0 \right)^2 = \$1,000,000 \qquad \text{(13A.7)}$$

Since $\sigma_2^2 > \sigma_1^2$, the second wager is riskier than the first.

STANDARD DEVIATION

An alternative way to express the riskiness of a set of random outcomes is the **standard deviation**. The standard deviation is simply the square root of the variance, that is:

$$\sigma = \sqrt{\sigma^2} \qquad \text{(13A.8)}$$

> **Standard deviation** The square root of the variance. The standard deviation is a commonly used measure of the riskiness of an investment.

For the above wagers the standard deviations are $\sigma_1 = \$10$ and $\sigma_2 = \$1,000$. Since the standard deviation is a monotonic transformation of the variance, the ordering of relative risks of the wagers is preserved. Thus, since $\sigma_2 > \sigma_1$ the second wager is riskier than the first.

―――――――――――――――― **Demonstration Problem 13A.3** ――――――――――――――――

Using the information provided in demonstration problem 13A.1, calculate the variance and the standard deviation of Bombay Emerald's expected profits.

Solution

From the above problem, expected profits are $640. The variance of Bombay Emerald's expected profits is:

$$\sigma^2 = E\left[\left(\pi - E(\pi) \right)^2 \right] = \sum_{i=1}^{2} \left[\pi_i - E(\pi) \right]^2 p_i$$

$$= 0.4 \left(\$100 - \$640 \right)^2 + 0.6 \left(\$1,000 - \$640 \right)^2 = \$194,400 \qquad \text{(D13A.3.1)}$$

The standard deviation is:

$$\sigma = \sqrt{\sigma^2} = \$440.91 \qquad \text{(D13A.3.2)}$$

―――――――――――――――― **Demonstration Problem 13A.4** ――――――――――――――――

From Demonstration Problem 13A.2, calculate the variance and standard deviation of Bob's expected payout.

Solution

Since the probability of any number between 1 and 6 is 1/6, Bob's expected payout is:

$$\sigma^2 = E\left[(v - E(v))^2\right] = \left(\frac{1}{n}\right)\sum_{i=1}^{6}\left[v - E(v)\right]^2$$

$$= \left(\frac{1}{6}\right)\left[(1 - 3.5)^2 + (2 - 3.5)^2 + (3 - 3.5)^3 + (4 - 3.5)^2 + (5 - 3.5)^2 + (6 - 3.5)^2\right]$$

$$= \$2.92 \tag{D13A.4.1}$$

The standard deviation is:

$$\sigma = \sqrt{\sigma^2} = \sqrt{2.92} = \$1.71 \tag{D13A.4.2}$$

COEFFICIENT OF VARIATION

Unfortunately, neither the variance nor the standard deviation can be used to compare the relative riskiness of two or more risky situations with different expected values. The reason for this is that neither measure is independent of its units of measurement. One way to measure the relative riskiness of two or more outcomes is to use the coefficient of variation, which may be calculated using Equation (13A.9). The coefficient of variation allows us to compare the riskiness of alternative investments by "normalizing" the standard deviation of each outcome by its expected value.

$$CV = \frac{\sigma}{\mu} \tag{13A.9}$$

> **Coefficient of variation** A dimensionless number that is used to compare the relative riskiness of two or more investments with different expected values. It is calculated as the ratio of the standard deviation of the investment to its the mean.

─────────────── **Demonstration Problem 13A.5** ───────────────

Suppose that capital investment project *A* has an expected value of $\mu_A = \$100{,}000$ and a standard deviation of $\sigma_A = \$30{,}000$. Additionally, suppose that project *B* has an expected value $\mu_B = \$150{,}000$ and a standard deviation of $\sigma_B = \$40{,}000$. Which is the relatively riskier project?

Solution

From Equation (13A.6) the relative riskiness of projects *A* and *B* are:

$$CV_A = \frac{\sigma_A}{\mu_A} = \frac{\$30{,}000}{\$100{,}000} = 0.300 \tag{D13A.5.1}$$

$$CV_B = \frac{\sigma_B}{\mu_B} = \frac{\$40{,}000}{\$150{,}000} = 0.267 \tag{D13A.5.2}$$

Although project *B* has the larger standard deviation, it is the relatively less risky project.

CHAPTER

14

TORTS AND CONTRACTS

In this chapter we will:

* *Introduce basic concepts of tort law, including liability, negligence, and damages;*
* *Demonstrate how static games with complete information and probabilistic payoffs can be used to illustrate the effects of different tort-law regimes on player incentives, allocation of damages, and social welfare;*
* *Illustrate the importance of legally enforceable contracts when payoffs are otherwise uncertain.*

INTRODUCTION

In the previous chapter we analyzed static and dynamic games with complete information under conditions of risk and uncertainty. In this chapter we will apply some of those concepts to an analysis of such fundamental legal concepts as negligence, liability, and damages in the Anglo-American system of torts and contracts.[1] We will demonstrate how the choice of a tort-law regime can affect payoffs from alternative strategy profiles. We will also see that if incentives are properly structured, the system of tort law can result in socially efficient Nash equilibria. This will be followed by a discussion of an area of the law that is of particular interest to market participants. Explicit and implicit contracts form the basis of all business transactions. A failure to perform a service or deliver a product with expected characteristics can result in grievous damages for either or all of the parties concerned.

TORTS

Tort law is a branch of the law dealing with torts. The French word *tort* is derived from the Latin *torquere,* meaning to twist. A tort is a private wrong or injury. It is an act committed by one person that causes injury to another. Torts may be intentional, or may result from **negligence**, which is a failure to exercise the same care that a reasonable or prudent individual would exercise under the same circumstances, or take an action that a reasonable person would not. Torts result in more civil litigation than any other field of the law, including contract, real property, and criminal law.

> **Tort law** A branch of the law dealing with torts, which is an act committed by one person that results in an injury to another.
>
> **Negligence** A failure to exercise the same care that a reasonable or prudent person would exercise under the same circumstances, or to take an action that a reasonable person would not.

Torts are central to the Anglo-American concept of the **common law**, which refers to the corpus of legal principles created largely by judges as a by-product of deciding cases. The evolution of common law is shaped by precedent rather than legislation. Common law is based on the legal principle of *stare decisis* (to stand by that which is decided). According to this legal principle, precedent is binding on lower court judges deciding similar subsequent cases.

> **Common law** The corpus of legal principles created largely by judges as the by-product of deciding cases. Common law is shaped by legal precedent rather than legislation.

In tort law, **liability** refers to the legal responsibility for acts of commission or omission. In this section we will analyze various liability rules that govern the distribution of accident costs between a motorist and a pedestrian as a static game with complete information. In this game, the likelihood that an accident will occur depends on the care exercised by a motorist while driving and a pedestrian while crossing the street. The behavior of the motorist and the pedestrian also depends on the legal rules of the game. Motorist and pedestrian are assumed to be aware of these rules, and the court's ability to enforce those rules. It has long been recognized by legal scholars that motorists will exercise greater caution if they are liable for injuries sustained by accident victims. In order to write effective laws governing behavior, however, we must understand how these laws affect a player's behavior.

> **Liability** Legal responsibility for an act of commission or omission.

Much of the literature in law and economics in recent decades has assumed that legal rules, including those relating to Anglo-American tort law, are designed to minimize the social costs associated with an accident, which include the costs of pain and suffering by the victim and avoidance costs incurred by both parties. Our concern is about which legal rule works best in achieving this social objective. We will demonstrate with the use of a one-time static game that there are several different legal rules that will induce rational players to behave in ways that may be both mutually and socially beneficial.

It should be emphasized that the choice of legal regimes is not predicated on some subjective determination as to which player is relatively more virtuous. We are only concerned with identifying the features of a tort-law regime that minimize the total cost to society from an accident. This, however, creates another problem. As we have noted in our discussion of strategic trade policy, it is not possible to engage in interpersonal comparisons of utility. Although economists generally assume diminishing marginal utility of money, there is no theorem or observation of economic behavior that will allow us to judge the relative merits of a transfer of wealth or income from one individual or group to another. As we have seen, one way that economists circumvent this problem is by assuming that the utility of a dollar transferred from one player to another is equivalent. The reader may recall that this is referred to as the one-dollar, one-vote rule. Adopting this rule implies a willingness to judge tort-law regimes in terms of the combined payoffs of both parties to an accident without regards to the distribution of interpersonal well-being.

Tort law is concerned with allocation of civil damages. It works by requiring one party to an accident (the injurer, wrongdoer, or tortfeasor) to pay damages to the other party (the victim) in certain circumstances, but not in others. Different tort-law regimes can result in different payoffs associated with each strategy profile. Different payoffs elicit different behavior. These strategy choices are continuous for both the pedestrian and the motorist. In the case of the motorist, for example, the choice of strategies may be between driving or not driving. If the choice is to drive, the decision is how carefully. For the pedestrian, the decision is whether or not to cross the street. If the decision is to cross, how much caution should the pedestrian exercise?

Each degree of caution (diligence) involves an opportunity cost, either explicit (out of pocket), implicit, or some combination of the two. For our purposes, it is only necessary to assume that each player must choose between a *negligent* or a *diligent* strategy. Adopting a *diligent* strategy

FIGURE 14.1
No-Liability Accident Game

Motorist

		Negligent	Diligent
Pedestrian	Negligent	**(−10, 0)**	(−10, −1)
	Diligent	(−11, **0**)	(**−2**, −1)

Payoffs: (Pedestrian, Motorist)

means that a player behaves in such a manner so as to minimize the social costs of an accident. Adopting a *negligent* strategy is to incur zero avoidance costs. The choice of strategy is based on a comparison of the marginal cost of an additional unit of caution and the marginal benefit of a reduction in expected accident costs.

No Liability

A legal rule in which the injured party (victim) in an accident has no right to recover damages from the injurer (tortfeasor) is referred to as a **no-liability** tort regime. The no-liability accident game is depicted in Figure 14.1. Under this legal rule, neither the tortfeasor nor the victim is held liable for damages in the event of an injury. The payoff matrix reflects the fact that the motorist will not be assessed **damages** in the event of an accident. Damages represent the amount of money or other compensation awarded by the court to the plaintiff in a lawsuit. In this game, the pedestrian has no right to collect damages from the motorist for injuries received. The payoff for a *diligent* pedestrian includes the cost of accident avoidance and/or the cost of the accident, which includes medical costs in the event of injury. If the pedestrian is *negligent,* the payoff includes only accident costs. The payoff for a *diligent* motorist's payoffs only includes accident avoidance costs. The *negligent* motorist incurs no cost whatsoever.

> **No-liability tort law** A legal rule in which neither the tortfeasor nor the victim is held liable for damages in the event of an injury.
>
> **Damages** An amount of money or other compensation awarded by the court to the plaintiff in a lawsuit.

In the game depicted in Figure 14.1, a {*Negligent, Negligent*} strategy profile results in a loss of $10 for the pedestrian and a $0 payoff for the motorist. The reason for this is that neither player incurs any costs associated with exercising due diligence, but the pedestrian incurs the full $10 cost of the accident. For simplicity, we assume that both players are risk neutral. We will further assume that the probability an accident will occur if both players are *negligent* is unity, and a 10 percent probability that both players are *diligent*. If the pedestrian is *diligent* while the motorist is *negligent,* the motorist's payoff is $0 while the pedestrian's expected loss is −$10 − 0.1($10) = −$11. Finally, if both players adopt a *diligent* strategy, each will incur a $1 avoidance cost, and the risk-neutral pedestrian will incur an expected injury cost of −0.1($10) = −$1. The strategy profile {*Diligent, Diligent*} in this game is socially optimal because it minimizes the combined payoffs of both players.

In the no-liability accident game, society is caught in a prisoner's dilemma. The Nash equilibrium strategy profile for this game is {*Negligent, Negligent*} since a rational motorist will adopt his or her strictly dominant *negligent* strategy. By contrast, the strategy adopted by the pedestrian depends on the strategy adopted by the motorist. Since a rational motorist will adopt a *negligent* strategy, it will be in the pedestrian's best interest to adopt a *negligent* strategy because of the better payoff. The no-liability accident game depicted in Figure 14.1 is an example of a prisoner's dilemma since it is clearly in society's best interest for both players to adopt a *diligent* strategy.

The Nash equilibrium strategy profile {*Negligent, Negligent*} may strike the reader as being somewhat counterintuitive, not because the motorist chooses to be *negligent,* but because the pedestrian chooses to be *negligent* as well. This implies that the motorist is indifferent to running down pedestrians. Under a no-liability tort regime, accidents are inevitable and pedestrians will be injured. Surely, a risk-averse pedestrian knows this and will probably exercise more, not less, diligence. On the other hand, the above example makes a rather significant prediction. When no player is assigned liability for the accident, motorists have no incentive to be *diligent* by incurring avoidance costs that minimize the expected social costs of accidents. Diligence is strictly dominated by negligence.

Pure Strict Liability

It should be clear from the above example that a legal regime in which neither player is held liable in the event of an accident is not socially optimal. This suggests that society can be made better off under a legal rule that compels both players to adopt a *diligent* strategy. The legal rule of no-liability will result in too many accidents, and pedestrians will incur uneconomically high avoidance costs.[2] This suggests that a tort-law regime that assigns liability is more socially efficient.

In this section, we will assume a legal regime in which the motorist is held strictly liable in the event of an accident and must pay damages to the victim, regardless of whether or not the motorist exercised due diligence. This legal rule, which is called **pure strict liability**, is the polar opposite of no-liability. The payoffs to the pedestrian include avoidance costs, if any, and the expected value of damages received from the motorist. The payoffs to the motorist are equivalently reduced by damages paid to the pedestrian in the event of an accident. The pure-strict-liability game is depicted in Figure 14.2.

> **Pure-strict-liability tort law** A legal rule in which the tortfeasor is held liable for damages, regardless of whether he or she exercised due diligence.

As in the no-liability accident game, exercising due diligence results in $1 of avoidance costs for each player. In the pure-strict-liability accident game, the payoffs for the strategy profile {*Negligent, Negligent*} are –$10 + $10 = $0 for the pedestrian and –$10 for the motorist. This is because neither has exercised due diligence and the motorist is wholly responsible for injuries incurred by the pedestrian. The payoffs from the strategy profile {*Diligent, Diligent*} are (–1, –2). Both players will incur $1 avoidance costs by adopting a *diligent* strategy, while the motorist incurs an additional –0.1($10) = –$1 expected injury cost incurred by the pedestrian. If the motorist is *diligent* and the pedestrian is *negligent,* the payoff to the motorist –$1 –$10 = –$11. That is, the motorist incurs a certain cost of $1 by exercising care and $10 injury costs incurred by the pedestrian. Lastly, if the pedestrian adopts a *diligent* strategy and the motorist is *negligent,* the payoff to the pedestrian is the cost of being careful, while the motorist pays for the pedestrian's injuries.

The change in the liability rule does not change the strategies open to each player, nor does it change the sum of the payoffs from each strategy profile. It does, however, reallocate the payoffs. Tort rules alter the players' behavior because they link actions to consequences. The reader should verify that the Nash equilibrium strategy profile for the game depicted in Figure 14.2 is again {*Negligent, Negligent*}. The difference with the no-liability case is that incentives have been completely turned around by the change in the structure of the payoffs. Since the motorist is now liable for accident costs, there is no incentive for the pedestrian to incur accident avoid-

FIGURE 14.2
Pure-Strict-Liability Accident Game

Motorist

Pedestrian		Negligent	Diligent
	Negligent	**(0, –10)**	(0, –11)
	Diligent	(–1, –10)	(–1, **–2**)

Payoffs: (Pedestrian, Motorist)

ance costs. Negligence is now a strictly dominant strategy for the pedestrian. In the no-liability case, negligence was a strictly dominant strategy for the motorist. In general, whenever a player is not liable for accident costs, there is no incentive to exercise due diligence. Compared with the no-liability tort regime, pure strict liability shifts the burden of the accident from the pedestrian to the motorist.

Although the strategy profile {*Negligent, Negligent*} also does not result in a socially optimal outcome, a pure-strict-liability tort regime should not be dismissed out of hand. This tort regime may be attractive in situations where it is not possible for the courts to determine whether the injured party exercised due diligence, or where it is not possible for the victim to take adequate precautions, such as in airplane crashes. On the other hand, pure strict liability would not be preferred in situations where it is necessary to have a legal rule that compels potential victims to take precautionary measures. In such cases, a no-liability tort-law regime would be preferred because it compels wrongdoers to exercise care.

Negligence with Contributory Negligence

The objective of the next three tort-law regimes is to provide both players with an incentive to choose a strategy that results in a socially efficient outcome. The first of these tort-law regimes is **negligence with contributory negligence**. This legal principle, which once dominated Anglo-American tort law, allows the injured party to recover damages in the courts only if the tortfeasor alone is negligent. The negligence-with-contributory negligence accident game is depicted in Figure 14.3.

As before, changing the legal rule changes the allocation of payoffs, but does not change the sum of the payoffs. The only difference between Figure 14.3 and Figure 14.1 is the allocation of payoffs for strategy profile {*Diligent, Negligent*}. In this case, the pedestrian's payoff is –$1 while the motorist's payoff is –$10. The pedestrian continues to bear the brunt of the accident under the remaining strategy profiles.

> **Negligence-with-contributory-negligence tort law** A legal rule that allows the victim to recover damages if the tortfeasor alone is negligent.

In this tort-law regime, the pedestrian has a strictly dominant *diligent* strategy. The motorist's strategy will depend on the strategy adopted by the pedestrian. A rational motorist will also adopt a *diligent* strategy because it results in a loss of only $1, compared with a loss of $10 by being *negligent*. The Nash equilibrium strategy profile for this game is {*Diligent, Diligent*}, which is socially optimal because it minimizes the sum of total accident costs. Regardless of how one feels about the manner in which liability has been assigned, negligence with contributory negligence results in a socially optimal outcome. Changing the legal rule not only alters player incentives and payoffs, it also results in a socially efficient Nash equilibrium strategy profile.

FIGURE 14.3

Negligence-with-Contributory-Negligence Accident Game

Motorist

		Negligent	Diligent
Pedestrian	Negligent	(−10, **0**)	(−10, −1)
	Diligent	(−1, −10)	(**−2, −1**)

Payoffs: (Pedestrian, Motorist)

FIGURE 14.4

Strict-Liability-with-Contributory-Negligence Accident Game

Motorist

		Negligent	Diligent
Pedestrian	Negligent	(−10, **0**)	(−10, −1)
	Diligent	(−1, −10)	(**−1, −2**)

Payoffs: (Pedestrian, Motorist)

Strict Liability with Contributory Negligence

The next tort-law regime that results in a socially efficient outcome is **strict liability with contributory negligence**. The strict-liability-with-contributory-negligence game is depicted in Figure 14.4. The difference between this and negligence with contributory negligence is that the motorist is responsible for damages when both players exercise due diligence. In general, the pedestrian bears the burden of an accident in negligence-based tort-law regimes, while the motorist bears the burden for liability-based tort-law regimes. The Nash equilibrium for the game in Figure 14.4 is again {*Diligent, Diligent*}. The pedestrian still has a strictly dominant *diligent* strategy. The motorist does not have a dominant strategy, but will adopt a *diligent* strategy because of the better payoff. Although the change in legal rules has altered the payoffs for the {*Diligent, Diligent*} strategy profile, this difference is not large enough to alter the players' behavior. As with negligence with contributory negligence, this legal rule results in a socially optimal outcome by minimizing the sum of total expected accident costs.

> **Strict-liability-with-contributory-negligence tort law** A legal rule in which the tortfeasor is responsible for damages when both the tortfeasor and victim exercise due diligence.

─────────────── **Demonstration Problem 14.1** ───────────────

Suppose that the cost of avoiding an accident if a motorist and pedestrian adopt a diligent strategy is $5. If an accident occurs, the pedestrian incurs accident costs of $100. If both players adopt a *negligent* strategy, the probability of an accident is 0.95. If either player adopts a *diligent* strategy, there is a 0.05 chance that an accident will occur. If both players adopt a diligent strategy, the probability that an accident will occur is 0.01. What is the Nash equilibrium strategy profile for each of the following tort-law regimes?

   a. No liability.
   b. Pure strict liability.

FIGURE D14.1.1
Accident Game in Demonstration Problem 14.1a

Motorist

Pedestrian		Negligent	Diligent
	Negligent	(–95, **0**)	(–5, –5)
	Diligent	(–10, **0**)	(–6, –5)

Payoffs: (Pedestrian, Motorist)

FIGURE D14.1.2
Accident Game in Demonstration Problem 14.1b

Motorist

Pedestrian		Negligent	Diligent
	Negligent	(**0**, –95)	(**0**, –10)
	Diligent	(–5, **–5**)	(–5, –6)

Payoffs: (Pedestrian, Motorist)

Figure D14.1.3
Accident Game in Demonstration Problem 14.1c

Motorist

Pedestrian		Negligent	Diligent
	Negligent	(–95, **0**)	(–5, –5)
	Diligent	(**–5, –10**)	(–6, –5)

Payoffs: (Pedestrian, Motorist)

c. Negligence with contributory negligence.
d. Strict liability with contributory negligence.

Solution

a. Under this legal rule, neither the motorist nor the injured party is held liable for damages in the event of an injury. The normal form of the no-liability accident game is depicted in Figure D14.1.1. The Nash equilibrium strategy profile for this game is {*Diligent, Negligent*}.
b. Under this legal rule, the motorist is strictly liable for damages in the event of an injury. The normal form of the no-liability accident game is depicted in Figure D14.1.2. The Nash equilibrium strategy profile for this game is {*Negligent, Diligent*}.
c. Under this legal rule, the pedestrian can recover damages only if the motorist is negligent. The normal form of the no-liability accident game is depicted in Figure D14.1.3. The Nash equilibrium strategy profile for this game is {*Diligent, Negligent*}.
d. Under this legal rule, the pedestrian can recover damages, even if both motorist and pedestrian are diligent. The normal form of the no-liability accident game is depicted

FIGURE D14.1.4
Accident Game in Demonstration Problem 14.1d

Motorist

		Negligent	Diligent
Pedestrian	Negligent	(−95, **0**)	(−5, −5)
	Diligent	(**−5, −5**)	(−5, −6)

Payoffs: (Pedestrian, Motorist)

in Figure D14.1.4. The Nash equilibrium strategy profile for this game is {*Diligent, Negligent*}.

Comparative Negligence

A problem with the preceding legal rules is that if one party is negligent while the other is diligent, one or the other will carry the burden of the accident costs. In the cases of no-liability and pure-strict liability, the objective is to affix blame. In the cases of negligence with contributory negligence and strict liability with contributory negligence, the objective is to minimize the social cost of the accident by altering payoffs and providing the proper incentives for both players to adopt a *diligent* strategy. This rather cold-hearted approach to assessing the relative merits of alternative tort-law regimes is normatively troubling because it ignores the pain and suffering of the injured party. It could be reasonably argued that when both parties are negligent, blame for the accident should be shared. Those who take this position would advocate the tort-law regime of **comparative negligence**, which requires both players shoulder some of the accident costs. In addition to sharing liability when both parties are negligent, this legal rule can also provide incentives for both parties to adopt a socially optimal *diligent* strategy.

> **Comparative-negligence tort law** A legal rule in which both tortfeasor and victim share liability if both are negligent.

The main difficulty with modeling comparative negligence is how to allocate liability when both players are negligent. In some jurisdictions, this is the responsibility of the jury after considering the facts of the case and the extent to which each player caused or contributed to injuries suffered. In other cases, liability may be allocated on the basis of the relative amount of due diligence exercised by both parties. Although the case of comparative negligence appears somewhat amorphous, it is still possible to develop insights into the rationale underlying comparative negligence by considering the extreme case in which the negligent party bears a disproportionately large share of the accident costs. As before, the cost of an accident is $10, which is certain if both parties are negligent. We will continue to assume that if both players adopt a *diligent* strategy there is a 10 percent likelihood that an accident will occur. The main difference between this and earlier tort-law regimes is the introduction of a *moderate* strategy in which a player exercises an amount of care that lies somewhere in between negligence and diligence. The comparative-negligence accident game is depicted in Figure 14.5.

In this game, the cost of exercising a *moderate* amount of care is $1, while the cost of being *diligent* increases from $1 to $2. If both players are negligent, neither will incur accident avoidance costs, but they will evenly divide accident costs. If both players purchase a moderate amount of prevention, each will incur $1 in prevention cost and evenly divide $10 in total accident costs.

FIGURE 14.5

Comparative-Negligence Accident Game

Motorist

		Negligent	Moderate	Diligent
	Negligent	(−5, −5)	(−9, −2)	(−10, −2)
Pedestrian	Moderate	(−2, −9)	(−6, −6)	(−11, −2)
	Diligent	(−2, −10)	(−2, −11)	(−3, −2)

Payoffs: (Pedestrian, Motorist)

When both players adopt a *diligent* strategy, each will incur $2 in avoidance costs and the pedestrian will pay another $1 in expected accident costs. If one player is *negligent* while the other player exercises a *moderate* amount of care, the *negligent* player incurs no avoidance costs but is responsible for 90 percent of the accident costs. The player exercising a *moderate* amount of care will incur $1 in prevention costs and 10 percent of the accident costs. Finally, if one player is *negligent* while the other player is *diligent,* the *negligent* player will pay all of the accident costs while the other player incurs $2 in avoidance costs.

The reader should verify that both players in Figure 14.5 have a dominant *diligent* strategy. Thus, the Nash equilibrium strategy profile for this game is {*Diligent, Diligent*} with payoffs of (−3, −2). In the comparative-negligence accident game with a sharing rule that skews damages toward the player who is more careless, each player has the correct incentive to incur avoidance costs by exercising due diligence. Even when strategies are continuous, this is the only strategy profile that survives iterated elimination of strictly dominated strategies (see, for example, Landes and Posner [1980]). A tort-law regime of comparative negligence not only allows for normative value judgments of the distribution of accident costs when both parties are negligent, but also provides players with the proper incentives to achieve socially efficient outcomes by minimizing total accident costs.

NASH EQUILIBRIUM AND ECONOMIC EFFICIENCY

The importance of tort law is that it influences players' behavior by allowing injured parties to sue for damages. A change in the tort-law regime reallocates payoffs between players. The above tort-regime games involved only two or three strategies. In fact, strategy choices are continuous, and Nash equilibria, if they exist, are found by successively eliminating dominated strategies. As noted above, the common law can be described as "efficient" if it provides the players with the correct incentives to adopt strategies that minimize the social costs.

We will define a strategy of *due diligence* as the amount of avoidance costs necessary to minimize the expected social cost of an accident. A *less than due diligence* strategy involves lower avoidance costs and higher expected accident costs. An *excess diligence* strategy involves higher avoidance costs, but does not lower expected accident costs. In principle, these strategies lie along a continuum. Thus, there are an infinite number of possible tort regimes that provide players with the correct incentives. Is there a common thread that ties together all efficient tort regimes under the same set of assumptions?

The tort-law regime's negligence with contributory negligence, strict liability with contributory negligence, and comparative negligence share three common features. First, they involve compensatory damages in which the players are responsible for their own accident avoidance costs, but do not require the tortfeasor to pay the victim for more than the cost of injuries received. Second, the tortfeasor who exercises at least *due diligence* is not financially responsible for the victim's injuries. Moreover, the victim who exercises *due diligence* is fully compensated for accident costs, but only if the tortfeasor does not exercise *due diligence*. Finally, when both the tortfeasor and the victim exercise at least due diligence, the cost of the accident is borne by one, or the other, or both. These tort-law regimes, which lie along a continuum of strategies, require players to exercise *due diligence* and may result in radically different distributional outcomes.

It is relatively easy to demonstrate that a tort-law regime that provides incentives for rational players to exercise due diligence is a Nash equilibrium. To see this, we begin by showing that *excess diligence* strategies are strictly dominated. The reason for this is that if one player adopts a *less than due diligence* strategy, the other player can avoid liability for accident costs by exercising just *due diligence*. Exercising *excess diligence* provides no additional benefit but increases avoidance costs. This will unambiguously reduce the player's net payoff because the expected accident costs will be borne by the player exercising *less than due diligence*. By analogous reasoning, *excess diligence* is a strictly dominated strategy for the other player as well. The only other strategies to consider are those involving *due diligence* and *less than due diligence*.

When both players exercise *due diligence*, it is necessary to consider the allocation of expected accident costs. Suppose that a player does not bear the full cost of the accident. In this case, exercising *less than due diligence* means bearing the full cost of the accident. For this to make sense, the gains from spending less on accident avoidance must exceed the increase in expected accident costs.

A player will always be better off by exercising *due diligence* and incurring a portion of expected accident costs than exercising less than *due diligence* and incurring the full amount of the accident costs. Since the increased cost of *due diligence* must be less than the reduction in the player's share of the accident costs, the player is more than compensated for the added avoidance costs than from a *less than due diligence* strategy. Thus, whenever a player does not incur the full cost of an accident, *due diligence* is the best response to a rival's *due diligence* strategy.

What if one player incurs the full costs of the accident? Since this player cannot shift accident costs, the objective is to minimize the possibility that the accident occurs. If this player exercises *due diligence*, it is in the best interest of the other player to exercise *due diligence* as well. Finally, if both players exercise *less than due diligence*, it will be in both players' best interest to switch strategies and adopt a *due diligence* strategy. Thus, *less than due diligence* is strictly dominated by *due diligence*.

It would appear that the only Nash equilibrium strategy profile that results in a socially efficient outcome is for all rational players to adopt a *due diligence* strategy. For this to be true, however, two conditions must be satisfied. First, the tortfeasor's share of accident plus avoidance costs must be strictly less than avoidance costs. Second, the portion of uncompensated accident costs plus the victim's avoidance costs must be less than the victim's avoidance costs. When both conditions are satisfied, the sum of avoidance and accident costs to both players (i.e., the social cost of the accident) must be less than the costs to the victim and the tortfeasor from exercising *due diligence*. But, this cannot be true since if both players exercise *due diligence*, it must also be the case that social cost is minimized. Thus, one player will always prefer a *due diligence* strategy rather than adopt a *less than due diligence* strategy when the other player adopts a less than *due diligence* strategy.

Anglo-American tort-law regimes of contributory negligence all share the common feature that each provides adequate incentives for players to adopt strategies that lead to socially efficient outcomes in games. A player who exercises due diligence, while the other player does not, is never responsible for the full cost of the accident. While different rules have different distributional outcomes, if the above conditions are satisfied, the players will always have an incentive to behave in a socially efficient manner.

CONTRACTS

A defining characteristic of free enterprise is the **contract**. As discussed in Chapter 12, a contract is a formal agreement that obligates the parties to perform, or refrain from performing, a specified act in exchange for something of value. The law provides for remedies if a party fails to abide by the terms and conditions of the contract. Because contracts lie at the heart of virtually all market transactions, contract law is among the most important areas of jurisprudence. Contract law is extremely complex and anything approaching a detailed discussion of its intricacies is clearly beyond the scope of this book and the expertise of the author. The purpose of this discussion is simply to illustrate how game theory can be used to enhance our understanding of certain elements that are common to all contracts.

Consider the following example of a simple loan contract. In this dynamic game, the lender must decide whether to extend a $100 loan in exchange for a promise by the borrower to repay the loan and 5 percent simple interest in one year. The borrower plans to use the proceeds of the loan to finance an investment with a certain 10 percent rate of return. If all goes well, the lender will earn $5 in loan interest and the borrower will earn a profit of $110 – $105 = $5. The lender has the ability to repay the loan with interest, but may not have the willingness to do so. In the literature dealing with asymmetric information, this is a *moral hazard* problem. It is the risk (hazard) that the borrower will engage in an activity (default) that is undesirable (immoral) from the lender's perspective.

We will begin by assuming that the lender has no legal recourse to compel the borrower to repay the principle and interest in the event of default. Is it possible for the lender to create a mechanism that will make it in the borrower's best interest to nominally service the debt? The extensive form of the lending game with certain payoffs is depicted in Figure 14.6.

The first stage of this game begins when the lender decides whether to *lend* or *not lend*. If the lender adopts a *not lend* strategy, the game is over and the payoff to both players is $0. If the lender decides to *lend,* the borrower must decide whether to *repay* or *default*. If the borrower repays, the payoff to both players is $5. If the borrower defaults, the lender will lose $100 while the borrower's payoff is $110 ($100 from the loan and $10 return on the investment). Using the method of backward induction, the subgame perfect equilibrium is for the lender to *not lend.* The reason for this is that if the loan contract cannot be enforced, it is in the borrower's best interest not to repay.

In the above game, the payoffs were assumed to be certain. In fact, there is always doubt in the mind of the lender as to whether the borrower will repay. To incorporate this element of risk, we will assume that the lender believes that there is an 80 percent chance that the borrower will *repay* ($p_R = 0.8$) and a 20 percent chance that the borrower will *default* ($p_D = 1 - p_R = 0.2$). From the lender's perspective, the expected payoffs for the lending game are summarized in Figure 14.7. Once again, the subgame perfect equilibrium for this game is for the lender to *not lend.* The payoffs are highlighted in boldface. Clearly, the lender is better off by not lending than by lending. It is left as an exercise for the reader to demonstrate the subgame perfect equilibrium if the lender is certain that the loan will be repaid ($p_R = 1.0$).

How can the loan contract in the lending game depicted in Figure 14.7 be modified to make the borrower's promise to repay credible? The importance of contract law is that it structures incentives to make promises or threats credible. Suppose that society institutes a legal rule that allows the lender to call upon the government to enforce loan contracts. Moreover, suppose that the contract is structured in such a way as to make the borrower responsible for all litigation (court and attorney) costs in the event of negative court ruling. Let us assume that litigation costs are $10, which is more than the borrower would earn if the loan is repaid. The amount of litigation costs and other penalties relative to expected payoffs from a loan default is important. Penalties that are too low will not provide an adequate incentive to keep the borrower from defaulting. With intervention

FIGURE 14.6
Lending Game with Certain Payoffs

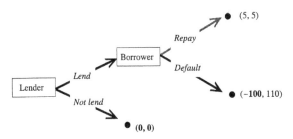

Payoffs: (Lender, Borrower)

FIGURE 14.7
Lending Game with Expected Payoffs

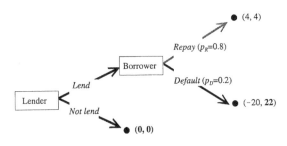

Payoffs: (Lender, Borrower)

FIGURE 14.8
Lending Game with Legal Enforcement

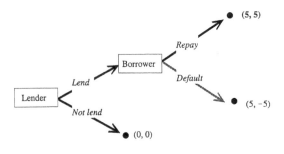

Payoffs: (Lender, Borrower)

by the courts, the payoffs in the lending game from alternative strategy profiles are known with certainty. The transformed lending game with legal enforcement is depicted in Figure 14.8.

The subgame perfect equilibrium for the lending game with legal enforcement is {*Lend* → *Repay*} with the payoff to both players of $5. The lender is better off by making the loan and the borrower is better off repaying the loan with interest. Default will expose the borrower to $10 in litigation costs, which will more than offset the payoff by nominally servicing the debt. Of course, this result depends on whether the case is litigated and whether the borrower loses the case in court (see Case Study 14.1).

The game depicted in Figure 14.8 underscores the important role played by the courts in adjudicating contracts. When the parties to a contract are reasonably certain that the victim of a breach has the legal right to recover damages, it becomes possible to structure the contract in such a way that rewards both parties for adopting strategies that are mutually and socially beneficial. Without legally enforceable contracts, social welfare from specialization and trade would be significantly, and negatively, affected. A contract would be reduced to the status of an unenforceable verbal agreement, which according to legendary film mogul Sam Goldwyn ". . . isn't worth the paper it's written on" (Berg 1989, p. 396)

CASE STUDY 14.1: ECONOMIC EFFICIENCY AND THE COMMON LAW

In this chapter we discussed the virtues of legally enforceable contracts. Laws that give the parties to a contract the ability to call upon the state to enforce a claim can transform a game with a socially inefficient outcome into a game that is socially efficient. Whether this actually happens depends on whether a dispute is litigated and whether the claimant is successful. Is this, in fact, what happens in reality? Do legal rules result in economically efficient outcomes? The evidence, or what there is of it, suggests otherwise.

Gordon Tullock (1997) noted that while the precise social cost of legal disputes is unknown, U.S. federal, state, and local spending on civil and criminal justice in 1983 amounted to almost $40 billion, or about $170 per capita, which was about 3 percent of total U.S. government expenditures. Of this amount, $37 per capita was expended on judicial services (Cooter and Ulen 1988, p. 478), which "amounts to only a small fraction of the social cost of resolving disputes through the courts, since most of these costs are borne by private parties" (Tullock 1997, p. 45). Cooter and Ulen estimated that the labor cost of a full trial was about $400 per hour, not including the cost of court facilities.

According to Tullock (1997, p. 52): "Since the early 1960s U.S. courts have systematically assaulted classical tort law by dismantling its twin historical pillars—deterrence and compensation—in favor of notions of societal insurance and risk-spreading, and undermining the concept of fault as a doctrinal mechanism for limiting tort liability to substantive tortfeasors. . . . The shift from negligence with contributory negligence to comparative negligence or strict liability has induced a sharp increase in moral hazard as plaintiffs lower their own standards of care and has stimulated a sharp increase in tertiary legal costs as the volume of lawsuits has exploded." Does this increase in legal costs resulting from changes in tort-law regimes imply a higher degree of social efficiency in U.S. court rulings?

Rubin (1977) and Posner (1979, 1992) have argued that there is a systematic tendency for common law rules to evolve toward economically efficient outcomes. Rubin has argued that when litigants have an ongoing interest in legal precedent, they are likely to resort to the courts whenever the relevant legal rules are inefficient and less likely to do so when the legal rules are efficient. According to Posner, once efficient rules have evolved, the incentive for further litigation is reduced. The resulting court rulings reduce the likelihood of future litigation and increase the probability that court rulings will persist.

Landes (1971), Gould (1973), and Tullock (1971, 1997) have determined that convergence of the common law to economically efficient outcomes is unlikely because it is in the litigant's best interest to settle out of court. Cooter and Kornhauser (1980) found that there is no automatic mechanism to guarantee that the common law converges to economic equilibrium, even when economically inefficient rules are litigated or when judges replace inefficient rules with efficient ones. By specifying the Rubin model as a two-person, non-cooperative, static game, Webster (2004) found that it may be in the litigant's best interest to negotiate an out-of-court settlement when legal rules are efficient if the expected net present value of accident and prevention costs is greater than the litigants' court costs, and litigate even when legal rules are efficient if the expected net present value of accident and prevention costs is greater than the sum of the litigant's court costs.

CHAPTER REVIEW

This chapter introduced the legal concepts of *negligence, liability,* and *damages* in the Anglo-American system of torts and contracts. *Negligence* refers to the failure to exercise the same care that a reasonable or prudent individual would exercise under the same circumstances, or taking an action that a reasonable person would not. *Liability* refers to legal responsibility for acts of commission or omission. *Damages* represent the amount of money or other compensation awarded by the court to the plaintiff in a lawsuit.

Tort law is a branch of the law dealing with torts, which refers to private wrongs or injuries. Torts may be intentional, or may result from negligence. Torts are central to the Anglo-American concept of the *common law,* which refers to the body of legal principles created largely by judges as the by-product of deciding cases. Common law is based on the legal principle of *stare decisis* in which precedent is binding on all subsequent cases.

This chapter provided a game theoretic analysis of the tort-law regimes of *no liability, pure strict liability, negligence with contributory negligence, strict liability with contributory negligence,* and *comparative negligence.* The importance of tort law is that it can be used to alter player incentives to achieve socially efficient outcomes.

The common law may be characterized as efficient if the resulting strategy profile minimizes the social cost of an accident. The only Nash equilibrium strategy profile that results in a socially optimal outcome is for rational players to exercise *due diligence*, which was defined as the amount of avoidance costs necessary to achieve this objective.

This chapter also examined *contracts,* which are formal agreements between two or more parties to perform, or refrain from performing, some specified act in exchange for something of value, valuable benefit, or consideration. The law provides for remedies if one or the other party fails to abide by the terms of the contract.

Contracts are at the heart of virtually all market transactions. The analysis presented in this chapter illustrated the important role of government in adjudicating contracts. When the parties to a contract are reasonably certain that the victim of a breach has the legal right to recover damages, it is possible to structure the payoffs in such a way that both parties have an incentive to adopt strategies that are mutually and socially beneficial.

CHAPTER QUESTIONS

14.1 Define and explain each of the following legal concepts:

   a. Common law
   b. *Stare decisis*
   c. Tort

    d. Tort law
    e. Negligence
    f. Liability
    g. Damages
    h. Contract
    i. Contract law

14.2 Define and explain the essential features of each of the following tort-law regimes:

    a. No liability
    b. Pure strict liability
    c. Negligence with contributory negligence
    d. Strict liability with contributory negligence
    e. Comparative negligence

14.3 Explain how changes in liability rules in tort-law regimes reallocate payoffs between plaintiffs and defendants.

14.4 Compared with a no-liability tort-law regime, pure strict liability shifts the burden of an accident from the motorist to the pedestrian. Do you agree with this statement? Explain.

14.5 What are the three common features of contributory negligence tort-law regimes?

14.6 In games of complete but imperfect information, prove that a tort-law regime that provides incentives for rational players to exercise due diligence is a Nash equilibrium.

14.7 A homeowner purchases a ladder from Home Depot to repair his roof. Directions for its proper use are clearly indicated on the ladder. Using the ladder improperly could result in a life-threatening injury. How much caution will the homeowner exercise under each of the following tort-law regimes?

    a. No liability
    b. Pure strict liability
    c. Negligence with contributory negligence
    d. Strict liability with contributory negligence
    e. Comparative negligence

14.8 Legally enforceable contracts are always necessary for a market transaction to take place. Do you agree with this statement? Explain.

14.9 When the parties to a contract are reasonably certain that the victim of a breach has legal recourse to recover damages, it is possible to structure a contract that gives players an incentive to adopt strategies that are mutually and socially beneficial. Why?

CHAPTER EXERCISES

14.1 Suppose that the cost of avoiding an accident if a motorist and pedestrian adopt a diligent strategy is $1. If an accident occurs, the pedestrian incurs accident costs of $10. If both players are *negligent,* the probability of an accident is 0.1. If both players adopt a *diligent* strategy, there is a 0.025 chance that an accident will occur. If either player adopts a diligent strategy, the probability that at accident will occur is 0.05. What is the Nash equilibrium strategy profile for each of the following tort-law regimes?

    a. No liability
    b. Pure strict liability

      c.   Negligence with contributory negligence

      d.   Strict liability with contributory negligence

14.2 Consider the lending game depicted in Figure 14.8. What is the subgame perfect equilibrium for this game if there is a certain probability of default?

14.3 Consider the lending game in Figure 14.7. What is the subgame perfect equilibrium for this game if $p_R = 1$?

14.4 In the lending game depicted in Figure 14.8, what probability of default will make the lender indifferent between lending and not lending to the borrower?

14.5 A lender has to decide whether to extend a $1,000 loan in exchange for a promise by the borrower to repay the loan and 10 percent simple interest in one year. The borrower plans to invest the proceeds of the loan with a certain 25 percent rate of return. The lender believes there is a 90 percent chance that the loan will be repaid. What is the subgame perfect equilibrium for this lending game?

ENDNOTES

    1. This chapter draws extensively from Douglas G. Baird, Robert H. Gertner, and Randal C. Picker, *Game Theory and the Law,* Cambridge, MA: Harvard University Press, 1994, Chapters 1 and 2.

    2. In the game depicted in Figure 14.1, the Nash equilibrium strategy profile {*Negligence, Negligence*} implies that pedestrians will incur no avoidance costs whatsoever. This is a by-product of the simplistic assumptions underlying this game. As we have noted, this strategy profile makes it inevitable that accidents will occur and that pedestrians will be injured. Knowing this, pedestrians will exercise even greater care than might otherwise be the case.

CHAPTER

15

AUCTIONS

In this chapter we will:

- *Discuss different types of auctions and analyze bidding strategies in auctions involving complete and incomplete information;*
- *Analyze optimal bidding strategies in auctions with independent and correlated private value estimates;*
- *Identify dominant bidding strategies of risk-neutral and risk-averse players;*
- *Examine common-value auctions and the winner's curse;*
- *Discuss auction mechanism design and the revelation theorem.*

INTRODUCTION

In Chapter 12 we analyzed the bargaining process as a multistage game in which buyers and sellers "haggle" over the terms of a contract or transaction. In this chapter we will examine the related concept of the **auction**. In an auction, individuals or institutions compete with each other for the right to purchase or sell something of value, such as goods, services, commodities, and financial assets.

> **Auction** A process whereby individuals or institutions compete with each other to buy or sell something of value.

Auctions to buy or sell items of value through price competition among bidders have been around for thousands of years. The Greek historian Herodotus described auctions involving the sale of brides in Babylonia over 2,500 years ago. Until nearly a century and a half ago, slave auctions were commonplace in the United States. Auctions have been used to sell livestock, tobacco, rare coins, antiques, used cars, and U.S. Treasury bills. The U.S. government has used auctions to award construction contracts, off-shore drilling rights, and licences for radio frequencies. State governments routinely use auctions to award construction, service, and repair contracts to private contractors. Online auction services such as eBay and Yahoo! account for annual transactions worth billions of dollars.

BUSINESS, AUCTIONS, AND THE INTERNET

One of the most dramatic developments in recent years is how Internet auction sites have transformed the way businesses acquire productive resources used to manufacture goods and services, and the manner in which output is priced, marketed, and sold to customers. For example, farmers have been able to cut growing and processing costs by using online exchanges to acquire seed, feed, and fertilizer, while receiving higher prices by selling their output in a global marketplace.

In addition to enabling businesses to acquire resources at lower prices, the Internet has dramatically reduced transaction costs. Instead of spending time and money on transportation, room and board, and other expenses to travel hundreds of miles to purchase raw materials from suppliers and locate buyers for their products, small-business owners can now do this, and much more, with a laptop computer in the comfort of their own homes. Large corporations such as General Motors, Daimler-Chrysler, and Ford have developed an online exchange to purchase large quantities of parts. Retailers have created the WorldWide Retail Exchange and Global Net Xchange. FreeMarkets has diversified from metal and plastic parts sales to online auctioning of tax-preparation, temporary employment, relocation, and other services, while Sears, Roebuck and Co., the second-largest retailer in the United States, and Carrefour SA of France, the world's second-largest retailer, have significantly reduced acquisition costs from nearly 50,000 suppliers around the world by using the Internet.

Online consumer auctions, such as eBay and Yahoo! Auctions, sell everything from fish sticks to fishing boats. MedicineOnline.com is an online auction site that enables plastic surgeons to bid against each other to sell face lifts, breast augmentations, rhinoplasties, and so on. Other Internet sites, such as Priceline.com for travel services, have a name-your-own-price feature. This option allows travelers to set a price that they are willing to pay, while airlines and hotels decide whether or not to accept the indicated price. In short, the Internet has revolutionized the global marketplace. While Internet auction sites may not be appropriate for every type of market transaction, it is now possible to buy and sell virtually anything you can think of.

CASE STUDY 15.1: FCC'S WIRELESS SPECTRUM AUCTION

Among the most significant contributions of game theory in recent years is the analysis and design of auctions. Its practical application became apparent in 1994 when the U.S. Federal Communications Commission (FCC) auctioned off ten nationwide narrow-band frequency licences to provide mobile telecommunication services for cellular phones, pagers, and other wireless personal communication service (PCS) devices (see McMillan 1994). What distinguished this from other government auctions was that, for the first time, the principles of game theory were used to design the auction rules. After consulting several game theorists and auction experts, the FCC adopted a simultaneous-move, multiple-round auction format. The auction involved multiple rounds during which bidders were allowed to submit simultaneous bids on all ten licenses. Each of four firms put up $3,500,000 for the right to purchase at most three licenses. The auction, which began on July 25, lasted five days and involved forty-six bidding rounds. The outcome of the FCC auction exceeded all expectations. In the first round alone, the ten highest bids totaled more than $100 million. The licenses sold for over $600 million—more than ten times the most optimistic prediction.

A distinguishing characteristic of the FCC auction was the frequent use of the **jump bid**, which is a bid that exceeds the previous bid by more than the minimum allowable incremental increase. Game theorists have demonstrated that jump bids are an effective way to signal a bidder's intention not to be outbid.

The FCC has since used similar auctions to allocate regional broadband licenses for more advanced PCS devices. Many of these bidders were interested in purchasing multiple licenses to create regional and nationwide networks. These auctions were quite complex and took several months to complete. In the end, these regional broadband frequencies sold for over $8 billion! Because of the success of these auctions, the federal government is considering similar auctions to allocate airport landing rights, pollution emission rights, and so on.

Jump bid A bid in a multiple-round auction that exceeds the previous bid by more than the minimum allowable increment.

CASE STUDY 15.2: EBAY

EBay, which is perhaps the best known and most popular online auction site, has proven to be fertile ground for economists, game theorists, market researchers, and others seeking to uncover underlying patterns in auction processes. The basic rules of an eBay auction are quite simple. A seller offers an item for sale and prospective buyers submit online bids. Whoever submits the highest bid at the end of a predetermined time period wins the auction.

Before eBay, game theorists had difficulty testing their theories because of a paucity of empirical data. Many researchers had to be satisfied with sifting through data on FCC narrow-band auctions. The advent of eBay and similar auction sites provided game theorists with a wealth of empirical data to sink their analytical teeth into. What has this research uncovered? For one thing, a seller's reputation significantly affects the price of the item placed at auction. EBay publishes anonymous ratings provided by auction participants. Studies indicate that reputation affects the final selling price and the quantity sold. Sellers with a single negative rating received significantly fewer bids, sold fewer items, and received sharply lower prices.

Another aspect of eBay auctions that has received a great deal of attention is the effectiveness of "sniping," which is possible because eBay auctions have fixed time limits. By waiting until just before the close of the auction to submit their bids, snipers hope to avoid being outbid. Many eBay users swear by the practice, and there are even online services that offer automated sniping. The evidence suggests, however, that in auctions involving standardized merchandise, such as consumer electronics, with a large number of bidders, snipers do not outperform anyone else in terms of their success rate or in the price paid. On the other hand, sniping appears to be successful in auctions with few bidders involving specialized merchandise because its value is difficult to determine.

Another area of research involves the effectiveness of a "secret reserve," which is a feature of eBay auctions. Sellers on eBay have the option of publicly setting a minimum acceptable bid, or rejecting bids below some undeclared minimum price. The evidence suggests that secret minimum prices dissuade prospective bidders from participating in such auctions. Fewer participants tend to result in lower selling prices.

TYPES OF AUCTIONS

The ubiquitous and growing importance of auctions in the global marketplace makes an understanding of auction processes and bidding strategies essential. This is especially true for managers who buy or sell goods, services, commodities, or factors of production in auctions involving multiple bidders.

There are two basic types of auctions: standard and procurement auctions. In a **standard auction**, multiple bidders compete to acquire an object of value from a single seller. The buyer who submits the highest bid wins the auction and acquires ownership of the object. Before the auction begins, both the bidder and the seller assign a value to the object being sold. The objective of the buyer and the seller is to maximize their positive net benefits. For the seller, this means selling the object for the highest possible price. For the bidder, this means buying the object at the lowest possible price.

Standard auction When multiple bidders compete to acquire something of value from a single seller.

Procurement (reverse) auction When multiple sellers bid for the right to sell something of value to a single buyer.

A **procurement (reverse) auction** reverses the roles of buyer and seller in a standard auction. Multiple sellers bid for the right to sell something of value to a single buyer. In a procurement auction, the lowest bid wins. Suppose, for example, that the U.S. Department of Defense (DoD) wants to auction off a contract for the construction of a military base. In this case, the objective of DoD is to maximize total net benefits by awarding the contract to the lowest bidder. By contrast, the objective of the construction companies (bidders) is to maximize total net benefits by procuring the contract at the highest possible price.

> *Principle:* In an auction, the objective of buyer and seller is to maximize the net benefit of acquiring an object of value.

In this chapter we will apply the basic tools of game theory to identify optimal strategies in the most common types of standard auctions. In general, risk preferences will affect bidder strategies, which in turn affect the seller's expected revenues. Although bidders generally exhibit risk-averse behavior, we will assume for the sake of simplicity that bidders are risk neutral. A discussion of the implications for auctions involving risk-averse bidders will be deferred until later in this chapter. We will begin by introducing the basic vocabulary of auctions. This will be followed with a short discussion of the most common types of auctions and the information set that is available to bidders about the object being auctioned.

Before discussing the most common types of auctions, it is necessary to distinguish between **auction rules** and an **auction environment**. Auction rules refer to the procedures that govern the bidding process. These procedures identify who may bid, which bids are acceptable, the manner in which bids are submitted, the nature of the information available to the players during the bidding process, when the auction ends, how the winner is determined, and the price paid. By contrast, the auction environment refers to the characteristics of the auction participants, including the number of bidders, risk preferences, how rival bidders value the object being auctioned, and so forth.

Auction rules The procedures that govern the bidding process, including who may bid, which bids are acceptable, the manner in which bids are submitted, the nature of the information available to the players during the bidding process, when the auction ends, how the winner is determined, and the price paid.

Auction environment The characteristics of the auction participants, including the number of bidders, their risk preferences, and information about how rival bidders value the good or service being auctioned.

Auctions may be an open-bid or closed-bid. In an **open-bid auction**, anyone may submit a bid. In a **closed-bid auction**, participation is by invitation only. In a closed-bid auction, bidders may be licensed, required to pay a nonrefundable entry fee, or post a performance bond to ensure timely payment.

Open-bid auction An auction in which anyone may submit a bid.

Closed-bid auction An auction in which participation is by invitation only.

Auctions may involve a **reservation price**. In a standard auction, this is the minimum price that a seller will accept. In ascending bid auctions, the reservation price is typically the opening bid specified by the **auctioneer**, who is the seller's agent. If the highest bid submitted is less than the reservation price, the auction is invalidated. Reservation prices are used to prevent bidder collusion ("bidding rings"), or when there are a small number of bidders. In the absence of a reservation price, low bidder turnout could result in a successful bid that is below the seller's valuation of the object being auctioned.

> **Reservation price** The minimum price that a seller will accept in a standard auction.
>
> **Auctioneer** The seller's agent at an auction.

An auction may be sealed-bid or oral. In a **sealed-bid auction**, competing buyers submit secret bids to the auctioneer. After all bids are received and opened, a winner is declared. Sealed-bid auctions are typical of many government auctions, such as for harvesting timber on public lands and off-shore drilling rights. In an **oral auction**, bids are announced publicly. The primary difference between sealed-bid and oral auctions is that players are more easily identifiable in an oral auction. The identities of rival bidders could convey important information to rival bidders about the value of the object being auctioned and the effectiveness of bidding strategies. Suppose, for example, it is known that one of the bidders is a rare art expert. Rival bidders may naturally assume that the expert has a better idea of an art object's true market value. Observing the expert's bidding strategies could alter the bidding strategies of rival bidders. Alternatively, suppose that it is known that one of the bidders, while not an acknowledged expert, is an avid collector and very wealthy. This knowledge may also cause rivals to become more aggressive or passive in their bidding strategies. Since this knowledge may affect the final auction price, it may be in the best interest of some bidders to hide their identities. Bidders can easily mask their identities by employing a surrogate, or by submitting bids by telephone or secure Internet connection.

> **Sealed-bid auction** An auction in which bids are secretly submitted.
>
> **Oral auction** An auction in which bids are announced publicly.

Auction rules may require that bids be submitted simultaneously, or may involve an auctioneer who announces prices in ascending or descending order. During an oral auction, bidders must decide whether to accept or pass on the announced price. In a **sealed-bid, first-price auction**, the winner pays the highest sealed bid submitted. In some auctions, the winner submits the highest sealed bid, but pays the second-highest bid. These auctions are called **second-price auctions**. Sealed-bid, second-price auctions are also called **Vickery auctions** in honor of Nobel laureate William Vickery, who discovered that these auctions have certain truth-revelation properties. In particular, Vickery auctions provide bidders with an incentive to reveal their valuations of the object being auctioned.

> **Sealed-bid, first-price auction** An auction in which the winner submits and pays the highest sealed bid.
>
> **Sealed-bid, second-price (Vickery) auction** An auction in which the winner submits the highest sealed bid, but pays the bid submitted by the second-highest bidder.

There are two main types of oral auctions. In a standard **English auction** (also called an **open ascending-bid auction**) the auctioneer may begin by announcing an opening bid (reservation price). Bidders must decide whether to accept, or pass, on the announced price. If two or more players accept, the auctioneer raises the price. The last player left standing wins the auction and pays the highest bid. In a standard Dutch auction (also referred to as an **open descending-bid auction**) the auctioneer begins by announcing a very high opening bid. As this price is incrementally lowered, players simultaneously decide whether to accept or pass. The first bidder to accept the price is declared the winner and the auction comes to an end.

> **English (open ascending-bid) auction** An auction in which an auctioneer announces prices in ascending order. Bidders decide to accept or pass on the announced price. The last remaining bidder wins the auction.

> **Dutch (open descending-bid) auction** An auction in which an auctioneer incrementally lowers an initially very high price. The first bidder to accept the current price wins the auction.

A variation of the Dutch auction is the **multiple-item Dutch auction**. In a multiple-item auction, which was named after the system of selling tulip bulbs in Holland, the seller accepts bids for blocks of identical items. Unlike a standard Dutch auction, a multiple-item Dutch auction can have many winners. After evaluating the range of sealed bids, the seller accepts the lowest price that will dispose of the entire amount placed at auction. Each bidder who wins an amount pays the same price. Multiple-item auctions have been used to sell large blocks of financial securities, including U.S. Treasury bills. They have also been used to sell equity shares in initial public offerings (see Case Study 15.3).

> **Multiple-item Dutch auction** A variation of a standard Dutch auction involving multiple bids for blocks of identical items. The seller accepts the lowest bid that will dispose of the entire amount placed at auction. Each player whose bid is accepted pays the same price.

AUCTIONS WITH COMPLETE AND PERFECT INFORMATION

We will begin our analysis of bidding strategies under alternative auction rules by assuming that bidders' private valuations of the object being auctioned are common knowledge. Private valuations of the object being auctioned reflect the demand determinants of the bidders, such as income, wealth, expertise, and so on. Clearly, this happy state of affairs is unrealistic. On the other hand, analyzing **complete information auctions** provides important insights into the formulation of optimal bidding strategies in situations where knowledge of bidders' private valuations are unknown.

> **Complete information auction** An auction in which the bidders' private valuations are common knowledge.

Sealed-Bid, First-Price Auction

Suppose that Molly and Robby are bidding for an antique pocket watch. Both players know how much the other bidder values the watch. Robby knows that Molly is willing to pay up to $562, while Molly knows that Robby is willing to pay $442. In a sealed-bid, first-price auction, the players submit a one-time, sealed bid. The winner of the pocket watch submits, and pays, the highest bid. In this type of auction, the winner pays the amount of his or her bid. In the event of a tie, the winner will be determined by the flip of a fair coin. We will assume that all bids must be in $50 increments, and that there is a reservation price (opening bid) of $300. The assumption of a reservation price is used to speed up the analysis of optimal bidding strategies, but has no effect on the outcome of the auction.

We will designate Molly's bid as b_M and Robby's bid as b_R. If Molly wins the auction, her net payoff is $562 - b_M$ and zero if she loses. If Robby wins the auction, his net payoff is $442 - b_R$

FIGURE 15.1

Sealed-Bid, First-Price Auction of an Antique Pocket Watch

		Molly					
		$300	*$350*	*$400*	*$450*	*$500*	*$550*
	$300	(71, 131)	(0, 212)	(0, 162)	(0, 112)	(0, 62)	(0, 12)
Robby	*$350*	(92, 0)	(46, 106)	(0, 162)	(0, 112)	(0, 62)	(0, 12)
	$400	(42, 0)	(42, 0)	(21, 81)	**(0, 112)**	(0, 62)	(0, 12)

Payoffs: (Robby, Molly)

and zero if he loses. Clearly, Molly will not submit a bid that is greater than $562 and Robby will not submit a bid that is greater than $442. In the jargon of game theory, bids greater than the players' valuation of the object being auctioned are strictly dominated strategies. If both players submit identical bids (b), Molly's expected payoff is $0.5(b_M - b)$ and Robby's expected payoff is $0.5(b_R - b)$. The normal form of this noncooperative, one-time static game is depicted in Figure 15.1. It is left as an exercise for the reader to demonstrate that the iterated weakly dominant Nash equilibrium strategy profile for this game is {*$400, $450*}. The net payoff for Molly is $562 – 450 = $112, while the net payoff for Robby is $0. These payoffs are highlighted in boldface.

The $50 spread between Molly's and Robby's bid is a by-product of the auction rule requiring that bids be submitted in $50 increments. It is easy to demonstrate that if the auction rules require that bids be in $10 increments, Robby would submit a bid of $440 and Molly would submit a bid of $450. If the bids are in $1 increments, the bids would be $442 and $443, respectively. Indeed, if the bids are in increments of $0.01, the Nash equilibrium strategy profile for game would be {*$442, $442.01*}. This leads to the following interesting result. In a sealed-bid, first-price auction with complete information, the loser will bid exactly his or her true private valuation of the object. The winner will submit a bid that is higher than the next-highest bid by the amount of the minimum allowable bidding increment. In principle, although not in practice, as the bidding increment shrinks to zero the winner will submit a bid that is equal to the bid submitted by the next-highest bidder.

Principle: In a sealed-bid, first-price auction with complete information, the winner is the player who values the object most and submits a bid that is higher by the minimum allowable bidding increment than the value placed on the object by the next-highest bidder. If there are no bidding increments, the winner pays what the object is valued by the loser.

Sealed-Bid, Second-Price Auction

How is this outcome changed if the highest bidder pays the price submitted by the next-highest bidder? As before, the players submit a one-time, sealed bid, and the highest bid wins. In the event of a tie, the flip of a fair coin determines the winner. There is a reservation price of $300, and all bids are in $50 increments. To illustrate how this auction differs from the game depicted in Figure 15.1, suppose that Molly submits a bid for $450 and Robby submits a bid for $400. In this case, Molly wins the auction, but instead of paying her winning bid of $450 she pays Robby's bid of $400. Molly's net payoff will be $562 – $400 = $162, instead of $562 – $450 = $112 previously. The normal form of this static game is depicted in Figure 15.2.

An examination of Figure 15.2 reveals something very interesting about sealed-bid, second-price auctions. Namely, every player—not just the loser as is the case in sealed-bid, first-price auctions—has

FIGURE 15.2

Sealed-Bid, Second-Price Auction of an Antique Pocket Watch

		Molly					
		$300	$350	$400	$450	$500	$550
	$300	(71, 131)	(0, 262)	(0, 262)	(0, 262)	(0, 262)	(0, 262)
Robby	$350	(142, 0)	(46, 106)	(0, 212)	(0, 212)	(0, 212)	(0, 212)
	$400	(142, 0)	(92, 0)	(21, 81)	(0, 162)	(0, 162)	**(0, 162)**

Payoffs: (Robby, Molly)

an incentive to submit bids that reflect their true value of the object. In fact, submitting a true value bid is a dominant strategy for every player, even when the other bidders' valuations of the object are unknown. The Nash equilibrium strategy profile for the game depicted in Figure 15.2 is {*$400, $550*}. Robby's $400 bid is the highest bid that he can submit without exceeding his valuation of the pocket watch ($442). Molly's $550 bid is the highest bid that she can submit without exceeding her valuation of the pocket watch ($562). When the bidding increments shrink to zero, each player will submit a bid that is exactly equal to his or her valuation of the object being auctioned.

Principle: In a sealed-bid, second-price auction with complete information, the winner is the player who values the object most, submits a bid that is lower by the minimum allowable bidding increment than his or her true value of the object, and pays the bid submitted by the next-highest bidder. If there are no bidding increments, the dominant strategy for each player is to submit a bid that is equal to his or her true value of the object.

English Auction

We shall now examine what happens when we replace a sealed-bid auction with an English auction. In a standard English auction, the auctioneer announces current bid prices in ascending order. When the price is announced, each of the bidders must decide whether to *accept* or *pass*. In a standard English auction, passing on the announced price does not preclude the bidder from accepting a higher price. In order to simplify matters, we will use the Japanese version of an English auction where passing on the current, announced price precludes the bidder from reentering the auction at a higher price. Once a bidder passes, he or she may no longer participate. In this type of auction, a player will continue to participate as long as the current price is less than or equal to his or her true value of the object. The value placed on the object by the player is called the **dropout price** because he or she will pass on all subsequent higher bids. If one player passes while the other accepts, the auction ends and the winner pays the current price.

Dropout price The price at which a player will drop out of an English auction. The dropout price exceeds the bidder's valuation of the object being auctioned.

We will now describe the dynamics of an English auction using the antique pocket watch example. As before, Molly is willing to pay up to $562 for the antique pocket watch while Robby is willing to pay $442. Bids are in $50 increments. To speed up the analysis, we will assume a reservation price (opening bid) of $400. In this multistage game, a new subgame commences whenever the auctioneer raises the current asking price. The game tree for this auction is depicted in Figure 15.3.

FIGURE 15.3
Subgame-Perfect Equilibrium in the Antique Pocket Watch English Auction

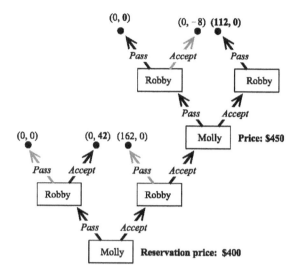

Payoffs: (Molly, Robby)

Despite appearances, the game depicted in Figure 15.3 is a two-stage, static game with different payoffs in each stage. In the first stage, Molly and Robby have a strictly dominant *accept* strategy since each enjoys net benefits of $162 and $42, respectively. If both bidders accept, the auctioneer raises the price to $450. In the second stage, Robby's dominant strategy is *pass* since accepting would result in a payoff of $442–$450 = –$8. Since Robby passes, Molly's dominant strategy is *accept* since her payoff is $562–$450 = $112. Since Robby passes and Molly accepts, the auction ends. The subgame perfect equilibrium strategy profile for this game when Molly moves first is {*Accept* → *Accept* → *Accept* → *Pass*} with payoffs of ($112, $0). Molly will win the auction and pay $450.

As with sealed-bid, first-price auctions, the winner of this auction is the bidder who values the object the most and submits a bid that is higher by the minimum allowable bidding increment than the value placed on the object by the next-highest bidder. If the bidding increment is made arbitrarily small, Molly will win the auction, pay $442 for the antique pocket watch, and enjoy a positive net payoff of $562 – $442 = $120.

Principle: In an English auction with complete and perfect information, the winner is the player who values the object most and submits a bid that is higher by the minimum allowable bidding increment than the value of the object to the next-highest bidder. If there are no bidding increments, the winner pays what the object is valued by the loser.

Dutch Auction

How do the above conclusions change with a Dutch auction format? Recall that in a Dutch auction, the auctioneer begins by announcing a very high opening bid. If no bidder accepts this price, the auctioneer incrementally lowers the price. At each announced price, the players must decide whether to *accept* or *pass.* The first bidder to *accept* wins and the auction comes to an end. If both bidders *pass,* the winner at the lower price is determined by the flip of a fair coin.

Once again, Molly and Robby value the pocket watch at $562 and $442, respectively, and these valuations are common knowledge. The auctioneer begins by announcing an opening bid of $550. If neither player accepts, the auctioneer lowers the price by $50. This game is modeled as a multistage game in which a new subgame begins whenever the auctioneer lowers the price by the bidding increment. The game tree for this auction is depicted in Figure 15.4.

Each bidder is expected to accept the current asking price if winning the auction yields a positive net payoff. At the reservation price, Molly will receive a positive net payoff of $562 – $550 = $12 if Robby passes, and an expected net payoff of 0.5($12) = $6 if Robby accepts. Robby will pass on the opening bid, however, since he will receive an expected net payoff of –$54 if Molly accepts, and a net payoff of –$108 if she passes. Even though Molly will receive a positive net payoff at an asking price of $550, however, she will *pass* because she knows how much Robby values the antique pocket watch. Moreover, Robby will continue to pass until the price falls below $450. Molly knows this and will continue to pass until the price reaches $450, at which point the auction comes to an end.

To see this, suppose that both Molly and Robby pass at a price of $450. In this case, the price is lowered to $400. If Molly and Robby accept, the net payoffs are $76 and $42, respectively. Molly will preempt this outcome by accepting a price of $450 because it will yield a higher payoff of $112. The subgame perfect equilibrium for this game in which Molly moves first is {*Pass →
Pass → Pass → Pass → Accept → Pass*}. Molly will win the auction, pay $450 for the antique pocket watch, and enjoy a positive net benefit of $112. It is left as an exercise for the reader to show that the payoffs for this Dutch auction using backward induction are (0, $112), which are highlighted in boldface.

As in the auctions already discussed, the winner of the auction is the player who values the object most. If the bidding increments shrink to zero, the winner will pay a price equal to the true value of the object to the loser, which is Robby's valuation of $442. Molly will again enjoy a positive net payoff of $120. Thus, a Dutch auction and a sealed-bid, first-price auction are strategically equivalent.

> ***Principle:*** In a Dutch auction with complete and perfect information, the winner is the player who values the object most and submits a bid that is higher by the minimum allowable bidding increment than the value placed on the object by the next-highest bidder. If the bidding increment is zero, the winner pays what the object is valued by the loser. A Dutch auction is strategically equivalent to a sealed-bid, first-price auction.

In each of the four auction formats examined (sealed-bid, first-price; sealed-bid, second-price; English and Dutch auctions) in which player valuations are common knowledge, the winner is the bidder who values the object being auctioned the most and pays what the object is valued by the next-highest bidder. Where the auctions differ is in their truth-revelation properties. In each of these auctions, the loser submits a bid that is equal to his or her true value of the object. Only in a sealed-bid, second-price auction will all players submit a bid that is equal to his or her true value of the object. In every other case, the winning bid submitted will be less than the true value.

AUCTIONS WITH INDEPENDENT PRIVATE VALUES

So what have we learned thus far? Not much, really. Our analysis has led to the rather mundane conclusion that in auctions where the participants have full knowledge of each other's valuations,

FIGURE 15.4
Dutch Auction of an Antique Pocket Watch

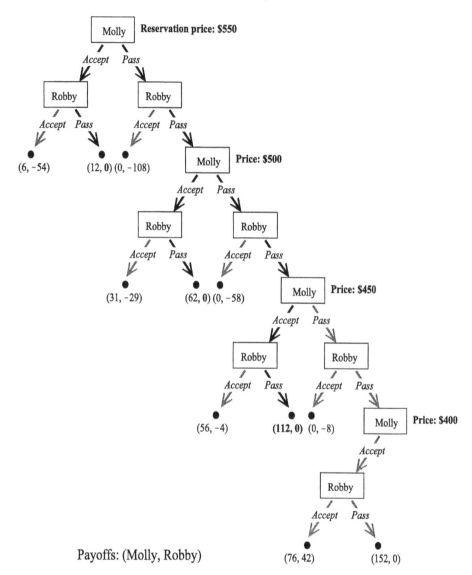

Payoffs: (Molly, Robby)

the bidder who wants the object most wins the auction. Moreover, the winner will pay an amount that is just sufficient to acquire the object being auctioned—no more, no less. In our example, we assumed that Robby valued the antique pocket watch at $442, and Molly knows this. If the increment between bids is 1 cent then Molly will bid $442.01.

Since bidders are not likely to have complete information, why all the fuss? The answer to this question is that we are now in a better position to understand more realistic auction scenarios. We begin by dropping the assumption that the players' valuations are common knowledge. Instead, we will assume that each player knows his or her own valuation of the object, but does not know the other bidders' valuations. In other words, we will assume that the bidders have **independent private values**.

> **Independent private values** When players know their own valuation of an object being auctioned, but not the other bidders' valuations.

Private values are determined by a player's tastes, preferences, income, wealth, and other demand determinants. These private values are assumed to be independent in the sense that each bidder's valuation neither affects, nor is affected by, the private valuations of the other bidders. Even if a player discovers how much the object is worth to a rival bidder, this will not affect his or her valuation of the object, although it might affect a player's bidding strategy.

We will use a Harsanyi transformation (see Chapter 13) to convert a game with incomplete information into a game with complete, but imperfect, information. We can do this by treating bidders with different valuations as distinct types. As before, Nature moves first by randomly determining the probability distribution of player types, which is common knowledge to all bidders. Bidding strategies are no longer based on maximizing certain payoffs, but on maximizing expected payoffs.

Sealed-Bid, First-Price Auction

In a sealed-bid, first-price auction with perfect information, the winner submits a bid that is incrementally higher than the value placed on the object by the second-highest bidder. This strategy is only possible when the bidders' private valuations are common knowledge. In the case of independent private values, however, the players' valuations are known only to themselves. Of course, each player has an incentive to submit a bid that is less than what he or she believes the object is worth. The lower the bid, the higher the expected net payoff, but the lower the probability of winning. The higher the bid, the greater the probability of winning the auction, but the lower the expected net payoff. So what is the bidder's optimal strategy when private valuations are not common knowledge? If we assume that players are risk neutral, the answer depends on the number of bidders and the distribution of private valuations. As we will see, the more competitive the auction (i.e., the greater the number of bidders), the closer each bid will be to the bidder's private valuation of the object.

To get a better idea of what is going on, consider the sealed-bid, first-price auction for the antique pocket watch discussed above. Recall that Molly values the antique pocket watch at $V_M = \$562$, while Robby values it at $V_R = \$442$. Although a rival's private valuation is unknown, all bidders assume these valuations are independent, random, and uniformly distributed. Moreover, in the antique pocket watch example, we will assume that each player believes that the lowest possible price is $L = \$200$ and that the highest possible price is $H = \$800$. This information may be arrived at in several ways, such as canvassing disinterested experts in antique pocket watches, or by surveying Internet sites to identify high and low selling prices. If both players are risk neutral, how much should Molly and Robby bid for the pocket watch? Each player's dominant bidding strategy is given by the equation:

$$b_i^* (V_i) = V_i - \frac{V_i - L}{n} \qquad (15.1)$$

In Equation (15.1), b_i^* is each player's optimal bid, V_i is the valuation placed on the object by bidder i, L is the lowest valuation by all bidders, and n is the number of bidders. This bidding strategy results in a pure strategy Nash equilibrium. Substituting the above information into Equation (15.1), Molly's optimal bid is:

$$b_M^* (V_M) = \$562 - \frac{\$562 - \$200}{2} = \$381 \qquad (15.2)$$

Robby's optimal bid is:

$$b_R^*(V_R) = \$442 - \frac{\$442 - \$200}{2} = \$321 \qquad (15.3)$$

In this example, Molly and Robby submit bids that are strictly less than what each believes to be the value of the antique pocket watch. Equation (15.1), and the above example, leads us to our first general principle of sealed-bid, first-price auctions with incomplete information.

Principle: In a sealed-bid, first-price auction with independent private values, risk-neutral bidders will submit bids that are strictly below what each believes the object is worth.

Suppose that a third bidder, Jeremy, were to participate in this auction. Assume that Jeremy's independent private valuation of the pocket watch is $V_J = \$386$. Each bidder's optimal bid is now:

$$b_M^*(V_M) = \$562 - \frac{\$562 - \$200}{3} = \$441.33 \qquad (15.4)$$

$$b_R^*(V_R) = \$442 - \frac{\$442 - \$200}{3} = \$361.33 \qquad (15.5)$$

$$b_J^*(V_J) = \$386 - \frac{\$386 - \$200}{3} = \$324.00 \qquad (15.6)$$

While each player submits a bid that is strictly below his or her private valuation of the pocket watch, each player's bid will increase with the number of bidders. From Equation (15.1), as the number of bidders approaches infinity ($n \to \infty$), each price will converge to the bidder's true value.

Principle: In a sealed-bid, first-price auction with independent private values, the bid of each risk-neutral player will increase and converge to his or her true value as the number of bidders increases.

——————————————— **Demonstration Problem 15.1** ———————————————

Suppose that three collectors, Tom, Dick, and Harry are bidding for a Wayne Gretsky rookie hockey card. Each bidder knows how much he values the card, but do not know the other bidders' valuations. The values placed on the card by Tom, Dick, and Harry are $V_T = \$250$, $V_D = \$170$, and $V_H = \$157.50$, respectively. Based on the opinion of disinterested experts, the lowest possible value of the card is $L = \$100$ and the highest possible value is $H = \$300$.

    a. If player valuations are independent, random, and uniformly distributed, how much should each risk-neutral player bid for the Wayne Gretsky rookie hockey card in a sealed-bid, first-price auction?

    b. If this is a Dutch auction, which player wins?

    c. Who wins if this is a sealed-bid, second-price auction?

Solution

a. The bidders' optimal bids, which may be calculated using Equation (15.1), are:

$$b_T^*(V_T) = \$250 - \frac{\$250 - \$100}{3} = \$200 \qquad (D15.1.1)$$

$$b_D^*(V_D) = \$170 - \frac{\$170 - \$100}{3} = \$146.67 \qquad (D15.1.2)$$

$$b_H^*(V_H) = \$157.50 - \frac{\$157.50 - \$100}{3} = \$138.33 \qquad (D15.1.3)$$

In a sealed-bid, first-price auction in which risk-neutral bidders have independent private values, the winner is the bidder who values the object most, but will submit a bid that is less than he or she each believes the object is worth. In this case, Tom values the object most at $250, but submits a winning bid of $200.

b. Since a Dutch auction is strategically equivalent to a sealed-bid, first-price auction, Tom still wins the Wayne Gretsky rookie card by submitting a bid of $200.

c. If this is a sealed-bid, second-price auction, each player should submit a bid that is equal to each player's true value of the rookie card. In this case, Tom will win with a bid of $250, but will pay $170.

Sealed-Bid, Second-Price Auction

Recall from our discussion of sealed-bid, second-price auctions with complete information that the winner is the player who values the object most, but pays what the object is worth to the next-highest bidder. The dominant bidding strategy for each player is to submit a bid that reflects his or her private valuation of the object being auctioned. How is this conclusion altered when we assume independent private values? Surprisingly, not at all! To see this, suppose that the antique pocket watch is sold using a sealed-bid, second-price auction with independent private values and that Molly submits a bid that is greater than her private value of the pocket watch. If Robby's bid is less than Molly's, this strategy yields no additional net benefit since she pays Robby's bid. Molly would have won anyway by bidding her higher private value. On the other hand, suppose that Robby's bid is greater than Molly's private value. If Molly wins by submitting a bid that is greater than her valuation of the pocket watch, her net payoff will be negative. Molly would have lost anyway by submitting a bid equal to her true valuation, and her net payoff would have been zero. Finally, suppose Robby's bid is equal to Molly's private value. In this case, it does not matter how Molly bids. If she bids high then she wins, but her net payoff is negative. If she bids low then she loses, and her net payoff is again zero. If she bids her private value, Molly's net payoff is zero, regardless of whether she wins or loses.

The above discussion makes it clear that Molly's dominant strategy is to submit a bid that is equal to her valuation of the pocket watch. The same is also true for Robby. As the above discussion makes clear, this result does not depend on the assumption of independent private valuations. What matters is that both players know their own valuations of the object being auctioned.

Principle: In a sealed-bid, second-price auction with independent private values, the dominant bidding strategy for a risk-neutral player is to submit a bid that is equal to his or her private valuation of the object.

English Auction

We shall now consider the effect that independent private values have on the outcome of an English auction. Recall that in an English auction, prices are announced in ascending order. A bidder will continue to participate until the announced price reaches his or her dropout price. The winner of the auction is the last bidder standing. The net payoff to the losing bidder is zero, while the net payoff to the winner is his or her valuation of the object less the second-highest bidder's dropout price. As in the case of an English auction with complete information discussed earlier, the dominant bidding strategy and net payoffs with independent private values are equivalent to those for a sealed-bid, second-price auction.

> *Principle:*   In an English auction with independent private values, the dominant bidding strategy for a risk-neutral player is to continue to bid until the price exceeds his or her valuation of the object.

Dutch Auction

Recall again that in a Dutch auction with complete information, the auctioneer begins with a very high price, which is incrementally lowered. The first bidder to accept the announced price is declared the winner, and the auction comes to an end. We saw that there was no strategic difference between this and a sealed-bid, first-price auction with complete information. As it turns out, the same is also true with uniformly distributed independent private values. The dominant bidding strategy for each risk-averse player is to submit a bid that satisfies Equation (15.1).

> *Principle:*   A Dutch auction with independent private values and risk-neutral bidders is strategically equivalent to a sealed-bid, first-price auction.

CASE STUDY 15.3: GOOGLE GOES DUTCH

On August 19, 2004, privately owned Google, the massive Internet search engine company, went public by selling 111.6 million shares at $85 per share using a multiple-item Dutch auction. The initial public offering (IPO) raised $1.67 billion, which was down from a projected $3.12 billion on sales of 25.6 million shares. The decision to go public using a multiple-item Dutch auction was meant to bypass the usual IPO process.

In a classic IPO, an investment bank estimates the market value of a company and determines investor interest at a recommended share price. At one time, this procedure was used to generate as much investment capital as possible for the company going public. Beginning in the mid-1990s, however, the success of IPOs began to be measured in terms of the increase in share prices on the first day of trading in the secondary market. By pricing the shares below market value, the swarm of investors would significantly drive up share prices. This procedure short-changed company owners and made big winners of investment bankers and privileged underwriters who got rich by flipping shares on the first day of public trading.

Dutch auctions curtail the ability of investment bankers to engage is such chicanery. In a multiple-item Dutch auction, prospective buyers submit sealed bids, usually within a specified price range for blocks of identical items. (The Google auction originally specified a price range of $108 to $135, which was later reduced to $85 to $95 per share as market conditions

TABLE C15.1
Multiple Item Dutch Auction for XYZ Shares

Investor	Bid per share	Shares	Shares remaining
A	$10,000	1000	999,000
B	$100	250,000	749,000
C	$50	150,000	599,000
D	$30	500,000	99,000
E	$25	100,000	−1,000
F	$24.99	50,000	0
⋮	⋮	⋮	⋮

weakened.) Unless the company specifies an acceptable price range, a bidder who submits an insanely high price for a block of shares will receive them, but will end up paying the lowest price that disposes of the entire offering. To illustrate, suppose that XYZ company announces that it will accept block bids on a million shares using a multiple-item Dutch auction. The highest bids and the number of shares desired are summarized in Table C15.1.

In the above example, the lowest bidder for whom shares were remaining is investor E, who offered to purchase 100,000 shares at $25 per share. Every investor who offered more than $25 is guaranteed to receive shares, but will pay the bid submitted by investor E. Investor E, on the other hand, does not receive the desired amount of shares, but rather the number of shares remaining (99,000) after all higher bidders have been satisfied. While bidding high will guarantee that an offer will be accepted, bidding too low runs the risk of not receiving the desired number of shares, or not receiving any shares at all. This was the case with investor F who bid just 1 cent below the bid submitted by investor E.

The reader may well ask whether submitting a very high bid constitutes an optimal bidding strategy. It does not. Bidding too high may result in substantial losses when the shares are sold in the secondary market. In principle, the optimal strategy in a multiple-item Dutch auction involving thousands of bidders is to establish a price range for the items placed at auction. This price range should reflect the reasoning of thousands of investors who have assessed the market value of XYZ company shares. To illustrate, suppose that this price range is $20 to $30 per share. If these valuations are independent, random, and uniformly distributed, Equation (15.1) can be used by each bidder to determine a dominant bidding strategy. If there is a very large number of bidders, an optimal bidding strategy is for each bidder to submit a bid equal to his or her private valuation of the shares. The final share price will equal the lowest valuation in this price range, unless the offer is fully subscribed before the lower limit is reached.

The above discussion suggests that Google's specified price range of $85 to $95 per share was not low enough to guarantee that the IPO would be fully subscribed. This begs the question: Why didn't Google lower its price range to rid itself of all 25.6 million shares? Perhaps Google believed it would earn more by selling the remaining 6 million shares in the secondary market because the demand for Google shares below $85 was price inelastic (i.e., a percentage increase in shares sold is less than the percentage decline in the share price). If so, lowering its share price would have resulted in a decline in the total value of Google's IPO.

AUCTIONS WITH CORRELATED VALUE ESTIMATES

Thus far, we have assumed that regardless of the auction environment, bidders were only concerned with their own valuation of the object. The value placed on the object by rival bidders played no role in a player's bidding strategy. It is very often the case, however, that private values of the object are correlated. **Correlated value estimates** means that a player's private value estimate reflects his or her information about the object being auctioned. Moreover, changes in estimated private value estimates affect, and are affected by, the changes in the private value estimates of rival bidders. In the antique pocket watch example, Molly may covet the antique pocket watch even more if she believes that Robby wants it more than she originally thought. In other words, Molly derives utility from the perceived prestige of owning something that somebody else wants.

> **Correlated value estimates** When a private value estimate reflects a player's information about an object being auctioned, and where changes in this estimate affect, and are affected by, changes in the private value estimates of rival bidders.

With independent private values, what information would cause Molly to raise or lower her private value of the object? Molly might acquire information from the bidding process about rival bidders' interest in the object. Molly might interpret a rival's high bid as a reflection of superior information about the quality of the watch. In an English auction, Molly might also interpret intense bidding activity as a sign that the antique pocket watch is more valuable than she initially believed. Alternatively, lackluster bidding activity might lead Molly to downgrade her valuation of the pocket watch.

How are bidding strategies affected when bids are correlated, or when bidders are uncertain about the true value of the object being auctioned? The answer depends upon the type of auction, and whether the bidder is able to act on any information received. New information about the value of the object being auctioned is of little or no value once the bidding has ended. Optimal bidding strategies require that new information not only be used to update private valuations, but that this information be acted upon during the bidding process.

COMMON-VALUE AUCTIONS AND THE WINNER'S CURSE

A special case of correlated value estimates is the **common-value auction**. In this case, differences in private valuations stem not from differences in player types, but from different estimates of the actual value of the object being auctioned. It is called a common-value auction because the payoff has a true, but unknown, value. To illustrate, suppose that the U.S. Department of the Interior announces its intention to use a sealed-bid, first-price auction to sell oil-drilling rights in Yosemite National Park. The problem is that the actual value of these oil reserves is unknown.

> **Common-value auction** An auction in which the true value of the object is the same for all bidders, but this value is unknown. Players' private value estimates reflect differences in information about the object's true value.

Before participating in this auction, the oil companies conduct independent seismic and geological tests to obtain an estimate of the quality and quantity of the oil reserves. Suppose that these estimates are independent and randomly distributed. In a sealed-bid, first-price auction, the oil company submitting the most optimistic estimate will be awarded drilling rights. This gives rise to a phenomenon of common-value auctions called the **winner's curse**.

> **Winner's curse** When a player wins a common-value auction with a bid that exceeds every other bidder's estimate of the value of the object being auctioned.

To illustrate the winner's curse, suppose that twenty oil companies participate in the auction for oil-drilling rights. Each company will submit a bid based on its individual estimate of the value of the oil in the ground. The company with the most optimistic estimate will submit the highest bid and win the auction. But, this means that the nineteen other bidders believe that the oil reserves are worth less than the winning bid. The winner of the auction must come to grips with the rather unpleasant realization that he or she probably paid more than the reserves are actually worth. How might the oil companies deal with the problem of the winner's curse?

> ***Principle:*** In a common-value auction, the winner has the most optimistic estimate of the object's value.

Suppose that the senior geologist for the Petroleum Exploration Company (PETROX) estimates that the crude oil reserve is worth $100 million. It is commonly believed that the minimum possible value of the drilling rights is $20 million. Rather than submit a bid equal to their estimate, the CEO of PETROX uses Equation (15.1) to submit a bid of $96 million. If PETROX wins the auction, the net payoff will be $100 – $96 = $4 million. Of course, PETROX may not win, but in a common-value auction, a bidder's dominant strategy is to avoid the winner's curse by submitting a bid lower than his or her private estimate of the object being auctioned.

> ***Principle:*** To avoid the winner's curse in a common-value auction, bidders must submit a bid that is lower than their private estimates of the value of the object being auctioned.

The winner's curse is most pronounced in sealed-bid, first-price auctions because it is not possible to learn anything about the other bidders' private valuations until after the bids have been submitted. By contrast, players may be able to infer something about their rivals' private value estimates during the course of a multistage auction or a static auction with multiple rounds. If so, the bidders will be able to modify their bidding strategies accordingly. During an English auction, for example, the player's optimal strategy is to continue bidding if the current price is less than his or her private value estimate, which is based on the bidder's private information and information gleaned from the behavior of rivals during the bidding process.

We will now consider a variation of the winner's curse that involves any situation in which the acquisition of additional information is critical to formulating strategies based on expected payoffs. First proposed by Bazerman and Samuelson (1983), this example is a variation of the so-called *lemons problem* (see Akerlof 1970). It involves the owner of a company who is planning to make a one-time, final offer to purchase a privately held company in a related industry. If the takeover is successful, the value of this company to the buyer will be worth at least 50 percent more than to the current owner because of substantial synergies (economies of scale). If the market value of the company's assets is common knowledge, both parties should be able to negotiate a mutually beneficial transaction price. The problem is that the buyer does not have complete information about the market value of the privately held company. This is an example of a situation involving *asymmetric information* because the seller knows more than the buyer about the value of the object being transacted.

To make this example more concrete, suppose that the buyer believes that the company is worth something between $0 and $100 million. By contrast, the current owner knows that the market value of the company is $50 million. Both parties agree, however, that the company's assets are worth 50 percent more to the prospective buyer. The expected value of the company to the buyer is 1.5[($0 + $100)/2] = $75 million. What is the buyer's best one-time offer?

Let us assume that the owner will accept any offer that is equal to or greater than the company's true market value. Suppose that the buyer makes an offer of $50 million. Since the owner will accept this offer, the buyer will reason that the owner values the company's assets at $50/1.5 = $33.33 million. Recall that it is common knowledge that the buyer values the assets 50 percent more than the owner. By accepting the offer, the buyer will reason that it has incurred an expected loss of $50 − $33.33 = $16.67 million. The buyer has fallen prey to the winner's curse. Using Equation (15.1), the only offer that the buyer can make to completely avoid the winner's curse is $0, although this is the result of a rather unrealistic assumption about the lowest market value of the company.

The foregoing discussion is an admittedly extreme case since the sale will not occur even though both parties benefit from the transaction. As with many noncooperative games, the outcome does not depend on the rules of this game, which in the above example is a take-it-or-leave-it offer. The possibility of a winner's curse does not necessarily rule out a mutually beneficial outcome. The importance of the above example is that it forces the players to consider a rival's incentives and to revise strategies as new information is gleaned during the course of the game.

AUCTIONS AND RISK-AVERSE BEHAVIOR

In each of the above auction scenarios, we assumed that players were risk neutral. In what way are the above conclusions affected if we assume that bidders are risk averse? It turns out that a player's attitude toward risk will affect some bidding strategies, but not others. In the case of an English auction with independent private values, attitudes toward risk play no role. A bidder will continue to participate until the announced price exceeds his or her private value of the object. Once the price exceeds this value, the bidder will drop out. As before, the winner values the object the most, and pays what the object is worth to the next-highest bidder. There is also no change in bidding strategies with a sealed-bid, second-price auction. A risk-averse player will submit a bid that is equal to his or her true value of the object being auctioned. The winner is the player who values the object most, and pays what the object is worth to the next-highest bidder.

By contrast, in a sealed-bid, first-price auction with independent private values, a risk-averse player will submit a bid that is higher than the bid submitted by a risk-neutral bidder. To see why, recall that a risk-neutral player will submit a bid that is somewhat lower than his or her true value of the object. How much lower depends on the lowest valuation of the object and the number of bidders. While this lower bid increases the possibility of losing the auction, it also increases the net payoff should the bidder win. A risk-averse player will want to lower the probability of losing by submitting a bid that is lower than his or her true value of the object, but higher than the bid submitted by a risk-neutral bidder. How much lower depends upon the degree of risk aversion. As a result, a seller's expected revenues from sealed-bid, first-price and Dutch auctions will be greater than for sealed-bid, second-price and English auctions.

What about auctions involving correlated value estimates? Recall that in these auctions the information revealed in an English auction reduces the winner's curse. Because of this, risk-averse players will submit bids that are somewhat higher in an English auction than in a second-price auction. Thus, the sellers' expected revenues will be greater in an English auction than in a sealed-bid, second-price auction, and may be greater than in sealed-bid, first-price and Dutch auctions.

REVELATION PRINCIPLE

Among the most significant contributions of game theory in recent years has been in the area of game design. This has been especially true of auctions where the seller would like to maximize expected payoffs. The practical significance of game design became abundantly clear in 1994 when the U.S. government consulted with game theorists and auction experts to design an auction to sell narrow-band frequency licenses to provide mobile telecommunication services. (See Case Study 15.1.) The resulting auction generated in excess of $600 million, which was ten times more than the most optimistic pre-auction estimate.

Mechanism design refers to that branch of game theory that deals with the design of games having certain desirable properties. In this section we shall briefly discuss one of the most important analytical tools for designing games—the **revelation principle**. The revelation principle, which can be traced to the pioneering work of Roger Myerson (1981), allows the game theorist to break down the problem of game design into two manageable steps. In the first step, the game theorist focuses on a simple subset of Bayesian games called **incentive-compatible direct mechanisms**, or ICDMs. A Bayesian game is a **direct mechanism** if players can report their types to a neutral "referee." This information is then used to determine the players' payoffs. This direct mechanism is **incentive compatible** if truthful reports of player types result in a Bayesian Nash equilibrium. From this smaller set of Bayesian games, the designer need only select those games that have certain desirable properties.

While the first step requires the participation of a neutral "referee," the second step is to find a game that does not require a referee, but that has the same Bayesian Nash equilibrium as an ICDM. The revelation principle asserts that for any player-type profile, the Bayesian Nash equilibrium of a static Bayesian game can be represented by a properly constructed ICDM. This result is known as the **revelation theorem**. The reader is referred to Myerson (1981) for a formal proof of this theorem.

> **Mechanism design** A branch of game theory that deals with the design of games having certain desirable properties.
>
> **Incentive-compatible direct mechanism** A Bayesian game is a *direct mechanism* if players can report their types to a neutral referee. This information is used to determine the players' payoffs. This direct mechanism is *incentive compatible* if truthful reports of player types result in a Bayesian Nash equilibrium.
>
> **Revelation theorem** For any player type, the Bayesian Nash equilibrium of a static Bayesian game can be represented by a properly constructed incentive-compatible direct mechanism.

CHAPTER REVIEW

An auction is a process whereby individuals or institutions compete for the right to purchase or sell anything of value. In a *standard auction,* multiple bidders attempt to acquire an object of value from a single seller. The bidder who submits the highest bid wins the auction and acquires ownership of the object. A *procurement (reverse) auction* reverses the roles of buyer and seller in a standard auction. In a procurement auction, the lowest bid wins.

Auction rules refer to the procedures that govern the bidding process, including who may bid, which bids are acceptable, the manner in which bids are submitted, the nature of the information available to the players during the bidding process, when the auction ends, how the winner is determined, and the price paid to the seller. An *auction environment* refers to the characteristics of the auction participants, including the number of bidders, risk preferences, how rival bidders value the object being auctioned, and so forth.

In an *open-bid auction,* anyone may submit a bid. In a *closed-bid auction,* participation is by invitation only. A *reservation price* is the minimum price that a seller will accept in a standard auction. In a *sealed-bid auction,* bids are submitted in secret to the auctioneer. After all bids are received, the auctioneer declares a winner. In an *oral auction,* bids are announced publicly. The primary difference is that players are more easily identifiable in an oral auction.

In a sealed-bid, first-price auction with complete information, the winner is the player who values the object most and submits a bid that is higher by the minimum allowable bidding increment than the value placed on the object by the next-highest bidder. If there are no bidding increments, the winner pays what the object is worth to the loser.

In a sealed-bid, second-price auction with complete information, the winner is the player who values the object most, submits a bid that is lower by the minimum allowable bidding increment than his or her value of the object, but pays the bid submitted by the next-highest bidder. If there are no bidding increments, the dominant strategy for each player is to submit a bid that is equal to his or her true value of the object.

There are two main types of oral auctions. In a standard *English auction* (also called an *open ascending-bid auction*) the auctioneer may begin by announcing an opening bid (reservation price). Bidders must decide whether to accept, or pass, on the announced price. If two or more players accept, the auctioneer raises the price. The last player left standing wins the auction and pays the highest bid. In a standard *Dutch auction* (also called an *open descending-bid auction*) the auctioneer begins by announcing a very high opening bid. As this price is incrementally lowered, players simultaneously decide whether to accept or pass. The first bidder to accept the price is declared the winner and the auction comes to an end.

In an English auction with perfect information, the winner is the player who values the object most and submits a bid that is higher by the minimum allowable bidding increment than the value of the object to the next-highest bidder. If there are no bidding increments, the winner pays what the object is valued by the loser.

In a Dutch auction with perfect information, the winner is the player who values the object most and submits a bid that is higher by the minimum allowable bidding increment than the value placed on the object by the next-highest bidder. If the bidding increment is zero, the winner pays what the object is valued by the loser. A Dutch auction is strategically equivalent to a sealed-bid, first-price auction.

In auctions in which the players know their own true value of the object, but not rival bidders' true value, are said to have *independent private values.* In a sealed-bid, first-price auction with independent private values, risk-neutral bidders will submit bids that are strictly below what each believes is the value of the object. These bids will increase and converge to their true values as the number of bidders increases. In a sealed-bid, second-price auction with independent private values, the dominant bidding strategy for a risk neutral player is to submit a bid that is equal to his or her value of the object.

In an English auction with independent private values, the dominant bidding strategy for a risk-neutral player is to continue bidding until the price exceeds his or her true value of the object. A Dutch auction with independent private values and risk-neutral bidders is strategically equivalent to a sealed-bid, first-price auction.

Correlated value estimates refers to auctions in which a bidder's private value estimate reflects his or her information about the object being auctioned, and where changes in estimated private value estimates affect, and are affected by, the changes in the private value estimates of rival bidders.

In a *common-value auction,* the actual value of the object is the same for all bidders, but this value is unknown. Players' private value estimates reflect differences in information about the object's actual value. Common-value auctions are plagued by the *winner's curse.* The bidder with the most optimistic estimate of the actual value of the object wins, but the winning bid exceeds the estimate of every other bidder.

A bidder's attitude toward risk affects some bidding strategies, but not others. In an English auction with independent private values, attitudes toward risk play no role. A bidder will continue to participate until the announced price exceeds his or her private value of the object. There is also no change in bidding strategies with a sealed-bid, second-price auction. A risk-averse player will submit a bid that is equal to his or her true value of the object being auctioned.

By contrast, in a sealed-bid, first-price auction with independent private values, a risk-averse bidder will submit a bid that is higher than the bid submitted by a risk-neutral bidder. How much lower depends on the lowest valuation of the object and the number of bidders. This lower bid increases the possibility of losing the auction, but increases the net payoff should the bidder win. How much lower depends upon the degree of risk averseness. In auctions with correlated value estimates, risk-averse players will submit bids that are somewhat higher in an English auction than in a second-price auction.

Among the most significant contributions of game theory in recent years has been in the area of auction design. *Mechanism design* refers to that branch of game theory that deals with the design of games having certain desirable properties. One of the most important analytical tools for designing games is the *revelation principle,* which asserts that for any player-type profile, the Bayesian Nash equilibrium of a static Bayesian game can be represented by properly constructed *incentive-compatible direct mechanisms.*

CHAPTER QUESTIONS

15.1   Explain each of the following types of auctions:

   a.   Standard auction
   b.   Procurement auction
   c.   Open bid auction
   d.   Closed bid auction
   e.   Sealed-bid auction
   f.   Oral auction
   g.   First-price auction
   h.   Second-price auction
   i.   Standard English auction
   j.   Standard Dutch auction
   k   Multiple item Dutch auction

15.2   Who is the winner in each of the following auctions involving perfect information? What price will the winner pay for the object being auctioned if the bidding increments are arbitrarily small?

   a.   Sealed-bid, first-price auction
   b.   Sealed-bid, second-price auction
   c.   Standard English auction
   d.   Standard Dutch auction

15.3   Who is the winner in each of the following auctions if bidders have independent private values? What price will the winner pay for the object being auctioned if bidding increments are arbitrarily small?

   a.   Sealed-bid, first-price auction
   b.   Sealed-bid, second-price auction
   c.   Standard English auction
   d.   Standard Dutch auction

15.4   In a sealed-bid, first-price auction with perfect information, the winner is the bidder who submits the highest bid and pays a price equal to his or her valuation of the object being auctioned. Do you agree? Explain.

15.5 In a sealed-bid, second-price auction with complete information, the winner is the bidder who submits the second-highest price, but pays the price submitted by the highest bidder. Do you agree? Explain.

15.6 Explain why a player in a sealed-bid, second-price auction would never submit a bid that exceeds his or her true value of the object being sold. (Hint: What if all players submitted bids greater than their valuations of the object?)

15.7 In a Dutch auction with perfect information, the winner is the bidder who values the object the most and pays his or her value of the object. Do you agree? Explain.

15.8 What is the difference between auctions with independent private value estimates and auctions with correlated value estimates?

15.9 In a sealed-bid, first-price auction with independent private values, risk-neutral bidders will submit bids that are higher than what each believes the object is worth. Do you agree? Explain.

15.10 In a sealed-bid, first-price auction with independent private values, as the number of risk-neutral bidders increases, each bid will be lowered, eventually converging to the next-highest bidder's valuation of the object being auctioned. Do you agree? Explain.

15.11 In a sealed-bid, second-price auction with independent private values, risk-neutral bidders should submit bids that are equal to what each believes the object is worth to the next-highest bidder. Do you agree? Explain.

15.12 A Dutch auction with independent private values and risk-neutral bidders is strategically equivalent to a sealed-bid, second-price auction. Do you agree? Explain.

15.13 What is a common-value auction? Explain what is meant by the "winner's curse."

15.14 In what way do auction outcomes differ if the winners are risk averse instead of risk neutral?

15.15 What does the revelation principle tell us about game design?

15.16 Why would a company going public using a multiple-item Dutch auction begin by specifying a price range for investor bids?

CHAPTER EXERCISES

15.1 At the Hemlock Bush tavern Jethro (Jellyroll) Bottom announces that he will auction off an envelope containing $35.00. Clem and Heathcliff are the only two bidders and each has $40 with which to bid. The rules of the auction are as follows:

(1) The bidders take turns. After a bid is made, the next bidder can make either another bid, or pass. The opening bid must be $10.00.
(2) After the opening bid, each bid must be in increments of $10.
(3) Bidders cannot bid against themselves.
(4) The bidding comes to an end when either bidder passes, except on the first bid. If the first bidder passes, the second bidder is given the option of accepting the bid.
(5) The highest bidder wins.
(6) All bidders must pay Jethro the amount of their last bid.
(7) Assume that Clem bids first.

    a. Illustrate this auction with a game tree.

    b. Determine the subgame perfect equilibrium for this game using the backward induction solution concept.

    c. What is the outcome of this auction?

15.2   Suppose that Ted and Bill are submitting sealed bids for a pair of tickets to a sold-out basketball game between the Orange of Syracuse University and the Wildcats of the University of Arizona. Earlier in the day, Ted overheard Bill tell his girlfriend that he is willing to pay as much as $212. The previous night, Bill, who is dating Ted's sister, learned that Ted would be willing to pay as much as $224. In this auction, there is a reservation price of $200 and bids must be submitted in $5 increments. Although the winner of this auction submits the highest bid, he will pay the bid submitted by the loser. If Ted and Bill submit the same bid, the winner will be determined by the flip of a fair coin.

    a. Illustrate this auction as a noncooperative, simultaneous-move, one-time game.

    b. Does Ted or Bill have a dominant strategy?

    c. What is the solution profile for this game? Who will win the auction, and how much will he pay?

    d. Suppose that the bidding increments were $0.01. Who will win the auction and how much will he pay?

15.3   The U.S. Department of the Interior has announced its intention to use a sealed-bid, first-price auction to sell petroleum-drilling rights on public land in Alaska. Four oil companies have indicated their intention to participate in this auction. The problem is that the actual value of the oil reserves is unknown with certainty. Before submitting their bids, each oil company conducted independent seismic and geological tests to obtain an estimate of the quality and quantity of the petroleum reserves. The petroleum-deposit value estimates in millions of dollars of the four companies are $V_1 = \$250$, $V_2 = \$170$, $V_3 = \$320$, and $V_4 = \$265$. Although the companies do not know their rivals' estimates, they believe that these valuations are independent and randomly distributed between a low value of $L = \$100$ and a high possible price of $H = \$400$.

    a. What is the optimal bid of each oil company?

    b. Who will win the auction and what price will the winner pay for the drilling rights?

    c. Do you believe that the winner of this auction paid too much for the drilling rights? Why?

15.4   Consider a sealed-bid, first-price auction in which the players have independent private values. The general consensus is that the true value of the object being sold is uniformly distributed between $1 and $10.

    a. If the true value of a risk-neutral player is $2, what is this player's optimal bidding strategy if there are only two bidders?

    b. If this is a Dutch auction, what is the player's optimal bidding strategy if there are three bidders?

    c. If this is a sealed-bid, second-price auction, what is a player's optimal bidding strategy?

15.5   Consider an independent private value auction with twenty risk-neutral players. The players' valuations are generally believed to be uniformly distributed between $0 and $50,000. Suppose that the true private value of one player is $40,000. For each of the following auction formats, what is the player's optimal bidding strategy?

a. Sealed-bid, first-price
b. Sealed-bid, second-price
c. English
d. Dutch

CHAPTER

16 DYNAMIC GAMES WITH IMPERFECT INFORMATION

In this chapter we will:

- *Review dynamic games with incomplete information;*
- *Introduce the concept of information sets and the use of Bayesian updating to find a perfect Bayesian equilibrium;*
- *Discuss the concepts of asymmetric information, adverse selection, and moral hazard;*
- *Introduce the concepts of signaling and screening;*
- *Discuss how separating and pooling strategies can be used to search for perfect Bayesian equilibria.*

INTRODUCTION

In dynamic games with incomplete information, players possess private information but are ignorant or ill informed about the information possessed by rivals. We have already seen how static games with incomplete information might be transformed into static games with complete but imperfect information using a Harsanyi transformation. Unfortunately, Harsanyi transformations are of limited usefulness when analyzing dynamic games in which the players possess private information. Without complete information, a player may not be able to distinguish between decision nodes. When this happens, we have a dynamic game with imperfect information—i.e., the player does not know for certain his or her position in the game tree. One way to deal with this problem is for players to form conditional expectations about payoffs from alternative strategies, and to update these expectations as more and better information is received. In this chapter we will consider how players can revise their expectations using Bayes' theorem—a procedure known as Bayesian updating.

INFORMATION SETS

In dynamic games with imperfect information, players are unaware of their positions in the game tree. When analyzing dynamic games in which the players possess private information, however, it is essential that we keep track of what the players know and when. Information sets allow us to keep track of decision nodes that appear the same to the player, but that are, in fact, different. Consider, for example, the dynamic game depicted in Figure 16.1. In this game, there are three decision nodes: $D_1, D_2,$ and D_3. Player A is at $D_1,$ the root of the game tree. Unlike dynamic games with complete and perfect information, however, in games with incomplete information, such as private value estimates in auctions, player B cannot tell how player A will move until after the fact. For this reason, player B cannot tell whether he or she is at D_2 or D_3.

Decision nodes D_2 and D_3 are said to be within the same information set. An information set is a collection of decision nodes that are indistinguishable to the decision maker. When a player

Figure 16.1
Valid Information Set

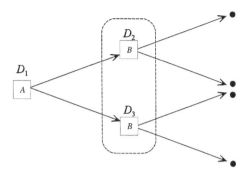

reaches a decision node that lies within an information set containing more than one decision node, the player is unable to determine his or her position in the game tree. Decision nodes in the same information set are identified with a rounded, dashed rectangle. In games involving multiple information sets, the collection of information sets constitutes an **information partition**. In the game depicted in Figure 16.1, the decision nodes of player B at stage two are in the same information set. A player's knowledge is the same at all decision nodes within this information set. Thus, a player will make the same move at each decision node.

> **Information partition** A collection of information sets.

The information set in Figure 16.1 is called a **valid information set** because player B does not know whether he or she is at D_2 or D_3. There are three conditions that must be satisfied for information sets to be valid. First, all decision nodes in an information set must belong to the same player. Second, no decision node in an information set can precede any other decision node in the same information set. Finally, in each information set the same collection of moves can be taken at each decision node.

> **Valid information set** A collection of decision nodes for the same player where no decision node precedes any other decision node in the same information set.

An example of an **invalid information set** is depicted in Figure 16.2 because it includes the decision nodes of players B and C. Player B knows that she or he is at decision node D_2 because player C is at decision node D_3.

> **Invalid information set** A collection of decision nodes for two or more players where a decision node by one player precedes the decision node of another player in the same information set.

Figure 16.3 is another example of an invalid information set because the decision node of player A at D_1 precedes the decision node of player B at D_2. The reason the information set is invalid is because player A knows that he or she is at D_1 because there were no prior moves. Player B knows he or she is at D_2 because player A, who is in the same information set, has already moved. Memory of player A's move allows player B to identify his or her position in the game tree.

The idea of a valid information set is not a new concept. In fact, we have been dealing with valid information sets throughout most of this text. Recall the prisoner's dilemma in Figure 1.1. Using the concept of an information set, Figure 16.4 illustrates the prisoner's dilemma in extensive format with A moving "first." Recall from our discussion of static games in Chapter 1 that the

FIGURE 16.2
Invalid Information Set

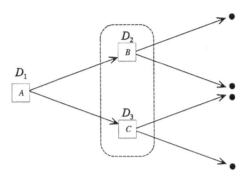

FIGURE 16.3
Invalid Information Set

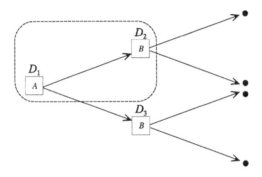

FIGURE 16.4
Extensive Form of the Prisoner's Dilemma

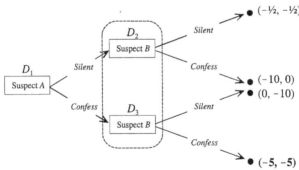

Payoffs in years: (Suspect A, Suspect B)

players are not actually required to move at the same time. It is only necessary for each player to move without knowledge of his or her rivals' moves. The essential element of static games is that players move with complete, but imperfect, information.

Expressing the prisoner's dilemma in extensive form seems problematic because it implies that suspect B's move is contingent on the move made by suspect A. Depicting the prisoner's dilemma

FIGURE 16.5
Extensive Form of the No-Liability Accident Game

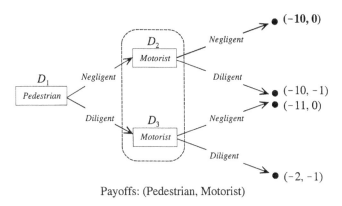

Payoffs: (Pedestrian, Motorist)

in extensive format does not appear to preserve the simultaneous nature of the game in which both players move without knowledge of the move made by the other player. To remedy this, decision nodes D_2 and D_3 are nested within the same valid information set. Suspect B is unable to infer his or her position in the game tree because of the simultaneous nature of the game. The choice of the "first" mover in Figure 16.4 is arbitrary. The same outcome results by positioning player B at the root of the game tree and by putting player A's decision nodes in an information set (although the order of the payoffs at the terminal nodes is reversed). Designating the order of moves only matters if the second mover can observe the first mover.

In Chapter 14 we discussed the Anglo-American tort no-liability law regime in which a victim cannot recover damages from the tortfeasor. This game was depicted in Figure 14.1. As with the prisoner's dilemma in Figure 16.4, the no-liability accident game is an example of a static game with complete, but imperfect, information. The players know the strategies and payoffs, but not how the other player plans to move. Figure 16.5 depicts the no-liability accident game in extensive format in which the pedestrian moves "first." By nesting decision nodes D_2 and D_3 within the same valid information set, the choice of the "first mover" is irrelevant.

We can use Figures 16.4 and 16.5 to clarify our definition of a proper subgame. Recall that a proper subgame begins at a subroot, but this is only true of dynamic games with perfect information. The reason for this is that all information sets in these types of games are invalid. The subgames beginning at decision nodes D_2 and D_3 in Figure 16.4, on the other hand, are not proper subgames because these subroots belong to a valid information set. Since it is not possible for suspect B to deduce his or her location in the game tree, suspect B cannot identify which subgame is being played. For dynamic games with incomplete information, a proper subgame can only begin at subroots not included within the same valid information set. The subgames beginning at decision nodes D_2 and D_3 in Figure 16.3 are examples of proper subgames, while the decision notes D_2 and D_3 in Figure 16.4 are not. Thus, an **improper subgame** is any subgame with a subroot located within a valid information set.

Improper subgame A subgame with a subroot located within a valid information set.

The validity of information sets also depends on the assumption of **total (perfect) recall**. Virtually all sequential-move games assume that players remember everything they ever learned about the game being played, including earlier moves made by other players. The implications of total recall are depicted in Figure 16.6. Consider the information sets $\{D_1\}$, $\{D_2\}$, and $\{D_3, D_4\}$. All of

FIGURE 16.6
Game Tree in Which Player *A* Has Total Recall

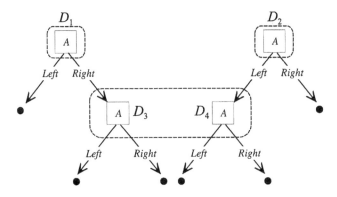

the decision nodes in this game belong to the same player. Although player *A* is able to distinguish between decision nodes D_1 and D_2, it would seem that he or she is unable to distinguish between decision nodes D_3 and D_4. But, this is incorrect since for player *A* to move to decision node D_3, he or she must have first moved *right*. Likewise, to get to decision node D_4, player *A* must have first moved *left*. Figure 16.6 implies that for decision nodes D_3 and D_4 to be in the same valid information set, player *A* must have forgotten his or her prior move, but this has been ruled out by the assumption of total recall, otherwise $\{D_3, D_4\}$ is invalid.

> **Total (perfect) recall** When players remember everything they ever learned about a game being played, including earlier moves made by other players.

The assumption of total recall also rules out situations like the one depicted in Figure 16.7. In this game, decision nodes D_4 and D_5 belong to the same information set. If we assume total recall, the information set $\{D_4, D_5\}$ must be invalid. The reason for this is not difficult to see. Since players are assumed to remember everything, including the moves made by other players, the only way that player *B* could have gotten to D_4 is if player *A* had first moved to the *left*. Likewise, the only way that player *B* could have gotten to D_5 is if player *A* had first moved to the *right*. The only way for decision nodes D_4 and D_5 to belong to a valid information set is to relax the assumption of total recall; otherwise, $\{D_4, D_5\}$ is invalid.

BAYESIAN UPDATING

In dynamic games with incomplete information, a player's prior moves may reveal private information to his or her opponent, and *vice versa*. Rational players will undoubtedly incorporate this new information into their future decisions. The process by which players revise their information set as the game is played is called Bayesian updating. Bayesian updating is based on Bayes' theorem, which deals with the statistical concept of conditional probabilities.

To see what is involved in Bayesian updating, consider the following variation of the oil-drilling game. GLOMAR must decide whether to purchase a two-year lease to drill on government land. The company believes the land lies directly above a four-million barrel deposit of crude oil, but does not know for certain. Independent geological surveys suggest that there is a 20 percent chance that the oil deposit exists. At such a low probability of success, high development and extraction costs renders the project infeasible. To reduce the risk, GLOMAR decides to conduct further seismic surveys. Unfortunately, seismic testing is not foolproof. The test can produce a

FIGURE 16.7
Game Tree in Which Player B Has Total Recall

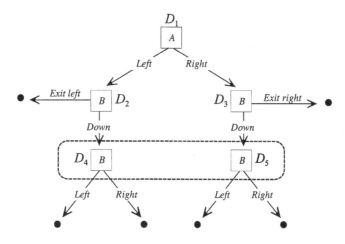

FIGURE 16.8
Oil-Drilling Game with Conditional Probabilities

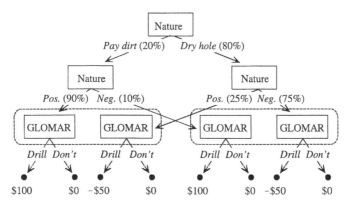

false positive (i.e., that oil exists when it does not) or a *false negative* (i.e., that oil does not exist when it does). Based on the results of seismic testing, should GLOMAR drill? The extensive form of this game is depicted in Figure 16.8.

Using a Harsanyi transformation, Nature moves first by assigning probabilities to the results of independent geological surveys as to the presence (*Pay dirt*) or absence (*Dry hole*) of crude oil, which are 20 percent and 90 percent, respectively. Nature then moves by assigning probabilities to determine the outcome of the seismic testing. The conditional probability of a false positive (*Dry hole*) is 25 percent. The conditional probability of a false negative (*Pay dirt*) is 10 percent. This assignment of probabilities is referred to as the players' **belief profile**. A **perfect Bayesian equilibrium** must be consistent with these beliefs. A player must be able to calculate the expected payoffs associated with a move from any given information set. Since these beliefs are likely to change during the course of game play, it must also be possible for players to revise their strategies and expected payoffs using Bayesian updating whenever possible. Should GLOMAR drill (*Drill*) or not drill (*Don't*)?

Belief profile A player's probability assessments at each nontrivial information set.

Perfect Bayesian equilibrium A Nash equilibrium for dynamic games that is consistent with the probability distribution of states of nature and player beliefs.

If GLOMAR drills and hits oil, the company will earn $100 million. If GLOMAR drills and does not hit oil, the company will lose $50 million. The expected payoff from drilling without seismic testing is 0.20($100) + 0.80(–$50) = –$20 million. Since the payoff from not drilling is $0, a risk-neutral GLOMAR will choose not to drill. How will seismic testing affect this decision? To answer this question, we must ask how seismic testing changes the probabilities used to estimate the expected payoff from drilling. More specifically, we need to calculate the conditional probability that oil exists given a positive seismic test result. To do this, we need to apply Bayes' theorem.

Bayes' theorem allows us to update the probability of an event (B_i) after learning that some other event (A) has occurred. Let us denote the unconditional probability that oil is present as $P(B_i)$. Since this probability comes first, we will refer to $P(B_i)$ as the *prior probability*. The probability that oil exists will be revised once GLOMAR receives the results of its seismic testing (A). We will refer to this as the *posterior probability* that oil exists. We will denote this revised probability that oil exists given the results of seismic testing as $P(B_i|A)$. The process of revising the prior probability that oil exists once the results of seismic testing are determined is an example of Bayesian updating. In general, Bayesian updating involves replacing prior probabilities with posterior probabilities once prior moves have been observed. In order to transform prior probabilities into posterior probabilities we need **Bayes' theorem**.

Bayes' theorem A statistical relationship that enables a player to update the probability of an event (B_i) after learning that some other event (A) has occurred. Let B_i be the *i*th of *k* mutually exclusive and collectively exhaustive events with a nonzero probability. If some other event A has a nonzero probability, then for each event *i*:

$$P\left(B_i\,|A\right) = \frac{P\left(A|B_i\right)\cdot P\left(B_i\right)}{\displaystyle\sum_{i=1}^{k} P\left(A|B_i\right)\cdot P\left(B_i\right)} \tag{16.1}$$

In Equation (16.1), $P(B_i|A)$ is the probability of event B_i conditional on event A having occurred.

Bayes' theorem allows us to combine prior information with the latest information to produce a revised probability, which is referred to as a posterior probability. To apply Bayes's theorem to the oil-drilling game, it is necessary to state our prior probabilities. Let $P(B_1)$ represent the prior probability that oil exists (*Pay dirt*) and $P(B_2)$ the prior probability that oil does not exist (*Dry hole*). Let $P(A|B_1)$ be the probability that the seismic test indicates the presence of oil exists when oil is present, and $P(A|B_2)$ the probability that oil does not exist when oil is not present. In our example, $P(A|B_1) = 0.90$, and $P(A|B_2) = 0.25$. Substituting these results into Equation (16.1) we obtain:

$$\begin{aligned} P\left(B_1\,|A\right) &= \frac{P\left(A|B_i\right)\cdot P\left(B_i\right)}{P\left(A|B_1\right)\cdot P\left(B_1\right)+P\left(A|B_2\right)\cdot P\left(B_2\right)} \\ &= \frac{0.90\cdot 0.20}{\left(0.90\cdot 0.20\right)+\left(0.25\cdot 0.80\right)} = 0.47 \end{aligned} \tag{16.2}$$

Equation (16.2) says that when the seismic test is positive, there is a 47 percent probability that GLOMAR will strike oil and a 53 percent probability of a dry well. We will use these posterior probabilities to recalculate the expected payoff of striking oil. The expected payoff of finding oil is now 0.47($100) + 0.53(–$50) = $20.5. Since the expected payoff from not drilling is $0, a risk-neutral GLOMAR will *drill*. By using Bayes' theorem and updating the prior probability of finding oil using seismic testing, the expected value of finding oil has increased from –$20 million to +$20.5 million.

The conditional probabilities using Bayes' theorem are very sensitive to changes in prior probabilities. To see this, suppose that the prior probability of finding oil is 10 percent instead of 20 percent. It is left as an exercise for the reader to show that the revised (posterior) probability of finding oil is only 28.6 percent. Using this result, the expected payoff of finding oil is –$7.1 million. Risk neutral and rational GLOMAR would choose not to drill. This result is somewhat surprising given that seismic testing in this example is highly accurate. It is left as an exercise for the reader to demonstrate that a less accurate test will result in an even larger expected loss from drilling.

ADVERSE SELECTION AND THE MARKET FOR LEMONS

There are many problems in economics in which some individuals have more or better information about a good or service being transacted than others. **Asymmetric information** may lead to problems of **adverse selection** and **moral hazard**. Adverse selection is an *ex ante* problem in that it arises before the transaction takes place. An often cited example of this arises in the loan market. Borrowers know more about their own creditworthiness than do lenders. Since bad credit risks are more likely to ask for loans, this market becomes crowded with bad credit risks. Lenders know this and may decide not to lend, even though there may be good credit risks who want to borrow. The problem is that lenders may not be able to distinguish good credit risks from bad credit risks.

> **Asymmetric information** When some individuals engaged in transactions have more or better information than others.

> **Adverse selection** An *ex ante* asymmetric information problem in which the market becomes crowded with individuals or products having undesirable characteristics.

Moral hazard is an *ex post* problem because it occurs after the transaction takes place. Moral hazard refers to the risk (hazard) that one party to a transaction behaves in undesirable (immoral) manner. Returning to the loan market example, suppose that the owner of a company obtains a loan from a bank to finance a capital investment project. The bank's decision to make the loan was based on the belief that the capital investment will generate a rate of return sufficient to nominally service the debt. Once the loan is made, however, the borrower has an incentive to use the loan for some other, riskier, purpose, such as spending a weekend in Las Vegas. Because moral hazard lowers the probability that the loan will be repaid, the bank may again decide not to make the loan without adequate safeguards.

> **Moral hazard** An *ex post* asymmetric information problem in which the risk (hazard) that a party to a transaction engages in activities that are undesirable (immoral) from the perspective of the other party to the same transaction.

Nobel laureate George Akerlof (1970) identified the problems associated with adverse selection in the used-car market. Professor Akerlof began by identifying two types of used cars: good used cars (peaches) and bad used cars (lemons). In this market, transactions take place directly between the buyers and sellers of used cars. The problem is one of asymmetric information since the sellers of used cars know the value of their cars, but buyers do not. Buyers of used cars are willing to pay a "low" price for lemons and a "high" price for peaches. Since they are unable to distinguish between them, the price will reflect the average quality of used cars. Owners of lemons are more than happy to sell at the higher, average price because it overstates the value of their used cars. Owners of peaches are not happy to sell at the lower, average price because it understates the value of their used cars. Unless the problem of asymmetric information is resolved, the market will become crowded with lemons. Potential buyers of used cars know this, so few used cars will be purchased. If the problem of asymmetric information is severe enough, the used-car market could disappear entirely.

The lemons problem has multiple applications. In the market for financial securities, for example, sellers of stocks and bonds have better information about a company's underlying financial strengths or weaknesses than buyers. Buyers of equities (stocks) are less able to distinguish high-earnings, low-risk companies from low-earnings, high-risk companies. As in the case of the used-car market, share prices tend to reflect the average financial strength of similar companies. Sellers of low-quality equities will be happy to receive the (higher) average price. Sellers of high-quality equities will not be happy to receive the (lower) average price. Similarly, buyers of debt (bonds) are less able to distinguish high-default-risk companies from low-default-risk companies. Because of the lemons problem, the interest rate paid will tend to reflect the average of high- and low-default-risk companies. Sellers of high-default-risk bonds will be more than happy to borrow at the (lower) average interest rate, while sellers of low-default-risk bonds will not be happy to borrow at the (higher) average interest rate. Once again, asymmetric information could result in inefficient equity and bond markets—or no securities markets at all. Of course, used-car, equity, and bond markets do exist, which means remedies have been devised to overcome adverse selection problems arising from asymmetric information.

SIGNALING

Among the many applications of game theory is an analysis of a process known as **signaling**. Signaling is one possible solution to adverse selection arising from the problem of asymmetric information. Signaling occurs when an informed player transmits a credible but unreproducible signal to an uninformed player. This signal is intended to convey important information that is otherwise not directly observable by other players. For example, companies often use signals to convey information about the benefits of a good or service to potential customers. Since these benefits are unobservable unless the firm's product is actually consumed, the signal is designed to convince potential customers to give the product a try. In many cases, the signals involve a "try-it-and-you'll-like-it" approach. Such signals include money-back guarantees, free trial periods, and so on. In some instances, the signals are linked to a firm's reputation in the market. These "you know me so you can trust me" signals often entail packaging that prominently displays the company's logo, or can be as simple as a neighborhood storefront that indicates the year in which the company was founded.

Signaling is very common in the labor market. Prospective employers typically require job applicants to submit a host of supporting documentation, such as letters of recommendation, college transcripts, résumés, etc. The résumé in particular is used by job applicants to transmit signals to the employer about his or her potential value to the firm. These signals include work experience, internships, educational background, schools attended, degrees earned, extracurricular activities, professional certifications, licenses, grade point average, professional and academic awards, fraternal and academic honor societies, and so forth. Each of these signals is intended to convince the prospective employer that the job applicant will be a valued asset and will positively contribute to the firm's "bottom line."

Signaling When an informed player transmits a credible and unreproducible signal to an uninformed player. This signal is intended to convey information that is important to the sender but is otherwise unobservable.

For a signal to be effective, it must not only be observable by uninformed players, it must also be reliable and not easily mimicked. Before formally discussing the use of signals in dynamic games with incomplete information, consider the case of a potential employer who wants to hire a worker from a pool of applicants. To keep things simple, suppose that the employer believes that there are only two types of prospective employees: Unproductive workers who contribute nothing to the firm's total revenues, and productive workers who add $100,000 annually to the firm's total revenues. In the terminology of economics, the value of an unproductive worker's marginal product is $0, while the value of an unproductive worker's marginal product is $100,000. In this example we have a problem of asymmetric information. The applicant knows more about his or her productivity than the employer. Suppose the employer believes that there is a 50-50 chance of hiring a productive worker. This information can be used to calculate the expected value of a worker's marginal product, which in this case is 0.5($0) + 0.5($100,000) = $50,000. If the prospective employer is risk neutral, the most that he or she is willing to offer the job applicant is $50,000. Regrettably, this creates the same kind of lemons problem that we observed in the used-car market. Unproductive workers will be very happy with a salary offer of $50,000, while productive workers will not. How might this affect the employer's hiring decision?

Suppose that productive workers have a choice. They can either accept the $50,000 offer, or start their own business and earn $75,000. Clearly, productive workers will refuse the offer and start their own business. Unproductive workers can accept the offer or start their own business and earn less. Clearly, unproductive workers will accept the offer. As a result, the market will become crowded with unproductive workers. If prospective employers are unable to distinguish productive from unproductive workers, an optimal strategy is not to hire anyone.

Is there a solution to this problem? Since productive workers are harmed by employers' inability to distinguish between worker types, it will be in the workers' best interest to transmit signals about their productivity. Of course, productive workers could just tell the prospective employers that they are productive. But what good is that? Unproductive workers will say the same thing. On the other hand, if productive workers are able to transmit credible signals that cannot be reproduced by unproductive workers, the employer is more likely to offer a salary that approximates the value of the applicant's marginal product.

At one time, having an undergraduate college degree served as an effective signal of worker productivity, as did a high school diploma until the mid-twentieth century. A worker either has a college degree, or does not. At one time, employers believed that only productive workers could earn an undergraduate college degree. As undergraduate college degrees became the norm, employers began to revise this belief, such as by concluding that there is a 90 percent probability a worker with a college degree is productive. In this case, the expected value of a worker's marginal product is 0.1($0) + 0.9($100,000) = $90,000. Alternatively, as the market became flooded with college graduates, employers began to look for different signals, such as graduates from prestigious universities and/or high grade point averages, applicants with graduate degrees from accredited professional programs, and so on. The employer will use these and other signals to weed out unproductive workers from the pool of job applicants.

SPENCE EDUCATION GAME

We will illustrate signaling games by analyzing a simple version of the Spence education game (Spence 1973).[1] This game involves a company (firm) that is considering hiring a job applicant (worker) at the prevailing competitive wage. This game involves asymmetric information because workers know whether they are high-productivity (HP) or low-productivity (LP) types, but the firm does not. Productivity is private information that the firm cannot directly observe. The probability that the worker is a high-productivity type is $p = P(HP)$, and the probability that the worker is a low-productivity type is $(1 - p) = P(LP)$. We will assume that the firm earns a per-worker gross profit of π_{HP} if the worker is high-productivity and a per-worker gross profit of π_{LP} if the worker is low-productivity, where $\pi_{HP} > \pi_{LP}$. Per-worker gross profit is defined as the per-worker total economic profit *less* the market-determined wage rate (w). Since the firm is unable to distinguish low-productivity from high-productivity workers, all workers are paid w.

The challenge confronting the firm is to infer a productivity type based on signals received from workers. In this example, we will assume that the firm uses education as its productivity signal. High-productivity workers have less difficulty obtaining an education than low-productivity workers. Thus, the opportunity cost of obtaining an education is less for high-productivity workers (c_{HP}) than for low-productivity workers (c_{LP}). If education were equally costly for both worker types, it could not be used as a productivity signal, in which case some other signal is necessary.

In this game, workers of both productivity types have college degrees (C) and do not have college degrees (HS). Since the firm cannot distinguish between productivity types, all workers are paid the same wage rate (w). A firm that hires a high-productivity worker earns $\pi_{HP} - w$. A firm that hires a low-productivity worker earns $\pi_{LP} - w$. If a high-productivity worker with a college degree is hired (H), that worker's payoff is $w - c_{HP}$. A high-productivity or low-productivity worker without a college degree earns w since the cost of a college education is not incurred. If a high-productivity worker with a college degree is not hired (NH), the payoff is $-c_{HP}$. If a low-productivity worker with a college degree is not hired, the worker's payoff is $-c_{LP}$. If a high-productivity or low-productivity worker without a college degree is not hired, the worker's payoff is $0. This version of the Spence education game is depicted in Figure 16.9. In this game, Nature moves first by randomly selecting a worker's productivity type. The worker knows his or her productivity type. The firm does not know the worker's productivity type, but knows whether the worker has a college degree.

We will begin our analysis by assuming that the firm does not use education as a signal for worker productivity. The firm is aware, however, of the probability distribution of worker productivity types, which is determined by Nature. Suppose that the probability of hiring a low-productivity worker is $p = 0.6$ and the probability of hiring a high-productivity worker is $(1 - p) = 0.4$. The firm's per-worker gross profit from high-productivity and low-productivity worker are $\pi_{HP} = \$12$ and $\pi_{LP} = \$8$, respectively. If the wage rate paid to all workers is $w = \$8$, the firm's net per-worker profit for high-productivity and low-productivity workers are $\pi_{HP} - w = \$12 - \$8 = \$4$ and $\pi_{LP} - w = \$8 - \$8 = \$0$, respectively. Finally, we assume that the cost of obtaining a college education for high-productivity and low-productivity workers are $c_{HP} = \$1$ and $c_{LP} = \$4$, respectively. This means that the payoff to a high-productivity or low-productivity worker without a college degree is $\$8 - \$0 = \$8$. The payoff to a high-productivity worker with a college degree is $\$8 - \$1 = \$7$, while the payoff to a low-productivity worker with a college degree is $\$8 - \$4 = \$4$. The payoff to workers of either type without a college degree who is not hired is $\$0$. The payoffs to high-productivity and low-productivity types with college degrees who are not hired are $-\$1$ and $-\$4$, respectively. Substituting these data into Figure 16.9, the Spence education game with certain payoffs is depicted in Figure 16.10.

The Spence education game depicted in Figure 16.10 uses a Harsanyi transformation in which Nature moves first by randomly selecting low-productivity or high-productivity workers. The worker observes the move made by Nature and moves next, followed by the firm, which observes the move made by the worker, but not the move made by Nature. The worker must decide whether to earn a college degree (C) or stop with a high school diploma (HS). The firm then decides whether to hire (H) or not hire (NH) the job applicant. Since the firm cannot distinguish between decision nodes within the same information set, its hiring strategy depends on expected payoffs. The Spence education game with expected payoffs is depicted in Figure 16.11.

FIGURE 16.9
Spence Education Game

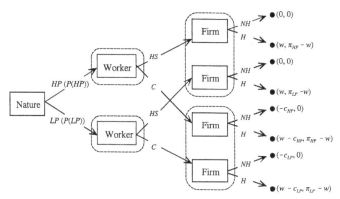

Payoffs: (Worker, Firm)

FIGURE 16.10
Spence Education Game with Certain Payoffs

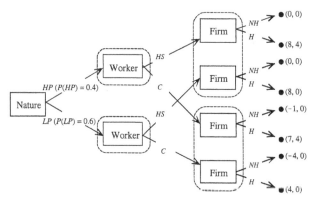

Payoffs: (Worker, Firm)

FIGURE 16.11
Spence Education Game with Expected Payoffs

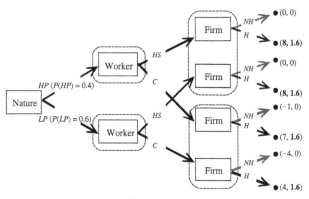

Payoffs: (Worker, Firm)

The reader should note that the game depicted in Figure 16.11 does not have a unique subgame perfect equilibrium. The reason for this is that there are no proper subgames. In fact, since the expected payoffs for the firm are the same, regardless of the worker's productivity type, this game appears to have four perfect Bayesian equilibrium strategy profiles: $\{LP\ (P(LP) = 0.6) \rightarrow C \rightarrow H\}$, $\{HP\ (P(HP) = 0.6) \rightarrow C \rightarrow H\}$, $\{LP\ (P(LP) = 0.4) \rightarrow HS \rightarrow H\}$, and $\{HP\ (P(HP) = 0.6) \rightarrow HS \rightarrow H\}$. Once again, the reader is cautioned that the actual payoffs for this game are known only after the firm discovers whether it has hired a high-productivity or low-productivity worker. If the firm hires a low-productivity high school graduate, the payoffs are (8, 4). If the firm hires a low-productivity college graduate, the payoffs are (4, 0). The four Nash equilibrium strategy profiles for this game suggest that the firm will hire anyone who applies for the job! Can the firm do better by taking into consideration information that is provided by job applicants? We will consider this question in the next two sections.

CASE STUDY 16.1: THE *U.S. NEWS & WORLD REPORT* COLLEGE RANKINGS

In recent years, the *U.S. News & World Report* (USNWR) tier rankings of colleges and universities have emerged as an important signal for and about colleges, universities, their students, and graduates. USNWR tier rankings provide important signals to graduating high school students seeking college admission because the information embedded in the rankings makes the search process more efficient, which reduces search costs. The tier rankings are important to college and university administrators because they help define the institution's market niche and influence perceptions of prospective students, which affect enrollments and operating budgets. The importance of the rankings as a market signal is underscored by the fact that administrators regularly meet with USNWR editors to discuss reasons why institutions are downgraded in the rankings. The USNWR tier rankings also serve as important productivity signals to prospective employers.

Despite their widespread use, the USNWR tier rankings have been the subject of severe criticism. Much of the debate revolves around the apparently arbitrary weighting scheme used in the ranking process (see, for example, Carter, 1998; Crissey, 1997; Garigliano, 1997; Gilley, 1992; Glass, 1997, Gleick, 1995; Graham and Diamond, 1999; Kirk and Corcoran, 1995; Machung, 1998; Schatz, 1993). The USNWR tier rankings affect the number and quality of admission applicants, which affects the overall profile of an institution's student body, the quality of an institution's academic and extracurricular programs, and the perceived value of an institution's degree. These perceptions influence retention rates, which affect tuition-based sources of revenues, financial resources, operating budgets, per-student expenditures, faculty/student ratios, etc. This, in turn, influences the institution's academic reputation, alumni contributions, foundation grants, and other non-tuition-based revenue sources. All of this suggests that the USNWR tier rankings have important feedback effects that reinforce existing positive and negative stereotypes about ranked institutions, their students, and graduates.

U.S. News & World Report bases its tier rankings on sixteen measures of academic quality that fall into seven broad categories, including academic reputation, student selectivity, faculty resources, student retention, financial resources, alumni contributions, and, graduation rates. Thomas Webster (2001) analyzed the relative contribution of eleven ranking criteria used by USNWR in its 1999 rankings. He found that the weighting scheme that is used by USNWR does not accurately reflect the actual contribution of each ranking criteria examined. The reason for this is the presence of widespread multicollinearity in the weighting scheme. According to Webster, while academic reputation is the most heavily weighted ranking criterion, it is fourth in terms of actual importance. The most important ranking criteria, according to Webster, is the SAT and ACT scores of enrolled students. Its importance stems not only from its direct effect on the tier rankings, but its indirect effect on several other ranking criteria, including actual graduation rates, predicted graduation rates, retention rates, alumni contributions, academic reputation, the percentage of enrolled students who ranked in the top 10 percent of their graduating high school class, and acceptance rates. If correct, Webster's results have significant implications for ranked institutions' admissions policies.

SEPARATING STRATEGY

Signaling games often have multiple Nash equilibria. In searching for these equilibria, we will investigate how the firm can use education as a productivity signal. There are two ways to proceed with our search. The first is for the firm to assume that high-productivity types always attend college and low-productivity types never go beyond high school. In the Spence education game, the firm uses education as a perfect signal of a worker's productivity type. This method of distinguishing (separating) high-productivity from low-productivity types is called the firm's **separating strategy**. Alternatively, the firm may choose to lump all productivity types into the same pool of workers. This is called the firm's **pooling strategy**. A worker's education reveals no definitive information about a worker's productivity. Rather, the firm knows the probability distribution of worker productivity types, which is determined by Nature. The firm also forms prior beliefs about how worker productivity types are distributed between college and high school graduates.

> **Separating strategy** When a different strategy is adopted for different player types. A separating strategy may be used as a perfect signal for other player types.

> **Pooling strategy** When the same strategy is adopted for different player types. A pooling strategy may not be used as a perfect signal for other player types.

We will begin our search for an equilibrium by assuming that the firm in the Spence education game adopts a separating strategy. That is, the firm assumes that only high-productivity workers go to college. Of course, the firm could proceed on the basis of a different separating strategy, such as assuming that low-productivity workers go to college and high-productivity workers do not, but this appears to be a less plausible separating strategy. Since the firm assumes that college graduates are high-productivity, these workers will be offered a higher wage rate (w_{HP}). Similarly, low-productivity high school graduates will be offered a lower wage rate (w_{LP}), where $w_{HP} > w_{LP}$. The firm's net per-worker profit from hiring a high-productivity and a low-productivity worker is now $\pi_{HP} - w_{HP}$ and $\pi_{LP} - w_{LP}$, respectively. The payoffs to low-productivity and high-productivity workers with a college degree who get hired are $w_{LP} - c_{LP}$ and $w_{HP} - c_{HP}$, respectively. This new game is depicted in Figure 16.12.

We will make this game more tangible by assuming that the wage for high-productivity workers is $w_{HP} = \$10$ and the wage for a low-productivity worker is $w_{LP} = \$7$. Substituting these values into Figure 16.12 and calculating conditional expected payoffs, we obtain Figure 16.13. The reader should verify that the perfect Bayesian equilibrium strategy profiles for this game with Nature moving first are $\{LP\ (P(LP) = 0.6) \rightarrow HS\ (P(HS|LP) = 1) \rightarrow H\}$ and $\{HP\ (P(HP) = 0.4) \rightarrow HS$

FIGURE 16.12
Spence Education Game in a Competitive Labor Market in Which the Firm Uses Education as a Perfect Signal for Productivity Type

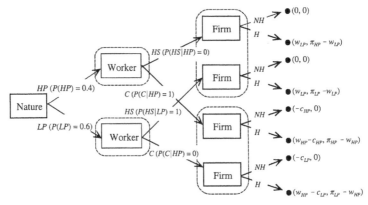

Payoffs: (Worker, Firm)

FIGURE 16.13
Spence Education Game with Conditional Expected Payoffs in Which the Firm Uses a Separating Strategy

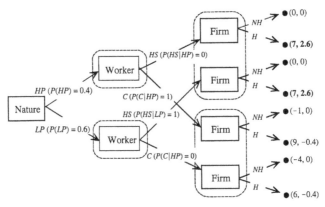

Payoffs: (Worker, Firm)

$(P(HS|HP) = 0) \rightarrow H\}$. That is, when the firm uses education as a perfect signal for productivity type, it will only hire high school graduates. This strategy of only hiring high school graduates is called a **separating equilibrium**. This is a tremendous improvement from the firm's earlier hiring strategy of hiring anyone who walks through the front door. Even though the firm takes education as a perfect signal for worker productivity type, payoffs depend on the type of worker actually hired. If the high school graduate is a low-productivity worker, the payoffs will be (8, 0). If the high school graduate is a high-productivity worker, the payoffs will be (8, 4).

> **Separating equilibrium** A Bayesian equilibrium that arises when a different strategy is adopted for players of different types.

POOLING STRATEGY

In the above section we assumed that the firm used the separating strategy of using education as a perfect signal for a worker's productivity type. This approach appears to be somewhat shortsighted since it is more likely that there are college and high school graduates of both productivity types. In this case, education is a less-than-perfect signal for a worker's productivity type. Because of this, the firm will adopt a pooling strategy by lumping college graduates of both productivity types into one group and high school graduates of both productivity types into another group. There are two issues that need to be addressed. Once again, productivity determines the wage rate paid by the firm. Since the firm cannot differentiate between productivity types, we will assume that all workers are paid the same wage rate (w). As in the Spence education game depicted in Figure 16.11, we will assume that this wage rate is $w = \$8$.

The second issue that needs to be resolved has to do with the firm's prior belief about the distribution of worker productivity types. Let us assume that in-house and independent studies suggest that 90 percent of college graduates are high-productivity types and 10 percent are low-productivity types. These studies also show that 30 percent of high school graduates are high-productivity and 70 percent are low-productivity types. We can summarize the firm's belief profiles as $\{C$: (HP: 0.90, LP: 0.10), HS: (HP: 0.30, LP: 0.70)$\}$. On the basis of these studies, the firm institutes a pre-employment screening program designed to identify a worker's productivity type. Figure 16.14 recreates Figure 16.11 with the inclusion of the firm's belief profile. The firm's belief profile and Bayes' theorem can be used to recalculate the firm's conditional probabilities. Substituting these data into Equation (16.1) the reader should verify that these conditional probabilities are $P(HP|C) = 0.86$; $P(LP|C) = 0.14$; $P(HP|HS) = 0.22$; $P(LP|HS) = 0.88$. These conditional probabilities were used to calculate the firm's conditional expected payoffs.

An analysis of the game depicted in Figure 16.14 reveals that when all workers receive the same wage rate, the perfect Bayesian equilibrium strategy profiles for this game are $\{LP$ ($P(LP) = 0.6) \rightarrow C$ ($P(C|LP) = 0.1) \rightarrow H\}$, $\{HP$ ($P(HP) = 0.4) \rightarrow C$ ($P(C|HP) = 0.9) \rightarrow H\}$. This strategy of hiring only college graduates is referred to as a **pooling equilibrium**. That is, when the firm adopts a pooling strategy, it will only hire college graduates regardless of productivity type. Of course, the actual payoffs depend on the productivity type of the worker hired. If the firm hires a high-productivity college graduate, the payoffs are (7, 4). If the firm hires a low-productivity college graduate, the payoffs are (4, 0). Once again, this outcome is an improvement over games in which the firm makes no effort to identify a worker's productivity type. As we saw earlier, when no signaling strategy is used, the firm will hire anyone who applies for the job.

> **Pooling equilibrium** A Bayesian equilibrium that arises when the same strategy is adopted for players of different types.

It may be somewhat disturbing that the existence of a perfect Bayesian equilibrium in the Spence education game depends crucially on the signaling strategy, the state of nature, and the firm's prior beliefs about worker productivity types. Unfortunately, this unhappy state of affairs is true of almost all signaling games, and unique Nash equilibria are the exception rather than the rule. The reason for this is that there are no proper subgames, which are only found in games with perfect information. Of course, this makes the search for a unique, perfect Bayesian equilibrium very problematic and extremely difficult. Unlike the method of backward induction, there is no commonly accepted methodology for finding a Bayesian equilibrium for games involving incomplete information.

For a game to have a perfect Bayesian equilibrium, it must satisfy three requirements. First, the collection of Nash equilibrium strategy profiles must be consistent with a player's beliefs about the state of nature and other players' strategies. Second, players attempt to maximize their conditional

FIGURE 16.14

Spence Education Game with Conditional Expected Payoffs in Which the Firm Uses a Pooling Strategy

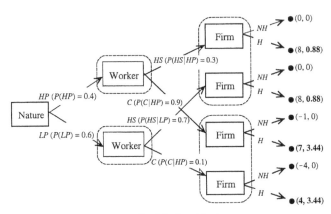

Payoffs: (Worker, Firm)

expected payoffs from each information set. Finally, players should revise conditional expected payoffs using Bayesian updating whenever possible.

Principle: For a strategy profile and a belief profile to constitute a perfect Bayesian equilibrium, it must satisfy three conditions: (i) Given the players' beliefs, the collection of strategy profiles must constitute a Nash equilibrium; (ii) players attempt to maximize their conditional expected payoffs from each information set, and (iii) conditional expected payoffs will be revised using Bayesian updating whenever possible.

Unfortunately, while it is easy to define the necessary conditions for a perfect Bayesian equilibrium, identifying them in practice can be very difficult. One possible, albeit crude, approach is suggested by Bierman and Fernandez (1998, p. 329). The first step is to propose a set of strategies and beliefs. Next, check whether these proposed strategies satisfy conditions (i) and (ii) in the above principle. Once this is done, check to see if these beliefs satisfy requirement (iii). According to Bierman and Fernandez, the main problem with this approach is that there are typically a very large number of possible strategies and belief profiles. Thus, finding a perfect Bayesian equilibrium can be very difficult and time consuming, and maybe even impossible.

———————————————— **Demonstration Problem 16.1** ————————————————

Consider the Spence education game in Figure 16.10. Suppose that the firm is considering using college degrees as a signal for worker productivity. Following an in-depth study of the performances and education levels of its current employees, the firm has estimated the conditional probabilities $P(HS|HP) = 0.9$ and $P(C|HP) = 0.88$.

   a. If the firm is risk neutral, what is its optimal hiring strategy?
   b. What does your answer to part a suggest about using education as a signal for worker productivity?
   c. What, if anything, do you believe is wrong with the firm's estimated conditional probabilities of worker productivity?

Solution

a. Applying Bayes' theorem, the conditional probabilities that high-productivity types are college and high school graduates are:

$$P(HP|HS) = \frac{P(HS|HP) \cdot P(HP)}{P(HS|HP) \cdot P(HP) + P(HS|LP) \cdot P(LP)}$$

$$= \frac{0.4(0.9)}{0.4(0.9) + 0.6(0.1)} = 0.86 \qquad (\text{D16.1.1})$$

$$P(HP|C) = \frac{P(C|HP) \cdot P(HP)}{P(C|HP) \cdot P(HP) + P(C|LP) \cdot P(LP)}$$

$$= \frac{0.4(0.88)}{0.4(0.88) + 0.12(0.6)} = 0.83 \qquad (\text{D16.1.2})$$

Using the result of Equation (D16.1.1), the conditional expected payoff from hiring a high school graduate is 0.86(4) + 0.14(0) = 3.44. Using the result of Equation (D16.1.2), the conditional expected payoff from hiring a college graduate is 0.83(4) + 0.17(0) = 3.32. On the basis of these conditional expected payoffs, the firm's optimal strategy is to hire only high school graduates.

b. The difference in the conditional expected payoffs calculated in part a appears to be statistically insignificant. Taken at face value, this suggests that having a college degree may not be a very good signal of worker productivity. Perhaps some other signal would perform better, such as previous work experience, community service, and so on.

c. The main problem with the firm's estimated conditional probabilities is that they were based on the performances and educational levels of *current* employees. It is reasonable to assume these employees are high-productivity. Low-productivity workers would probably have lost their jobs with the firm.

SCREENING

A phenomenon that is related to signaling is **screening**. Screening is an attempt by an uninformed player to sort out other players according to their characteristics. In the loan market, for example, the problem of adverse selection requires that lenders screen out bad credit risks. When an individual applies for a mortgage loan, for example, the bank or mortgage broker will request copious information about the applicant's salary, income tax returns, checking and savings account balances, other financial assets, real assets (such as automobiles and real estate), credit history, and employment history. If the prospective borrower is a business, the lender will require financial statements, such as income statements, balance sheets, feasibility studies, etc. It is the responsibility of the lending officer to assess credit risk and decide whether the loan should be made. Some banks specialize in the types of loans that they make. For example, banks may specialize in lending to small businesses, or to firms in specific industries or geographic regions. By specializing, banks develop the expertise necessary to assess the creditworthiness of prospective borrowers.

Screening When an uninformed individual attempts to sort players according to defined characteristics.

Screening is also used by prospective employers to sort out applicants according to their productivity characteristics. One type of screening mechanism that is used in the labor market is

self-selection. Job applicants are informed by the prospective employer that there are prerequisites to employment, such as minimum level of education, minimum grade point average, specific kinds of work experience, etc. Only the applicants know whether they have these characteristics. Those who do will continue the application process. Those who do not may decide to seek employment elsewhere.

> **Self-selection** When players voluntarily sort themselves into different groups according to defined characteristics.

To illustrate how screening works, consider the situation in which Adam is considering applying for a management position that pays $50,000 or a sales position that pays $40,000. As it turns out, Adam knows that he is a lousy manager and is only able to add $30,000 of value to the firm. On the other hand, Adam knows that he is a self-starter with great interpersonal skills, and that he is able to add $60,000 of value to the firm as a salesperson. Although Adam would make a better salesperson, he would prefer to work as a manager. By contrast, the personnel manager does not have enough information to decide whether to offer Adam a position as a manager or a salesperson. To avoid this problem, the personnel manager comes up with the following scheme. The personnel manager offers Adam his choice of either job at a guaranteed $20,000 annually, and a bonus of $5,000 for each $10,000 of value added to the firm. With this compensation scheme, Adam can earn $20,000 + $5,000($30,000/$10,000) = $35,000 as a manager, and $20,000 + $5,000($60,000/$10,000) = $50,000 as a salesperson. Faced with these choices, if income is his most important consideration then Adam will accept the job as a salesperson. Thus, even though the personnel manager has incomplete information, this self-selection mechanism has directed Adam to the job that yields the greatest benefits to both players.

CHAPTER REVIEW

In dynamic games with incomplete information players possess private information. When analyzing these games, it is important to keep track of what players know, and when. One way to do this is with the use of *information sets*. As new information becomes available, players revise their conditional expected payoffs using the procedure of *Bayesian updating*.

A *perfect Bayesian equilibrium* is a Nash equilibrium for dynamic games that is consistent with the probability distribution of states of nature and player beliefs. It is typical for dynamic games with incomplete information to have multiple Nash equilibria. The reason for this is that there are no proper subgames, which are only found in games with complete information.

Among the most important contributions of game theory is the process known as *signaling*, which occurs when an informed player transmits a credible and unreproducible signal to an uninformed player. For a signal to be effective, it must not only be observable by uninformed players, it must also be reliable and not easily mimicked.

Signaling is commonly used in the labor market, where job résumés are used to convey information about the applicant's potential value to the firm to an otherwise uninformed employer. The signals included in résumés are intended to convince prospective employers of an applicant's ability to contribute to the firm's "bottom line."

One way to organize our search for a Bayesian equilibrium is to use *separating* and *pooling strategies*. With a separating strategy, different strategies are used for other players of different types. A separating strategy may be used as a perfect signal for the player's type. With a pooling strategy, the same strategy is used for other players of different types. A pooling strategy may not be used as a perfect signal for the player's type.

The existence of a perfect Bayesian equilibrium in the Spence education game depends on the firm's signaling strategy, states of nature, and the firm's prior beliefs about worker productivity types. This is true of almost all signaling games. This, of course, makes the search for a unique, perfect Bayesian equilibrium very problematic and extremely difficult. Unlike the method of backward induction, there is no commonly accepted methodology for finding a Bayesian equilibrium for games involving incomplete information.

For a game to have a perfect Bayesian equilibrium, it must satisfy three requirements: (i) The collection of Nash equilibrium strategy profiles must be consistent with a player's beliefs about the state of nature and other players' strategies; (ii) players attempt to maximize their conditional expected payoffs from each information set; and (iii), players should revise conditional expected payoffs using Bayesian updating whenever possible. Unfortunately, dynamic games with incomplete information that satisfy these requirements results in a very large number of possible strategies and belief profiles. In practice, finding a perfect Bayesian equilibrium can be very difficult and time consuming, and maybe even impossible.

A phenomenon that is related to signaling is *screening*. With screening, prospective employers attempt to sort out job applicants according to their characteristics. One type of screening mechanism is the process of *self-selection*. In this case, job applicants are informed by the prospective employer that certain characteristics are required. Only the applicants know whether they have these characteristics. Those who do will continue the application process.

CHAPTER QUESTIONS

16.1 Explain the difference between a valid and an invalid information set.

16.2 Explain how a Harsanyi transformation transforms a static game with incomplete information into a static game with imperfect information. Is it possible to find a Bayesian Nash equilibrium for such static games? Why?

16.3 Explain how a Harsanyi transformation transforms a dynamic game with incomplete information into a static game with imperfect information. Is it possible to find a perfect Bayesian equilibrium for such games? Why?

16.4 Dynamic games with incomplete information that have been transformed into games with imperfect information using a Harsanyi transformation do not have proper subgames. How does this affect our ability to find an equilibrium strategy profile using the backward induction solution algorithm?

16.5 In dynamic games with incomplete information, the pattern of a player's prior moves may reveal private information to his or her opponent, and *vice versa*. Describe the process by which a rational player will revise his or her information set based on this revealed information.

16.6 Dynamic games with incomplete information typically have only one subgame. Explain.

16.7 Explain why implausible Nash equilibria in dynamic games with imperfect information are subgame perfect.

16.8 Explain why it is difficult to find an equilibrium strategy profile for dynamic games with incomplete information.

16.9 Asymmetric information gives rise to the problems of adverse selection and moral hazard. Explain.

16.10 What is the market for lemons? In what way do you think used-car dealers help reduce the problem of adverse selection?

16.11   How does the lemons problem manifest itself in the labor market?

16.12   How might government regulation help remedy the problem of asymmetric information in financial markets?

16.13   Would you be more or less willing to make a business loan to a friend who had invested his or her life savings in a business? Why?

16.14   Do you think that the lemons problem is more or less severe for financial securities traded in the New York Stock Exchange or in the over-the-counter market? Explain.

16.15   In terms of signaling, explain why graduates of Ivy League universities receive higher starting salaries than do graduates from less prestigious institutions. Why is this a problem of adverse selection, and do you believe that this outcome is fair? Explain.

16.16   In your opinion, what is a better signal for worker productivity: A so-so grade point average from a prestigious research university, or a high grade point average from a mediocre private or public college or university? Regardless of your answer, what other signals should job applicants provide to improve their chances of employment?

CHAPTER EXERCISES

16.1   Illustrate the strict-liability-with-contributory-negligence accident game depicted in Figure 14.4 in extensive form using a valid information set.

16.2   Illustrate the comparative-negligence accident game depicted in Figure 14.5 in extensive form using a valid information set.

16.3   Consider the oil-drilling game depicted in Figure 16.8.

   a. Without seismic testing, should GLOMAR drill for oil if the company believes that there is a 30 percent chance of finding oil?
   b. Suppose GLOMAR conducts seismic testing and determines that the conditional probability of a positive test $P(A|B_1) = 0.75$, and the conditional probability that the test is negative when no oil exists is $P(A|B_2) = 0.40$. Should the company drill for oil?

16.4   Consider the Spence education game depicted in Figure 16.10. Suppose that the firm's belief profile is $\{C: (HP: 0.80, LP: 0.20), HS: (HP: 0.50, LP: 0.50)\}$.

   a. Use Bayes' theory to calculate conditional expected payoffs? Illustrate the extensive form of this game.
   b. Using a pooling strategy, what is the firm's optimal hiring strategy?
   c. Using a separating strategy, what is the firm's optimal hiring strategy?
   d. Compare your answers to parts b and c. What, if anything, can you say about the existence of a unique perfect Bayesian equilibrium in signaling games? Be sure to mention the importance of the firm's beliefs, states of nature, and the signaling strategy used.

16.5   Consider the Spence education game depicted in Figure 16.10. Suppose that the firm's belief profile is $\{C: (HP: 0.20, LP: 0.80), HS: (HP: 0.80, LP: 0.20)\}$.

   a. Use Bayes' theory to calculate conditional expected payoffs. Illustrate the extensive form of this game.
   b. Using a pooling strategy, what is the firm's optimal hiring strategy?
   c. Using a separating strategy, what is the firm's optimal hiring strategy?

    d. Compare your answers to parts b and c. What, if anything, can you say about the existence of a unique perfect Bayesian equilibrium in signaling games? Be sure to mention the importance of the firm's beliefs, states of nature, and the signaling strategy used.

ENDNOTE

1. Largely on the basis of his pioneering work on signaling, Michael Spence was awarded the 1981 John Bates Clark Medal by the American Economic Association. First awarded in 1947 to Paul Samuelson, the John Bates Clark medal is awarded every other year to the most promising economist under the age of forty. Michael Spence was further recognized for his work in signaling, which spawned an enormous growth in the literature on contract theory, when he shared the 2001 Nobel Prize in economics with George Akerlof and Joseph Stiglitz.

GLOSSARY OF TERMS AND CONCEPTS

Adverse selection An *ex ante* asymmetric information problem in which the market becomes crowded with individuals or products having undesirable characteristics.

Asymmetric impatience The unequal reduction in the gains to all players by failing to quickly reach a bargaining agreement. The players have different discount factors.

Asymmetric information When some individuals engaged in transactions have more or better information than others.

Auction A process whereby individuals or institutions compete with each other to buy or sell something of value.

Auctioneer The seller's agent at an auction.

Auction environment The characteristics of the auction participants, including the number of bidders, their risk preferences, and information about how rival bidders value the good or service being sold.

Auction rules The procedures that govern the bidding process, including who may bid, which bids are acceptable, the manner in which bids are submitted, the nature of the information available to the players during the bidding process, when the auction ends, how the winner is determined, and the price paid.

Autarky When a country does not engage in international trade.

Backward induction A method for finding the outcome of a game by projecting forward and reasoning backward.

Bargaining A process whereby individuals or groups of individuals negotiate over the terms of a contract.

Bayes' theorem A statistical relationship that enables a player to update the probability of an event after learning that some other event has occurred.

Bayesian Nash equilibrium A Nash equilibrium for a static Bayesian game.

Bayesian updating A process whereby players revise their expectations based on a rival's prior moves.

Belief profile A player's probability assessment at each nontrivial information set.

Bertrand model A static price-setting game in which firms in the same industry cannot subsequently switch strategies without great cost. In the Bertrand model, industry output adjusts to firms' pricing decisions.

Bertrand Nash equilibrium A Nash equilibrium for a Bertrand price-setting game.

Bertrand paradox The Nash equilibrium in which duopolies that produce a homogeneous product and have symmetric marginal cost charge a price equal to marginal cost and earns zero economic profit. This is the same outcome as in perfect competition.

Best-response function The same thing as a reaction function.

Bluffing An attempt by a player to gain a strategic advantage over a rival through a display of bravado that has no basis in fact.

Branches Arrows from decision nodes in game trees indicating a player's possible moves.

Buyer's reservation price The maximum price that a buyer in a bargaining scenario is willing to pay.

Cartel A formal agreement among firms in an industry to allocate market shares and/or increase industry profits.

Cheating rule for a finitely repeated game with an uncertain end A rule that predicts the stability of a cartel in games that are finitely repeated, but in which the end of the game is uncertain. This rule is based on the present value of the stream of future payoffs from not cooperating, cooperating, and defecting, and the probability that the game will end in each stage of the game.

Cheating rule for infinitely repeated games A rule that predicts the stability of a cartel in games that are infinitely repeated. This rule is based on the present value of the stream of future payoffs from not cooperating, cooperating, and defecting.

Closed-bid auction An auction in which participation is by invitation only.

Coase theorem The assignment of well-defined private-property rights will result in a socially efficient allocation of productive resources and a socially optimal, market-determined level of goods and services.

Coefficient of variation A dimensionless number that is used to compare the relative riskiness of two or more investments involving different expected values. It is calculated as the ratio of the standard deviation of the investment to its mean.

Collusion When firms in an industry coordinate their activities to restrict competition to increase market power and profits.

Commercial policy Policies adopted by a government to influence the quantity and composition of a country's international trade.

Common law The corpus of legal principles created largely by judges as the by-product of deciding cases. Common law is shaped by legal precedent rather than legislation.

Common-value auction An auction in which the true value of the object is the same for all bidders, but this value is unknown. Players' private value estimates reflect differences in information about the object's true value.

Comparative-negligence tort law A legal rule in which both tortfeasor and victim share liability if both are negligent.

Complete information When all players' strategies and payoffs are common knowledge.

Complete information auction An auction in which the bidders' private valuations are common knowledge.

Constant returns to scale When a proportional increase in the use of all inputs results in the same proportional increase in total output.

Consumer surplus The difference between what a consumer is willing to pay for a given quantity of a good or service and the amount that he or she actually pays.

Contestable market An industry with an unlimited number of potential competitors producing a homogeneous product using identical production technologies, have no significant sunk costs, and are Bertrand competitors.

Contestable monopoly A contestable market consisting of a single firm.

Contract A formal agreement that obligates the parties to perform, or refrain from performing, some specified act in exchange for something of value.

Cooperative game A strategic situation in which the players agree to coordinate strategies, usually to achieve a mutually beneficial outcome.

Coordination game A game in which rational players attempt to coordinate their strategies to achieve a mutually beneficial outcome.

Correlated value estimate When a private value estimate reflects a player's information about an object being auctioned, and where changes in this estimate affect, and are affected by, the changes in private value estimates of rival bidders.

Cournot-Nash equilibrium A Nash equilibrium for a Cournot output-setting game.

Cournot model A static output-setting game in which firms in the same industry cannot subsequently switch strategies without great cost. In the Cournot model, prices adjust to firms' output decisions to clear the market.

Credible threat When it is in a player's best interest to follow through with his or her threat.

Damages An amount of money or other compensation awarded by the court to the plaintiff in a lawsuit.

Decision node The location in a game where the designated player decides his or her next move.

Decision tree A diagram used to determine an optimal course of action in a situation that does not involve strategic interaction.

Differentiated products When there are real or perceived differences in products produced by different firms in the same industry. Differentiated goods are close, but not perfect, substitutes.

Discount factor The rate used for finding the present value of the gains from a bargaining agreement. The greater the players' impatience (the higher the discount rate) the less advantageous will be the gains from bargaining.

Discount rate The rate used to discount to the present the flow of future payoffs.

Dominant strategy A strictly or weakly dominant strategy.

Dropout price The price at which a player will drop out of an English auction. The dropout price exceeds his or her valuation of the object being auctioned.

Duopoly An industry consisting of two firms producing homogeneous or differentiated products.

Dutch auction An auction in which an auctioneer incrementally lowers an initially very high price. The first bidder to accept the current price wins the auction.

Dynamic game The same thing as a sequential-move game.

Economies of scale A decline in a firm's per unit cost of production following a proportional increase in the use of all factors of production.

End-of-game problem For finitely repeated games with a certain end, each stage effectively becomes the final stage, in which case the game reduces to a series of noncooperative, one-time games.

English auction An auction in which the auctioneer announces prices in ascending order. Bidders decide to accept or pass on the announced price. The last remaining bidder wins the auction.

Equilibrium A situation from which there is no tendency to change, unless acted upon by an outside force.

Evolutionary equilibrium A population-dynamic equilibrium strategy profile that results when successful strategies replace unsuccessful strategies. In the end, only successful strategies will be selected.

Evolutionary game theory A branch of game theory which posits that animal behavior can be explained in terms of instinctual strategies that are genetically passed along from generation to generation.

Export subsidy A payment made by government to a firm or industry to promote overseas sales.

Extensive-form game A diagrammatic representation of a dynamic game that summarizes the players, the stages of the game, the information available to each player at each stage, player strategies, the order of the moves, and the payoffs from alternative strategies.

Externality A cost or benefit resulting from a market transaction that is imposed upon third parties.

Fair gamble When the expected value of winning a wager is equal to the expected value of losing.

Finitely repeated game A game that is played a limited number of times. Finitely repeated games may have a certain or an uncertain end.

First-mover advantage When a player who can commit to a strategy first enjoys a payoff that is no worse than if all players move simultaneously.

Focal-point equilibrium When a single strategy profile stands out from among multiple Nash equilibria because the players share a common understanding of the environment in which the game is being played.

Fold-back method The same thing as backward induction.

Framing When players make decisions based on a familiar frame of reference.

Game theory The study of how rivals make decisions in situations involving strategic interaction, i.e., move and countermove.

Game tree The complete collection of decision nodes and branches in an extensive-form game.

Government franchise A publicly authorized monopoly.

Grim strategy A trigger strategy that involves the most punitive response to a rival's defection from a cooperative agreement.

Harsanyi transformation Transforms games with incomplete information into games with complete but imperfect information by having Nature move first to determine states of nature and/or a player type.

Homogeneous products When there are no real or perceived differences in products produced by different firms in the same industry. Differentiated goods are perfect substitutes.

Imperfect information When players move without knowledge of the moves by other players.

Improper subgame A subgame with a subroot located within a valid information set.

Incentive-compatible direct mechanism A Bayesian game is a *direct mechanism* if players can report their types to a neutral referee. This information is used to determine the players' payoffs. This direct mechanism is incentive compatible if truthful reports of player types result in a Bayesian Nash equilibrium.

Incomplete information When players' strategies and payoffs are not common knowledge.

Increasing returns to scale When a proportional increase in the use of all inputs results in a more than proportional increase in total output.

Independent investments When acceptance of one investment project does not preclude acceptance of other investment projects.

Independent private values When players know their own valuation of an object being auctioned, but not the bidders' valuations.

Infinitely repeated game A game that is played over and over without end.

Information partition A collection of information sets.

Information set A collection of decision nodes that are indistinguishable to a player.

Instant messaging The exchange of real-time text messages between and among users of the Internet.

Intraindustry trade When a country exports and imports the same, or very similar, goods and services.

Invalid information set A collection of decision nodes for two or more players where a decision node by one player precedes the decision node by another player in the same information set.

Investor indifference curve Summarizes the combinations of risk and expected rates of return for which an investor is indifferent between a risky and a risk-free investment.

Iterated strictly dominant strategy equilibrium A strictly dominant strategy equilibrium that is obtained after all strictly dominated strategies have been eliminated.

Jump bid A bid in a multiple-round auction that exceeds the previous bid by more than the minimum allowable increment.

"Kinked" demand curve model The same thing as the Sweezy model.

Last-mover advantage When a player is able to dictate the final terms of a negotiated agreement.

Liability Legal responsibility for an act of commission or omission.

Marginal utility of money The additional satisfaction received from an additional dollar of money income or wealth

Market failure When the forces of supply and demand fail to generate socially efficient levels of consumption and production.

Market power The ability of a firm to influence the market-clearing price of its product by significantly changing industry output.

Market structure The nature and degree of competition in an industry.

Maximin strategy A strategy that selects the best payoff from among the worst payoffs. By adopting a maximin strategy, a player avoids the worst possible outcome.

Mean The weighted average of a set of random outcomes, where the weights are the probabilities of each outcome.

Mechanism design The branch of game theory that deals with the design of games having certain desirable properties.

Minimax regret strategy A strategy that minimizes the maximum possible opportunity loss of an incorrect decision.

Minimax theorem In a zero-sum, noncooperative game, a player will attempt to minimize a rival's maximum (minimax) payoff, while maximizing his or her own minimum (maximin) payoff. The minimax and the maximin payoffs of all players will be equal.

Mixed strategy A game plan that involves randomly mixing pure strategies.

Mixed-strategy Nash equilibrium A unique Nash equilibrium that occurs when players adopt their optimal mixing rules.

Monopolistic competition A market structure characterized by a large number of firms producing differentiated products in which entry into and exit from the industry is relatively easy.

Monopoly An industry consisting of a single producer of a good for which there are no substitutes, and where entry into the industry is impossible.

Moral hazard An *ex post* asymmetric information problem in which the risk (hazard) that a party to a transaction engages in activities that are undesirable (immoral) from the perspective of the other party to the same transaction.

Multiple-item Dutch auction A variation of a standard Dutch auction involving multiple bids for blocks of identical items. The seller accepts the lowest bid that will dispose of the entire amount placed at auction. Each player whose bid is accepted pays the same price.

Multistage game The same thing as a sequential-move game.

Mutual interdependence in pricing and output decisions When the pricing and output decisions of a firm affects, and are affected by, the pricing and output decisions of firms in the same or different industries.

Mutually exclusive investments When acceptance of one investment project precludes acceptance of other investment projects.

Nash bargaining A noncooperative static game in which the players bargain over the distribution of a divisible object of value.

Nash equilibrium When each player adopts a strategy that is the best response to the strategies adopted by rivals. A strategy profile is a Nash equilibrium when no player can improve his or her payoff by switching strategies.

Nash equilibrium existence theorem Randomly mixing pure strategies using an optimal mixing rule is sufficient to guarantee that every game with a finite number of players and pure strategies will have a unique Nash equilibrium.

Natural monopoly When a firm that can satisfy total market demand at lower per unit cost than an industry consisting of two or more firms.

Negligence A failure to exercise the same care that a reasonable or prudent person would exercise under the same circumstances, or to take an action that a reasonable person would not.

Negligence-with-contributory-negligence tort law A legal rule that allows the victim to recover damages if the tortfeasor alone is negligent.

Network When the activities of groups of players sharing a common technology have lower costs and greater benefits than if individual players use different technologies.

No-liability tort law A legal rule in which neither the tortfeasor nor the victim is held liable for damages in the event of an injury.

Noncooperative game A strategic situation in which players do not agree to coordinate their strategies.

Nondominant strategy A strategy that is neither strictly nor weakly dominant. A player's best strategy depends on what he or she believes is the strategy adopted by a rival.

Non-zero-sum game When one player's gain does not equal another player's loss. In a non-zero-sum game, both players may be made better or worse off.

Normal-form game Summarizes the players and payoffs from alternative strategies in a static game.

Oligopoly An industry dominated by a few large firms producing identical or closely related products in which the pricing and output decisions of rivals are interdependent.

One-dollar, one-vote rule The arbitrary value judgment that a dollar transferred to one group is equivalent in value to a dollar transferred from another group.

One-time game A game that is played just once.

Open ascending bid auction The same thing as an English auction.

Open-bid auction An auction in which anyone may submit a bid.

Open descending bid auction The same thing as a Dutch auction.

Opportunity loss The difference between a given payoff and the best possible payoff.

Optimal mixing rule A randomized mix of pure strategies that maximizes a player's minimum payoff, while minimizing a rival's maximum payoff.

Oral auction An auction in which bids are announced publicly.

Patent An exclusive right granted to an inventor by government to a product or a process.

Payoff The gain or loss to a player at the conclusion of a game.

Perfect Bayesian equilibrium A Nash equilibrium for dynamic games that is consistent with the probability distribution of states of nature and player beliefs.

Perfect competition A market structure consisting of a large number of utility-maximizing buyers and profit-maximizing sellers of a homogeneous good or service in which factors of production are perfectly mobile, buyers and sellers have perfect information, and entry into and exit from the industry is very easy.

Perfect information When each player is aware of a rival's prior moves. A player with perfect information knows his or her location in a game tree.

Perfect recall The same thing as total recall.

Player A decision maker in a game.

Political rent seeking When one group attempts to gain special benefits from the government at the expense of taxpayers or some other group.

Pooling equilibrium A Bayesian equilibrium that arises when the same strategy is adopted for other players of different types.

Pooling strategy When the same strategy is adopted for players of different types. A pooling strategy may not be used as a perfect signal for other player types.

Positive feedback effects When the benefits received by a player joining a network are shared by existing members of the same network. The value of network membership rises with an increase in the number of members.

Positive network externalities The same thing as positive feedback effects.

Preemption An example of a strategic move in which a player moves first in an attempt to limit a rival's strategy options.

Price discrimination The practice of charging different consumers, or groups of consumers, different prices for the same good or service.

Price leadership A form of price collusion in which a dominant firm initiates a price change that is matched by the rest of the industry.

Price maker A firm with market power.

Price taker A firm that does not have market power.

Prisoner's dilemma A game in which it is in the best interest of all players to cooperate, but where each player has an incentive to adopt his or her dominant strategy.

Procurement auction When multiple sellers bid for the right to sell something of value to a single buyer.

Producer surplus The difference between the total revenues earned from the production and sale of a given quantity of output and the minimum that the firm would accept to produce that output.

Proper subgame A subgame that begins at a decision node other than the root of the game tree.

Pure strategy A complete and nonrandom game plan.

Pure-strict-liability tort law A legal rule in which the tortfeasor is held liable for damages, regardless of whether he or she exercised due diligence.

Rational behavior Players endeavor to optimize their payoffs.

Reaction function A relationship that expresses the best response by a player to the moves of a rival.

Repeated game A game that is played more than once.

Reproductive success A weaker assumption than rationality which asserts that successful strategies replace unsuccessful strategies over time.

Reservation price The minimum price that a seller will accept in a standard auction.

Residual demand In the Cournot model, it is the possible price and output combinations for a firm given the output levels of rivals.

Revelation theorem For any player type, the Bayesian Nash equilibrium of a static Bayesian game can be represented by a properly constructed incentive-compatible direct mechanism.

Reverse auction The same thing as a procurement auction.

Risk Multiple payoffs from alternative states of nature with known probabilities.

Risk aversion When an individual prefers the certain value of a wager to its expected value. In general, a risk-averse individual would not accept a fair gamble.

Risk loving When an individual prefers the expected value of a wager to its certain value.

Risk neutral When an individual is indifferent between the certain value of a wager to its expected value.

Risk premium The difference between the expected rate of return on a risky investment and the expected rate of return on a risk-free investment.

Root of a game tree The first decision node in a game tree.

Savage strategy The same thing as the minimax regret strategy.

Schelling-point equilibrium The same thing as a focal-point equilibrium.

Scorched-earth policy A preemptive strategy in which a player commits to destroying his or her assets in an effort to alter a rival's behavior.

Screening When an uninformed individual attempts to sort players according to defined characteristics.

Sealed-bid auction An auction in which bids are secretly submitted.

Sealed-bid, first-price auction An auction in which the winner submits and pays the highest sealed bid.

Sealed-bid, second-price auction An auction in which the winner submits the highest sealed bid, but pays the bid submitted by the second-highest bidder.

Secure strategy The same thing as a maximin strategy.

Self-fulfilling strategy profile When the players believe it is in their collective best interest to adopt a particular strategy profile.

Self-selection When players voluntarily sort themselves into different groups according to defined characteristics.

Seller's reservation price The minimum price that a seller in a bargaining scenario is willing to accept.

Separating equilibrium A Bayesian equilibrium that arises when a different strategy is adopted for players of different types.

Separating strategy When a different strategy is adopted for different player types. A separating strategy may be used as a perfect signal for other player types.

Sequential-move game A game in which the players take turns.

Signaling When an informed player transmits a credible and unreproducible signal to an uninformed player. This signal is intended to convey information that is important to the sender but is otherwise unobservable.

Simultaneous-move game A game in which all players move at the same time.

Stackelberg leader A player who has the ability to commit to a strategy in a dynamic, output-setting game.

Stackelberg model A theory of strategic interaction in which one firm, the Stackelberg leader, believes that its rival, the Stackelberg follower, will not alter its level of output. The production decisions of the Stackelberg leader will exploit the anticipated behavior of the Stackelberg follower.

Standard auction When multiple bidders compete to acquire an object of value from a single seller.

Standard deviation The square root of the variance. The standard deviation is a commonly used measure of the riskiness of an investment.

Static game A game in which the players are ignorant of their rivals' decisions until all moves have been made. A simultaneous move game is an example of a static game.

Static Bayesian game A static game with incomplete information.

Strategic behavior When the actions of an individual or group affect, and are affected by, the actions of other individuals or groups.

Strategic form of a sequential-move game Summarizes the payoffs to each player from every possible strategy response profile.

Strategic move An attempt by a player to gain a strategic advantage by altering the behavior of a rival.

Strategic trade policy Measures adopted by government to influence the decisions of consumers and producers engaged in international trade.

Strategy A decision rule that defines a player's moves. It is a complete description of a player's decisions at each stage of a game.

Strategy profile The collection of all players' strategies.

Strict-liability-with-contributory-negligence tort law A legal rule in which the tortfeasor is responsible for damages when both the tortfeasor and the victim exercise due diligence.

Strictly dominant strategy A strategy that strictly dominates every other strategy. It is a strategy that results in the best payoff given the strategies adopted by the other players.

Strictly dominant strategy equilibrium A Nash equilibrium that results when each player has, and adopts, a strictly dominant strategy.

Strictly dominated strategy A strategy that is dominated by every other strategy.

Subgame A subset of branches and decision nodes in a game tree.

Subgame equilibrium The Nash equilibrium of a proper subgame.

Subgame perfect equilibrium The Nash equilibrium for a dynamic game, which found among the subgame equilibria.

Subroot The root of a subgame.

Sunk cost A cost that is not recoverable once incurred.

Sweezy model A model of firm behavior that attempts to explain infrequent price changes in oligopolistic industries. The model postulates that a firm will not raise its price because this will not be matched by rivals and will result in a loss of market share. Neither will it lower its price since this will be matched by rivals, resulting in no gain in market share and a loss of revenues and profits.

Symmetric impatience The equal reduction in the gains to all players by failing to quickly reach a bargaining agreement. Each player has the same discount factor.

Terminal node The point on a game tree where a game ends.

Tit-for-tat strategy An enforcement mechanism used in which a player does not knowingly violate an agreement, but neither will he or she allow defection to go unpunished nor quietly accept punishment.

Tort law A branch of the law dealing with torts, which is an act committed by one person that results in an injury to another.

Total recall When players remember everything they ever learned about a game being played, including earlier moves made by other players.

Trigger strategy A strategy adopted by one player in response to an unanticipated move by a rival. A trigger strategy will continue to be used until a rival makes another unanticipated move.

Trivial subgame A subgame that begins at the root of the game tree.

Type profile A complete listing of player types.

Uncertainty Multiple payoffs from alternative states of nature with unknown or meaningless probabilities.

Unfair gamble When the expected value of winning a wager is less than the expected value of losing.

Valid information set A collection of decision nodes for the same player where no decision node precedes any other decision node in the same information.

Variance A measure of the dispersion of a set of random outcomes. It is the sum of the products of the squared deviations of each outcome from its mean and the probability of each outcome.

Vickery auction The same thing as a sealed-bid, second-price auction.

Weakly dominant strategy A strategy that results in a payoff that is no lower than any other payoff regardless of the strategy adopted by the other player.

Weakly dominant strategy equilibrium A Nash equilibrium that results when both players adopt a weakly dominant strategy.

Winner's curse When a player wins a common-value auction with a bid that exceeds every other bidder's the estimate of the value of the object being auctioned.

Zero-sum game When one player's gain is another player's loss.

REFERENCES AND SUGGESTIONS FOR FURTHER READING

Akerlof, G. 1970. "The Market for 'Lemons': Quality, Uncertainty, and the Market Mechanism." *Quarterly Journal of Economics* 84: 488–500.

Aumann, R. 1987. "Correlated Equilibrium as an Extension of Bayesian Rationality." *Econometrica* 55(1): 1–18.

Axelrod, R. 1984. *The Evolution of Cooperation*. New York: Basic Books.

Baghwati J., V. Ramaswami, and T. Srinivasam. 1969. "Domestic Distortions, Tariffs and the Theory of Optimum Subsidy: Some Further Results." *Journal of Political Economy* 77: 1005–10.

Baigent, M., R. Leigh, and H. Lincoln. 1982. *Holy Blood, Holy Grail*. New York: Delacorte Press.

Bain, J. S. 1956. *Barriers to New Competition*. Cambridge, MA: Harvard University Press.

Baird, D. G., R. H. Gertner, and R. C. Picker. 1994. *Game Theory and the Law*. Cambridge, MA: Harvard University Press.

Bajari, P. and A. Hortacsu. 2004. "Economic Insights from Internet Auctions." *Journal of Economic Literature* 42(2): 457–486.

Barry, S., J. Levinsohn, and A. Pakes. 1995. "Voluntary Export Restraints on Automobiles: Evaluating a Strategic Trade Policy." *National Bureau of Economic Research Working Paper No. 5235* (August).

———. 1995. "Voluntary Export Restraints in Automobiles." *American Economic Review* 89(3): 400–430.

Baumol, W. J., J. C. Panzar, and R. D. Willig. 1982. *Contestable Markets and the Theory of Industry Structure*. New York: Harcourt Brace Jovanovich.

Baye, M. R. 2003. *Managerial Economics and Business Strategy*. Boston: McGraw-Hill Irwin.

Bazerman, M. and W. Samuelson. 1983. "I Won the Auction But Don't Want the Prize." *Journal of Conflict Resolution* 27(4): 618–34.

———. 1983. "The Winner's Curse: An Empirical Investigation." In *Aspiration Levels in Bargaining and Economic Decision Making*, ed. Reinhard Tietz, 186–200. New York: Springer-Verlag.

Berg, A. S. 1989. *Goldwyn: A Biography*. New York: Alfred A. Knopf.

Bernoulli, D. 1738. "Specimen Theoriae Novae de Mensura Sortis." *Commentarii Academiae Scientiarum Imperialis Petropolitannae*. Translated by L. Sommer as "Exposition of a New Theory on the Measurement of Risk," *Econometrica*, 22, (1954): 23–36.

Bertrand, J. 1883. "Théorie Mathématique de la Richesse Sociale." *Journal des Savants* 499–508.

Besanko, D., D. Dranove, and M. Shanley. 2000. *Economics of Strategy*. 2d ed. New York: John Wiley & Sons.

Bierman, H. S. and L. Fernandez. 1998. *Game Theory with Economic Applications*, 2d ed. New York: Addison-Wesley.

Binmore, K. 1991. *Fun and Games: A Text on Game Theory*. Lexington, MA: D. C. Heath.

Binmore, K. and P. Dasgupta. 1986. *Economic Organizations as Games*. Oxford: Basil Blackwell.

Borel, E. 1921. "La Theorie du Jeu et les Equations Integrales a Noyen Symetrique Gauche." *Comptes Rendus Hebdomadaires des Seances de l'Academie Sciences* 173: 1304–1308. Translated by Leonard J. Savage as "The Theory of Play and Integral Equations with Skew Symmetric Kernels," *Econometrica* 21 (1953): 97–100.

———. 1923. "Sur les Jeux ou Interviennent L'Hasard et L'Habilete des Joueurs." *Association Francaise pour l'Advancement des Sciences*: 79–85.

———. 1924. *Elements de la Theorie des Probabilites*. 3d ed. Paris: Librairie Scientifique, J. Hermann: 204–221. Translated by Leonard J. Savage as "On Games that Involve Chance and the Skill of the Players," *Econometrica* 21 (1953): 101–115.

———. 1924. "Sur les Jeux ou Interviennent L'Hasard et L'Habilete des Joueurs." *Theorie des Probabilites*. Paris: Librairie Scientifique, J. Hermann: 204–224.

———. 1926. "Un Theoreme sur les Systemes de Formes Lineaires a Determinant Symetrique Gauche." *Comptes Rendus Academie des Sciences* 183: 925–927, Avec Erratum: 996.

———. 1927. "Sur les Systemes de Formes Lineaires a Determinant Symetrique Gauche et la Theorie Generale du Jeu." *Comptes Rendus Hebdomadaires des Seances de l'Academie de Sciences* 184: 52–53. Translated by Leonard J. Savage as "On Systems of Linear Forms of Skew Symmetric Determinant and the General Theory of Play," *Econometrica* 21 (1953): 116–117.

———. 1938. "Traite du Calcul des Probabilites et de Ses Applications." *Applications des jeux de hasard* (Lectures by Emile Borel, transcribed by Jean Ville). Paris: Gauthier-Villars, Tome. IV, Fascicule 2.

———. 1938. "Jeux ou la Psychologie Joue un Role Fondamental." *Applications Des Jeux de Hasard* (Lectures by Emile Borel, transcribed by Jean Ville). Paris: Gauthier-Villars, Tome. IV, Fascicule 2.

Boyes, W. 2004. *The New Managerial Economics*. Boston: Houghton Mifflin.

Brams, S. and M. Kilgour. 1988. *Game Theory and National Security*. Oxford: Basil Blackwell.

Brander, J. 1981. "Intra-Industry Trade in Identical Commodities." *Journal of International Economics* 11: 1–14. Reprinted in *Intra-Industry Trade,* ed. P. Lloyd and H. Grubel. Camberly, UK: Edward Elgar, 2002.

Brander, J. A. and B. J. Spencer. 1981. "Tariffs and the Extraction of Foreign Monopoly Rent Under Potential Entry." *Canadian Journal of Economics* 14: 371–389. Reprinted in *Readings in International Trade,* ed. J. Bhagwati. Cambridge, MA: MIT Press, 1987.

———. 1983. "International R&D Rivalry and Industrial Strategy," *Review of Economic Studies* 50: 707–722.

———. 1985. "Export Subsidies and International Market Share Rivalry." *Journal of International Economics* 18: 83–100. Reprinted in *International Trade, Volume 1. Welfare and Trade Policy*, ed. P. Neary, Aldershot, UK and Brookfield, VT: Edward Elgar, 1995.

Brown, D. 2003. *The Da Vinci Code*. New York: Doubleday.

Burnham, T and J. Phelan. 2000. *Mean Genes: From Sex to Money to Food: Taming Our Primal Instincts*. New York: Penguin Books.

Cabral, L. M. B. 2000. *Industrial Organization*. Cambridge, MA: MIT Press.

Carter, T. 1998. "Rankled by the Rankings." *ABA Journal* 84: 46–53.

Chamberlin, E. H. 1933. *The Theory of Monopolistic Competition*. Cambridge, MA: Harvard University Press.

Chiappori, P. A., S. Levitt, and T. Groseclose. 2002. "Testing Mixed-Strategy Equilibria When Players Are Heterogeneous: The Case of Penalty Kicks in Soccer." *American Economic Review* 92(4): 1138–1151.

Chiang, A. 1974. *Fundamental Methods of Mathematical Economics*. 2d ed. New York: McGraw-Hill.

Coase, R. H. 1960. "The Problem of Social Costs." *Journal of Law and Economics* 3 (October): 1–44.

Commack, E. B. 1991. "Evidence on Bidding Strategies and the Information in Treasury Bill Auctions." *Journal of Political Economy* 99(1): 100–130.

Conway, E. 2007. "Stern Backs Global Carbon Tax to Avoid 'Biggest Market Failure'." *Telegraph. co.uk* (January 26). <http://www.telegraph.co.uk/money/main.jhtml?xml=/money/2007/01/25/cndavosstern125.xml>.

Cooper, R., D. V. DeJong, R. Forsythe, and T. W. Ross. 1996. "Cooperation without reputation: Experimental evidence from Prisoner's Dilemma games." *Games and Economic Behavior* 12 (February): 187–218.

Cooter, R. and K. Kornhauser. 1980. "Can Litigation Improve the Law Without the Help of Judges." *Journal of Legal Studies* 9(1): 139–163.

Cooter, R. and T. Ulen. 1988. *Law and Economics*. Glenview: Scott, Foresman & Co.

Cournot, A. 1838. *Recherches Sur les Principes Mathematiques de la Theorie des Richesses*. Paris: Hachette. Translated by Nathaniel T. Bacon as *Researches into the Mathematical Principles of the Theory of Wealth*. New York: Macmillan, 1897.

Crissey, M. 1997. "Changes in Annual College Guides Fail to Quell Criticisms on Their Validity." *The Chronicle of Higher Education* 44(2) (September 5): A67.

Darwin, C. 1859. *On the Origin of Species by Means of Natural Selection*. London: John Murray.

Davis, D. D. and C. A. Holt. 1993. *Experimental Economics*. Princeton: Princeton University Press.

Davis, M. D. 1997. *Game Theory: A Nontechnical Introduction*. Mineola, NY: Dover Publications.

Davis, O. and A. Whinston. 1962. "Externalities, Welfare, and the Theory of Games." *Journal of Political Economy* 70 (June): 241–262.

de Fraja, G. and F. Delbono. 1990. "Game Theoretic Models of Mixed Oligopoly." *Economic Surveys* 4: 1–17.

Dimand, R. W. and M. Dimand. 1992. "The Early History of the Theory of Strategic Games from Waldegrave to Borel." In *Toward a History of Game Theory*, ed. E. R. Weintraub. Durham: Duke University Press.

Dixit, A. 1979. "A Model of Duopoly Suggesting a Theory of Entry-Barriers." *Bell Journal of Economics* 10: 20–32.

———. 1984. "International Trade Policy for Oligopolistic Industries." *Economic Journal* 94 (supplement): 1–16.

Dixit, A. and B. Nalebuff. 1991. *Thinking Strategically: The Competitive Edge in Business, Politics, and Everyday Life*. New York: W. W. Norton.

Dixit, A. and S. Skeath. 1999. *Games and Strategy*. New York: W. W. Norton.

Doyle, A. C. 1890. *The Sign of the Four*. London: Lippencott's Monthly Magazine.

Eaton, J. and G. Grossman. 1986. "Optimal Trade and Industrial Policy Under Oligopoly." *Quarterly Journal of Economics* 101: 383–406. Reprinted in *Readings in International Trade*, ed. J. Bhagwati. Cambridge, MA: MIT Press, 1987.

Friedman, J. W. 1990. *Game Theory with Applications to Economics*, 2d ed. New York: Oxford University Press.

Fudenberg, D. and D. K. Levine. 1998. *The Theory of Learning in Games*. Cambridge, MA: MIT Press.

Fudenberg, D. and J. Tirole. 1991. *Game Theory*. Cambridge, MA: MIT Press.

Fuller, J. 1962. *The Gentlemen Conspirators: The Story of the Price-Fixers in the Electrical Industry*. New York: Grove Press.

Garigliano, J. 1997. "U.S. News College Rankings Rankle Critics." *Folio*, 26: 4.

Gibbons, R. 1992. *Game Theory for Applied Economists*. Princeton, NJ: Princeton University Press.

Gilley, J. W. 1992. "Faust Goes to College." *Academe: Bulletin of the AAUP*, 78(3) (May/June): 9–11. Reprinted as "Best college' lists: The ranking game." *Current* 348 (December): 8–10.

Glass, S. 1997. "The College Ranking Scam." *Rolling Stone* (October 16): 93–94.

Gleick, E. 1995. "Playing the Numbers." *Time* 145(16) (April 17): 52.

Gould, J. P. 1973. "The Economics of Legal Conflicts." *Journal of Legal Studies* 2(2): 279–300.

Graham, H. D. and N. Diamond. 1999. "Academic Departments and the Rating Game." *The Chronicle of Higher Education* 45(41) (June 18): B6.

Hagenmayer, S. J. 1995. "Albert W. Tucker, 89, Famed Mathematician," *Philadelphia Inquirer* (February 2): B7.

Hansen, R. G. and W. F. Samuelson. 1988. "Evolution in Economic Games." *Journal of Economic Behavior and Organization* 10(3) (October): 315–338.

Hardin, G. 1968. "Tragedy of the Commons." *Science* 162(3859) (December 13): 1243–1248.

Harsanyi, J. C. 1968. "Games with Incomplete Information Played by 'Bayesian' Players. Parts I–III." *Management Science* 14(3): 159–182, 320–334, 486–502.

———. 1995. "Games with Incomplete Information." *American Economic Review* 85(3): 291–303.

Harsanyi, J. C. and R. Selten. 1972. "A Generalized Nash Solution for Two-Person Bargaining Games with Incomplete Information." *Management Science* 18(5): 80–106.

Heckscher, E. 1919. "The Effects of Foreign Trade on the Distribution of Income." Originally published in *Readings in the Theory of International Trade*, ed. Howard S. Ellis and Lloyd M. Metzler. Philadelphia: Blackiston, 1949.

Heywood, J. S. and X. Wei. 2004. "Education and Signaling: Evidence from a Highly Competitive Labor Market." *Education Economics* 12(1): 1–16.

Hofbauer, J. and K. Sigmund. 1998. *Evolutionary Games and Population Dynamics*. Cambridge: Cambridge University Press.

Horowitz, I. 1991. "On the Effects of Cournot Rivalry Between Entrepreneurial and Cooperative Firms." *Journal of Comparative Economics* 15 (March): 115–121.

Ivanova-Stenzel, R. and T. C. Salmon. 2004. "Bidder Preferences Among Auction Institutions." *Economic Inquiry* 42(2) (April): 223–236.

Kirk, S. A. and K. Corcoran. 1995. "School Rankings: Mindless Narcissism or Do They Tell Us Something?" *Journal of Social Work Education* 31(3): 408–414.

Kreps, D. M. 1987. "Nash Equilibrium." In *The New Palgrave: A Dictionary of Economics*, ed. J. Eatwell, M. Milgate, and P. Newman. London: Macmillan.

———. 1990. "Corporate Cultures." In J. E. Alt and K. A. Shepsle *Perspectives in Political Economy*. Cambridge: Cambridge University Press.

———. 1990. *A Course in Microeconomic Theory* (Part III). Princeton, NJ: Princeton University Press.

Krugman, P. 1984. "Import Protection as Export Promotion: International Competition in the Presence of Oligopoly and Economies of Scale." In *Monopolistic Competition and Product Differentiation and International Trade*, ed. H. Kierzkowski. Oxford: Oxford University Press.

Landes, W. M. 1971. "An Economic Analysis of the Courts." *Journal of Law and Economics* 61(14): 61–108.

Landes, W. M. and R. A. Posner. 1980. "Joint and Multiple Tortfeasors: An Economic Analysis." *Journal of Legal Studies* 9: 517–56.

Lapan, H. E. and T. Sandler. 1988. "To Bargain or Not to Bargain: That Is the Question." *American Economic Review, Papers and Proceedings* 78(2): 16–20.

Lewontin, R. C. 1961. "Evolution and the Theory of Games." In *Toward a History of Game Theory*, ed. E. R. Weintraub. Durham: Duke University Press.

Lind, B. and C. Plott. 1991. "The Winner's Curse: Expectations with Buyers and with Sellers." *American Economic Review* 81 (March): 225–346.

Lloyd, W. F. 1833. *Two Lectures on the Checks to Population*. Oxford: Oxford University. Press. Reprinted (in part) in *Population, Evolution, and Birth Control*, ed. G. Hardin. Freeman, San Francisco: Freeman 1964, p. 37.

Luce, R. D. and H. Raiffa. 1957. *Games and Decisions*. New York: John Wiley & Sons.

Machung, A. 1998. "Playing the Rankings Game." *Change*, 30(4) (July/August): 12–16.

Maynard Smith, J. 1982. Evolution and the Theory of Games. Cambridge: Cambridge University Press.

McAfee, R. P. and J. McMillan. 1987. "Auctions and Bidding." *Journal of Economic Literature* 25(2) (June): 699–783.

McDonald, J. 1950. *Strategy in Poker, Business, and War*. New York: W. W. Norton.

McMillan, J. 1994. "Selling Spectrum Rights." *Journal of Economic Perspectives* 8(3) (Summer): 145–162.

———. 1992. *Games, Strategies, & Managers: How Managers Can Use Game Theory to Make Better Decisions*. London: Oxford University Press.

Milgrom, P. R. and R. J. Weber. 1982. "A Theory of Auctions and Competitive Bidding." *Econometrica* (50): 1089–1122.

———. 1989. "Auctions and Bidding." *Journal of Economic Perspectives* 3(3) (Summer): 3–22.

Moulin, H. 1982. *Game Theory for the Social Sciences*. New York: New York University Press.

Myerson, R. B. 1981. "Optimal Auction Design." *Mathematics of Operations Research* 6(1): 58–73.

———. 1991. *Game Theory: Analysis of Conflict*. Cambridge, MA: Harvard University Press.

Myerson, R. B. and M. A. Satterthwaite. 1983. "Efficient Mechanisms for Bilateral Trading." *Journal of Economic Theory* 29: 265–81.

Nasar, S. 1998. *A Beautiful Mind*. New York: Simon & Schuster.

Nash, J. 1950a. "The Bargaining Problem." *Econometrica* 18. 155–162.

———. 1950b. "Equilibrium Points in *n*-Person Games." *Proceedings of the National Academy of Sciences, USA* 36: 48–49.

———. 1950c. "A Simple Three-Person Poker Game" (with Lloyd S. Shapley). *Annals of Mathematics Study* 24.

———. 1951. "Noncooperative Games." *Annals of Mathematics* 51: 286–295.

———. 1953a. "A Comparison of Treatments of a Duopoly Situation" (with J. P. Mayberry and M. Shubik). *Econometrica* 21: 141–154.

———. 1953b. "Two-Person Cooperative Games." *Econometrica* 21: 405–421.

Noussair, C. 1995. "Equilibrium in a Multi-Object Uniform Price Sealed-Bid Auction with Multi-Unit Demands." *Economic Theory* 5(2) (March): 337–351.

Ohlin, B. 1933. *International and Interregional Trade*. Cambridge, MA: Harvard University Press.

Osborne, M. J. 2006. *An Introduction to Game Theory*. London: Oxford University Press.

Osborne, M. J. and A. Rubinstein. 1994. *A Course in Game Theory*. Cambridge, MA: MIT Press.

Oster, C. V. and J. S. Strong. 2001. "Predatory Practices in the U.S. Airline Industry," (January 21). <http://ostpxweb.dot.gov/aviation/domestic-competition/predpractices.pdf>.

Porter, M. 1980. "General Electric vs. Westinghouse in Large Turbine Generators." *Harvard Business School Case No. 9-380-129.*

Porter, M. and A. M. Spence. 1982. "The Capacity Expansion Decision in a Growing Oligopoly: The Case of Corn Wet Milling." In *The Economics of Information of Uncertainty*, ed. J. J. McCall, 259–316. Chicago: University of Chicago Press.

Posner, R. A. 1979. "Utilitarianism, Economics, and Legal Theory." *Journal of Legal Studies* 8(1): 103–40.

———. 1992. *Economic Analysis of Law*, 4th edition. Boston: Little Brown.

Poundstone, W. 1992. *Prisoners' Dilemma: John von Neuman, Game Theory, and the Puzzle of the Bomb.* New York: Doubleday.

Pruett-Jones, S. and M. Pruett-Jones. 1994. "Sexual Competition and Courtship Disruptions: Why Do Bowerbirds Destroy Each Other's Bowers?" *Animal Behavior* 47: 607–20.

Rasmusen, E. 1989. *Games and Information: An Introduction to Game Theory.* New York: Basil Blackwell.

Ricardo, D. 1817. *On the Principles of Political Economy and Taxation.* London: John Murray.

Riley, J. G. 1989. "Expected Revenues from Open and Sealed-Bid Auctions." *Journal of Economic Perspectives* 3(3) (Summer): 41–50.

Rosenthal, R. W. 1982. "Games of Perfect Information, Predatory Pricing, and the Chain Store Paradox." *Journal of Economic Theory* 25: 92–100

———. 1993. "Rules of Thumb in Games." *Journal of Economic Behavior and Organization* 22(1) (September): 1–13.

Roth, A. E., J. K. Murnighan, and F. Schoumaker. 1988. "The Deadline Effect in Bargaining: Some Experimental Evidence." *American Economic Review* 78(4): 806–823.

Rubenstein, A. 1982. "Perfect Equilibrium in a Bargaining Model." *Econometrica* 61 (1): 97–109.

Rubin, P. 1977. "Why Is the Common Law Efficient?" *Journal of Legal Studies* 6 (1): 51–61.

Samuelson, L. 1998. *Evolutionary Games and Equilibrium Selection.* Cambridge, MA: MIT Press.

Schatz, M. D. 1993. "What's Wrong with MBA Ranking Surveys?" *Management Research News* 16(7): 15–18.

Schelling, T. 1960. *The Strategy of Conflict.* London: Oxford University Press.

———. 1989. "Strategy and Self-Command." *Negotiation Journal* (October): 343–347.

———. 2007. "Climate Change: The Uncertainties, the Certainties and What They Imply about Action." *The Economist's Voice* 4(3). <http://www.bepress.com/ev/vol4/iss3/art3>.

Schotter, A. 1985. *Free Market Economics: A Critical Appraisal.* New York: St. Martin's Press.

———. 1998. *Microeconomics: A Modern Approach.* New York: Addison-Wesley.

Selten, R. 1965. "Spieltheoretische Behandlung Eines Oligopolmodells Mit Nachfragetragheit." *Zeitschrift fur die gesamte Staatswissenschaft* 121: 301–324.

———. 1975. "Reexamination of the Perfectness Concept for Equilibrium Points in Extensive Games." *International Journal of Game Theory* 4: 25–55.

Silberberg, E. 1990. *The Structure of Economics: A Mathematical Analysis*, 2d ed. New York: McGraw-Hill.

Smith, J. M. 1982. *Evolution and the Theory of Games.* Cambridge: Cambridge University Press.

Spence, M. 1973. "Job Market Signaling." *Quarterly Journal of Economics* 87(3): 355–374.

Stackleberg, H. 1934. *Marktform und Gleichgewicht*, Vienna: Julius Springer.

Stigler, G. J. 1947. "The Kinky Oligopoly Demand Curve and Rigid Prices," *Journal of Political Economy* (October): 432–449.

Sweezy, P. 1939. "Demand Conditions Under Oligopoly." *Journal of Political Economy* (August): 568–573.

Tellis, G. and P. Golder. 1991. "First to Market, First to Fail: Real Causes of Enduring Market Leadership." *Sloan Management Review* 37(2) (Winter): 65–75.

Thaler, R. H. 1988. "Anomalies: The Winner's Curse." *Journal of Economic Perspectives* 2(1): 191–202.

Tucker, A. W. 1955. *Game Theory and Programming.* Stillwater, OK: Department of Mathematics, Oklahoma Agricultural and Mechanical College.

Tullock, G. 1971. *The Logic of the Law.* New York: Basic Books.

———. 1997. "The Case Against the Common Law." *The Blackstone Commentaries, No. 1.* Durham, NC: The Locke Institute.

UNFCCC. 1992. *The United Nations Framework Convention on Climate Change.* <http://unfccc. int/essential_background/convention/background/items/1353.php>.

Vickery, W. 1961. "Counterspeculation, Auctions, and Competitive Sealed Tenders." *Journal of Finance* 16: 8–37.

Von Neumann, J. 1928. "Zur Theorie der Gesellschaftsspiele." *Mathematische Annalen* 100: 295–320. Translated as "On the theory of games of strategy" in *Contributions to the Theory of Games*, IV (*Annals of Mathematics Studies*), ed. A. Tucker and R. D. Luce. Princeton: Princeton University Press, 1959.

Von Neumann, J. and O. Morgenstern. 1944. *Theory of Games and Economic Behavior.* New York: John Wiley & Sons.

Von Stackleberg, H. 1934. *Marktform und Gleichgewicht.* Vienna: Julius Springer, 1934. Reprinted in *The Theory of the Market Economy*, translated by A. T. Peacock. London: William Hodge, 1952.

Walker, M. and J. Wooders. 2001. "Minimax Play at Wimbledon." *American Economic Review* 91(5) (December): 1521–1538.

Webster, T. J. 2001. "A Principle Component Analysis of the *U.S. News & World Report* Tier Rankings of College and Universities." *Economics of Education Review* 20: 235–244.

———. 2004. "Economic Efficiency and the Common Law." *Atlantic Economic Journal* 32(1) (March): 39–48.

Williams, J. D. 1966. *The Compleat Strategyst.* New York: McGraw-Hill.

Index

Thomas J. Webster is a professor in the Department of Finance and Economics of Pace University's Lubin School of Business. He previously held positions as an international economist with the Central Intelligence Agency, Continental Illinois National Bank and Trust Company, and Manufacturer's Hanover Trust Company.

In addition to his teaching duties at the Lubin School, he has served as graduate and undergraduate finance program chair, and as faculty adviser to Beta Gamma Sigma, the international honor society for collegiate schools of business. He has received the Lubin School of Business Scholarly Research Award for Basic Scholarship, the Lubin School of Business Outstanding Faculty Service Award, the Pace University Award for Distinguished Service, and the Beta Gamma Sigma Commitment to Excellence Award.

Professor Webster received his BA from American University's School of International Service, and his MA, MPhil, and PhD from the City University of New York.